Springer Series in Materials Science

Volume 354

The Springer Series in Materials Science covers the complete spectrum of materials research and technology, including fundamental principles, physical properties, materials theory and design. Recognizing the increasing importance of materials science in future device technologies, the book titles in this series reflect the state-of-the-art in understanding and controlling the structure and properties of all important classes of materials.

Suprakas Sinha Ray · Ritima Banerjee

Polymer Blend Nanocomposites

Springer

Suprakas Sinha Ray
Centre for Nanostructures and Advanced Materials, DSTI-CSIR Nanotechnology Innovation Centre
Council for Scientific and Industrial Research
Pretoria, South Africa

Department of Chemical Sciences
University of Johannesburg
Johannesburg, South Africa

Ritima Banerjee
Department of Chemical Sciences
University of Johannesburg
Johannesburg, South Africa

Centre for Nanostructures and Advanced Materials, DSTI-CSIR Nanotechnology Innovation Centre
Council for Scientific and Industrial Research
Pretoria, South Africa

ISSN 0933-033X ISSN 2196-2812 (electronic)
Springer Series in Materials Science
ISBN 978-3-032-00308-9 ISBN 978-3-032-00309-6 (eBook)
https://doi.org/10.1007/978-3-032-00309-6

This Springer imprint is published by the registered company Springer Nature Switzerland AG
The registered company address is: Gewerbestrasse 11, 6330 Cham, Switzerland

Dedicated to Our Parents

Preface

Polymer blends offer the possibility of combining the characteristics of two or more polymers with different properties to produce materials that meet requirements that cannot be fulfilled by any single polymer. Most polymers are thermodynamically immiscible, and the properties of an immiscible polymer blend are guided by its morphological characteristics. The addition of nanofillers is an effective means of enhancing the properties of polymer blends. In addition to imparting special properties, such as electrical conductivity provided by carbon nanotubes and graphene, nanofillers may also induce morphological changes in polymer blends, resulting in property enhancement. It has been reported that nanofillers refine and stabilize blend morphology, thus reducing the need to use compatibilizers, which are expensive and sometimes lower the properties of blends.

In a nanocomposite of an immiscible polymer blend, nanoparticles may be localized in either phase or at the interface between the two phases, which profoundly influences the morphological characteristics of the ternary composition. The migration and localization of nanoparticles in immiscible polymer blends are primarily guided by thermodynamic equilibrium. However, the importance of kinetics cannot be overlooked. By controlling various process parameters during melt processing, one can exploit the effects of kinetics to tailor the localization of nanoparticles. Moreover, by modifying the surface of the filler particles and inducing interfacial reactions, these particles can be localized at the interface. Such control over the migration of nanoparticles is crucial for achieving desired properties, such as high electrical conductivity and a low percolation threshold in conductive polymer blend nanocomposites.

The book begins with the fundamentals of the morphological development of these multiphase polymeric materials, providing a holistic understanding of all critical aspects related to controlling the migration and localization of commonly used nanoparticles, along with their subsequent effects on properties. There is a chapter on processing technologies, offering an in-depth discussion on how a combination of different processing techniques and tailored processing conditions can achieve the desired morphology and properties of polymer blend nanocomposites. This is followed by a chapter on structural and morphological characterization techniques, in which each technique is explained with case studies specific to polymer blend nanocomposites. Another chapter addresses interface modification through the interfacial localization of nanoparticles, detailing the methods for achieving such localization and its subsequent impact on blend morphology and properties. Following this, a chapter on properties discusses the characteristics achieved with commonly used nanofillers, both alone and in combination. Case studies covering a wide spectrum of polymers are included in this chapter. Additionally, a chapter on degradation studies provides insight into the durability and degradability of these multiphase materials, with a special focus on biodegradability. Finally, the book presents an in-depth discussion on advanced applications of polymer blend nanocomposites, such as packaging, EMI shielding, and biomedical applications.

The following are the key features of this book:

- *The fundamentals of morphology development of polymer blends and blend nanocomposites, with special focus on nanoparticle migration, interface crossing and localization*—This includes an in-depth discussion on the various thermodynamic and kinetic factors governing particle migration and localization.
- *Processing technologies used for manufacturing polymer blend nanocomposites*—Though the basic processing techniques of polymer blends, nanocomposites and blend nanocomposites are the same, the processing of polymer blend nanocomposites is more complex. This is because often various processing techniques are used in combination and by tailoring this combination, one can control the morphological characteristics of these multiphase materials.
- *Structural and morphological characterisation of polymer blend nanocomposites*—This includes an in-depth discussion on how various characterisation techniques can be used for gaining an understanding of blend morphology, the location of the filler particles, their quality of dispersion both qualitatively and quantitatively and the interfacial adhesion between the blend components and the filler particles.
- *Nanoparticles as interface modifiers*—Interface modification of immiscible polymer blends can be achieved by the selective localisation of nanoparticles at the blend interface. Such interfacially localised nanoparticles can interact with both blend components, either through reactions (reactive compatibilization) or through other modes, such as hydrogen bonding or physical entanglements with the macromolecular chains. The molecular structure (the presence of functional groups, grafted chains, etc.) of the nanoparticles play a pivotal role in such interactions.
- *An overview of the structure-property*—processing relationship of polymer blend nanocomposites.
- *Degradation studies on polymer blend nanocomposites.*
- *Applications of polymer blend nanocomposites*—This includes an in-depth discussion of advanced applications of polymer blend nanocomposites, such as, packaging, EMI shielding and biomedical applications. Each of these applications is explained with case studies, wherein researchers have achieved the desired properties by proper choice of materials and morphology control.

A single book of this nature, covering the fundamentals of polymer blend nanocomposites and extending to its applications in advanced technologies, will be a very useful reference for undergraduate and postgraduate students, academic researchers, engineers, scientists, and other professionals interested in this exciting field of research. This book will also help industrial researchers and R&D managers who want to bring nanoclay-based advanced products into the market.

Pretoria, South Africa Suprakas Sinha Ray
Johannesburg, South Africa Ritima Banerjee

Acknowledgments The authors thank the Department of Science, Technology and Innovation, the Council for Scientific and Industrial Research, and University of Johannesburg, South Africa, for financial support. We express our sincerest appreciation to all colleagues, postdoctoral fellows, and students for their valuable contributions as well as the reviewers for their critical evaluation of the proposal and chapters. This book is based on the invited review article (Polymer 2025, 316, 127844, open access) published by the authors of this book. We also thank the authors and publishers for their permission to reproduce their published works. Our special thanks go to Dr. Zachary Evenson at Springer Nature for his encouragement, cooperation, suggestions, and advice during various phases of preparation, organization, and production of this book.

Pretoria, South Africa — Suprakas Sinha Ray
Johannesburg, South Africa — Ritima Banerjee

Competing Interests The authors have no competing interests to declare that are relevant to the content of this manuscript.

Contents

About the Authors

Suprakas Sinha Ray is a Chief Research Scientist and Manager of the Centre for Nanostructures and Advanced Materials, DSTI-CSIR Nanotechnology Innovation Centre, Council for Scientific and Industrial Research, Pretoria, South Africa. He received his Ph.D. degree in Physical Chemistry at the University of Calcutta, India, in 2001 and was a recipient of the "Sir P. C. Ray Research Award" for the best Ph.D. work. Professor Ray current research focuses on the applications of advanced nanostructured and polymeric materials. He is one of the most active and highly cited authors in the field of polymer nanocomposite materials, and he has recently been rated by Thomson Reuters as being one of the Top 1% most impactful and influential scientists and Top 50 high impact chemists.

Professor Ray is the author of 14 authored-books, co-author of 13 edited books, 65 book chapters on various aspects of polymer-based nanostructured materials and their applications, and author and co-author of 507 articles in high-impact international journals and 79 book chapters. He also has 11 patents (registered or filled) and 10 new demonstrated technologies (commercialized) shared with colleagues, collaborators, and industrial partners. So far, his team commercialized 19 different products. Over the last 15 years, he has trained 23 postdoctoral fellows and produced 41 Ph.D. and 17 Master students under his direct supervision. He is one of the most active and highly cited (43400 citations; h factor 92, GS) authors in nanostructured materials. So far, his team has commercialized 19 different products. His honors and awards include South Africa's most *Prestigious 2016 National Science and Technology Award* (NSTF); Prestigious 2014 CSIR-wide Leadership award; Prestigious 2014 CSIR Human Capital development award; *Prestigious 2013 Morand Lambla Awardee* (top award in the field of polymer processing worldwide), International Polymer Processing Society, USA. He is also appointed as Extraordinary Professor, University of Pretoria and Distinguished Professor of Chemistry, University of Johannesburg.

Dr. Ritima Banerjee did her Masters in Polymer Science and Technology from the Indian Institute of Technology, Delhi (IITD). She returned to academia after working in the polymer industry (GE Plastics and SABIC) for seven years. She taught in Delhi Technological University for two years. Subsequently, she did her Ph.D. at the Department of Material Science and Engineering, IITD, where her work was microcellular processing of thermoplastic elastomer-based blends and nanocomposites. She is a researcher at the Department of Chemical Sciences, University of Johannesburg, Johannesburg South Africa. She is also associated with the Centre for Nanostructures and Advanced Materials, Council for Scientific and Industrial Research, Pretoria 0001, South Africa. Her research interests include microcellular processing and the structure-property-processing relationship of polymeric materials.

Chapter 1
Introduction

Abstract This chapter is divided into two sections. The first section offers a concise overview of polymer blend nanocomposite technology. The second section outlines the scope and advantages of this book.

1.1 Introduction to Polymer Blend Nanocomposite Technology

A polymer blend consists of two or more polymers or copolymers mixed together. These blends form a category of materials aimed at achieving better properties than the individual polymers [1]. They are characterised from both economic and scientific perspectives. For example, polymer blending serves as an effective approach for recycling waste plastics into usable materials [2, 3]. However, the majority of chemically dissimilar polymers do not mix well, leading to materials with weak interfacial adhesion and, as a result, poor mechanical performance [3]. Such unfavourable interactions typically cause polymer blends to display a rough morphology, elevated interfacial tension, distinct interfaces, and low adhesion between the phases. This phase-separated structure can adversely affect the final properties of the blend, including thermal and mechanical characteristics, rather than delivering the expected enhancements [3]. Consequently, controlling the morphology of immiscible polymer blends is crucial, as their mechanical properties are strongly influenced by their morphology [4].

In the case of immiscible polymer blends, the control of processing parameters during the melt-blending process plays an important role in the evolution of the phase morphology, which subsequently impacts the properties [4, 5]. It is critical to know the rheological properties of the polymer to understand the morphological development, especially in cases where the contributions of thermodynamics to uniform morphologies are low [4].

It is widely recognised that the structural evolution induced by flow in immiscible polymer blends is influenced by various factors, including the mixing conditions, interfacial tension between the phases, volume fraction, and viscosity ratio of the

S. S. Ray and R. Banerjee, *Polymer Blend Nanocomposites*, Springer Series in Materials Science 354, https://doi.org/10.1007/978-3-032-00309-6_1

constituent polymers. A review by Briscoe et al. [6] on immiscible fluid mixing, stemming from Taylor's foundational work in 1934 [7], outlines the factors affecting the shear strain necessary for break-up in a steady laminar flow as: (i) the viscosities of the dispersed and continuous phases, respectively; (ii) the initial radius, R; and (iii) the interfacial tension. In the case of Newtonian flow, droplet deformation is determined by two dimensionless numbers: the viscosity ratio $p = (\eta_d/\eta_m)$, and the capillary number C_a, which represents the ratio of deforming viscous stress to the restoring stress from interfacial tension in the following expression, equation (1.1) [6]:

$$C_a = \frac{\eta_m \dot{\gamma}}{\sigma/R} \tag{1.1}$$

where is the viscosity of the primary matrix, is the shear strain rate, is the radius of the dispersed phase, and is the interfacial tension. The numerator represents the deforming stresses, whereas the denominator signifies the restoring interfacial stress.

Koning et al. [8] illustrated the processes involved in the melt blending of two polymers, as shown in Fig. 1.1. Initially, the C_a is small, allowing the interfacial stress to endure the shear stress, which keeps the drop in a persistent ellipsoidal shape.

When the concentration exceeds a critical value, C_a, particularly during the initial mixing phase with larger dispersed domains, the shear stress overpowers the interfacial stress, causing drops to elongate into fine threads due to their affinity with the matrix. If the local radius of a thread becomes sufficiently minimal, interfacial disturbances—known as "Rayleigh" disturbances—emerge on the thread, leading to its fragmentation into smaller drops. Once exceeding a specific diameter, these smaller drops can undergo further stretching and breakage. For very tiny drops, the ratio δ/R is significant enough to hinder additional stretching and breakup.

Empirical correlations between the C_a and viscosity ratio, p, have been established, allowing prediction of droplet size based on processing parameters for low and constant concentrations of the dispersed phase [8]. However, the morphology of

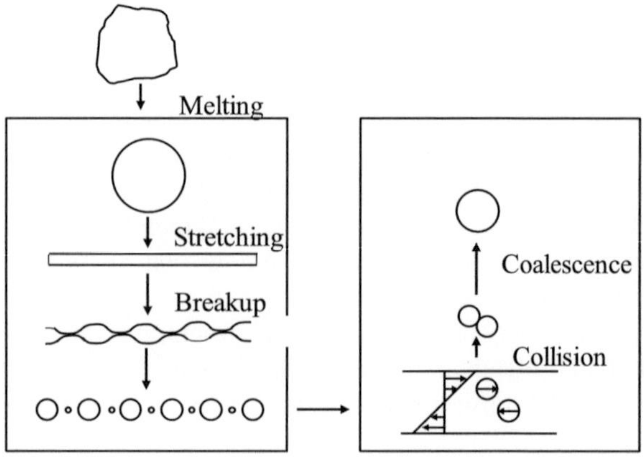

Fig. 1.1 Schematic representation of the processes occurring during the melt-blending of two polymers [8]. Copyright 1998. Reproduced with permission from Elsevier Science Ltd.

specific blends may not reach equilibrium and can vary with post-processing treatments. Thus, to fully leverage the advantages of the two polymers, the dispersed phase should be small in size, while also maximising the droplet surface area per unit volume of the blend.

Transforming the immiscible mixture into a functional polymer product with specific properties requires manipulation of the interface. A traditional approach to enhance adhesion between the phases is to introduce a third component, known as a compatibilizer, that is compatible or miscible with both phases [9–11].

Compatibilizers play a crucial role in the blending process by first reducing the interfacial tension, which helps delay the formation of Rayleigh disturbances on the polymer threads. This reduction allows for greater stretching by deforming stresses, leading to smaller sizes of the dispersed phase droplets. Additionally, compatibilizers on the surfaces of these droplets prevent coalescence, ensuring they remain as small and stabilized as possible. Finally, thanks to the strong interfacial adhesion provided by compatibilizers, stress transfer between phases is enhanced, improving overall efficiency.

Historically, compatibilizers have been viewed as macromolecular entities with interfacial roles in heterogeneous polymer blends. These compatibilizers can take the form of a homopolymer [12], or block, graft, or star copolymers [13, 14]. Among macromolecules, typical compatibilizers consist of block-structured copolymers, where one block is compatible with one component of the blend, while another block is compatible with the opposing component. Such block structures can be synthesised beforehand and introduced into the immiscible polymer blend; alternatively, they may be produced in situ during the blending process. The latter method, known as reactive compatibilization, necessitates that both blend components have mutual reactivity. Reactive polymers can also facilitate compatibilization through the formation of copolymers, which result from the addition of low molecular weight substances like peroxides and bifunctional agents, as well as through interchange reactions such as transesterification, the inclusion of selective crosslinking agents, and ionomers [8].

Compatibilization produces the desired blend morphology by regulating the interfacial properties, which in turn control the size of the dispersed droplets of the minor phase. This process stabilizes the morphology against coalescence during later processing stages and ensures adhesion between the phases in the solid state, thereby enhancing the mechanical properties. The outcome is a blend characterized by a finer and more stable morphology, along with improved performance. These traditional compatibilization methods have been extensively employed to create various industrial polymer blends with diverse properties [15].

In the past two decades, extensive research has focused on materials featuring a nanometre-sized microstructure. These materials consist of crystalline building blocks at the nanometre scale, which may vary in atomic structure, chemical composition, or crystallographic structure [16, 17]. Depending on these factors, the nanoscale building blocks can form incoherent or coherent interfaces. This heterogeneous nanometre-scale structure is essential for the distinct properties that set them apart from conventional homogeneous microstructures, such as glass and gels. This new category of materials, characterised by their heterogeneous microstructure and nanometre-sized building blocks, is known as polymer blend nanocomposites or polymer blend nanostructures [16, 17].

In recent years, activities related to nanoparticle-modified polymer blend composites have significantly increased. Research output, including journal articles, conference proceedings, book chapters, and trade publications, has steadily grown since 2000, as depicted in Fig. 1.2a. Figure 1.2b illustrates applications in various fields. With ongoing funding and heightened interest from academia, along with the growing use of polymer blend nanocomposites in products, research output is projected to keep rising across different sectors.

These materials consist of bulk solids made from either partially or entirely nanoscale building blocks, having historical significance that dates back many

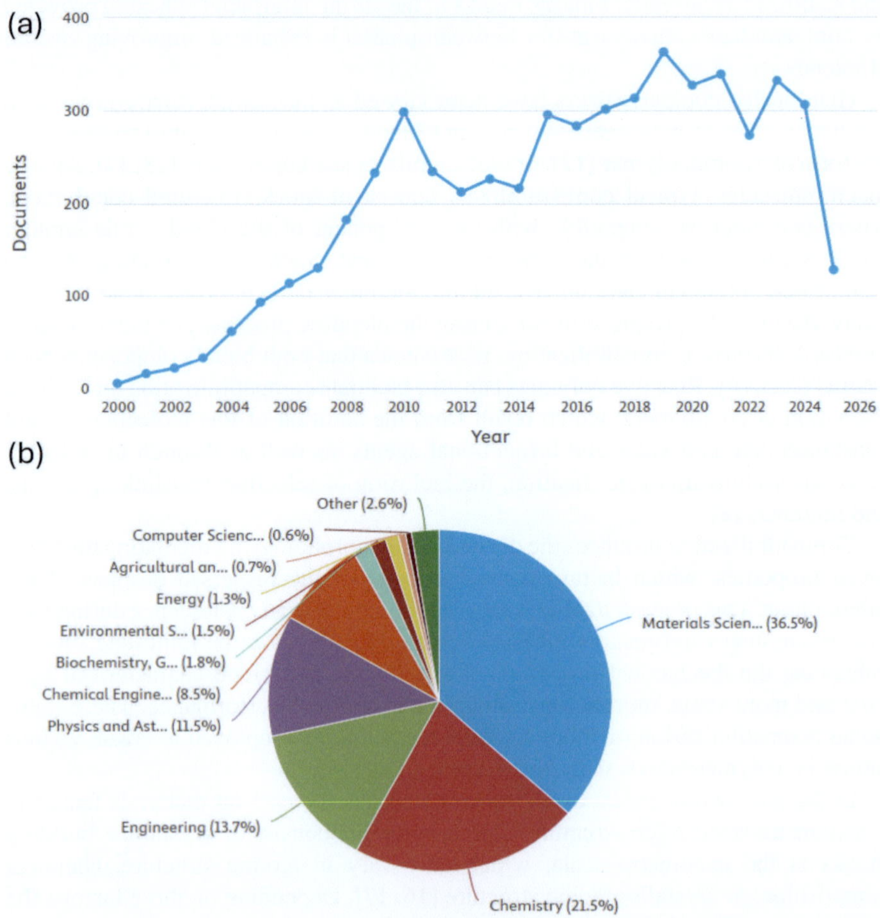

Fig. 1.2 **a** Annual research output documents and **b** applications across various fields. *Source* Scopus, with the search keyword "Polymer blend and nanocomposites" (March 24, 2025)

decades [18]. Recently, however, bulk nanostructured materials have gained prominence in nanomaterials research, attributed to their unique properties found in various forms of bulk nanostructured materials [18]. The most prevalent types of bulk nanostructured materials include those with nanoscale components, such as nanostructured films, nanocomposites, nontextured surfaces, and nanocrystalline materials. Polymer nanocomposites have become an essential part of bulk nanostructured materials, with additional details available in Chap. 2.

Recently, in the realm of immiscible polymer blend systems, a notably examined compatibilization method involves using inorganic solid particles. Lipatov [19] initially observed that in a three-component system with polymers A and B, the free energy of mixing is expressed as follows, equation (1.2):

$$\Delta G_m = \Delta G_{AS} + \Delta G_{BS} - \Delta G_{AB} \tag{1.2}$$

The subscripts denote the interacting pairs. Thermodynamically, the system achieves greater stability when the blend is immiscible with a positive DGAB than when it is miscible with DGAB values less than zero. Therefore, incorporating S into the A-B blend enhances its stability, functioning as a compatibilizer by adsorbing A and/or B polymers onto its surface. The energy gain from stabilization clearly results from the adsorption of polymeric materials onto the solid surface. For this purpose, the inorganic phase should possess the maximum surface area per unit weight. This need is met by nanoparticles such as nanoclay (with montmorillonite exhibiting a specific surface area of approximately So = 700–800 m^2/g) [20–30] or other nanoparticles (like carbon nanotubes, which have specific surface areas of So = 250–1000 m^2/g) [31–38] that can be well-dispersed in a two-phase matrix. For example, nanoclays can significantly improve morphology by altering the viscosity ratio when particles preferentially localize in one phase, reducing interfacial tension through conventional intercalation or tethering, and preventing the coalescence of droplets. Additionally, adding a nanoclay platelet that can adsorb macromolecules can also achieve compatibilization in an immiscible polymer blend.

The high adsorption of organic species by crystalline solids is well established by mathematical simulations [39] or experimentally using neutron scattering and surface force analysis [40, 41]. However, there is a fundamental principle of additive incorporation into a multiphase polymeric system. It must be inserted into a specific phase—if in the polymer blend, one phase is rigid and the other ductile, increasing the volume of each of these with the same additive will have the opposite effects. For example, incorporation of polystyrene (PS) into the rubbery phase of high-impact PS increases the ductility of the system, while incorporation of glass fiber into polyamide 66 in its blend with acrylonitrile-butadiene acrylate increases the modulus.

Since the early 1990s, immiscible polymer blends have been utilised as matrices in polymer nanocomposites, which are bulk nanostructured materials [42]. Many patents related to the preparation of nanocomposites assert that the methods can be applied to a range of polymers and their blends. Additionally, some emphasise using these blends to modify performance, such as enhancing stiffness, controlling permeability, or achieving the desired performance balance. However, other studies have highlighted the compatibilizing effect of organoclay on immiscible polymer blends. These findings indicate that organically modified layered silicates may function as compatibilisers between immiscible polymers.

The impact of solid particles, which vary from a few hundred nanometers to several microns, in immiscible polymer blends has been thoroughly investigated for their roles as compatibilizers, morphology stabilizers, and reinforcement agents. The significance of particle localisation in the structural properties of these blends is well understood. The wetting coefficient parameter ω_{AB} is the main thermodynamically controlling parameter for particle localization. Kinetic effects significantly influence where particles ultimately settle. This book discusses a variety of nanoparticles, focusing on the mechanisms and factors that affect the movement of nanofillers within immiscible polymer blends. The migration process entails bulk movement within the matrix or minor phase, interaction with the interface, and eventual penetration into the minor phase or matrix. Research indicates that a number of kinetic effects, including the mixing protocol (sequence), intensity, and duration, profoundly impact particle migration and localization. Additionally, the migration rate is affected by the viscosity ratio of the blend, as well as the particles' size, shape, and surface chemistry. It has been observed that particles in more viscous phases migrate more slowly compared to those with high aspect ratios and surface polarities that align with the opposite phase, which migrate more rapidly. This book provides an insightful overview of how nanoparticle migration affects localization and, consequently, the development of morphology, mechanical properties, electrical conductivity, and met-rheological features of blended composite materials.

1.2 Scope of the Book

This book is based on the invited review article (Polymer 2025, 316, 127844, open access) published by the authors of this book. The book opens by discussing the fundamentals of morphology development in multiphase polymeric materials, offering a comprehensive overview of critical aspects related to the migration and localisation control of commonly used nanoparticles, along with their subsequent effects on properties. A chapter on processing technologies delves deeply into how various processing techniques and tailored conditions can achieve desired morphologies and properties in polymer blend nanocomposites. Following this, a chapter on structural and morphological characterisation techniques explains each method with case studies focused on polymer blend nanocomposites. Another chapter discusses interface modification through the interfacial localisation of nanoparticles, detailing methods for achieving this localisation and its effects on blend morphology and properties. This is succeeded by a chapter addressing properties, which examines the characteristics attained with commonly used nanofillers, whether individually or in combination, supported by case studies spanning a variety of polymers. Additionally, there is a chapter on degradation studies, providing insights into the durability and degradability of these multiphase materials, with particular emphasis on biodegradability. Finally, the book presents an extensive discussion on the advanced applications of polymer blend nanocomposites, including uses in packaging, EMI shielding, and biomedical fields.

Though there are books related to polymer blends and polymer nanocomposites, there is no book dedicated to polymer blend nanocomposites, which covers all aforementioned points, without remaining restricted to any specific type of nanofiller or polymer systems. There are some reviews on some aspects of this subject, but none which holistically covers all the above-mentioned aspects. Thus, there is a need for a comprehensive, authored book on this topic. Such a book would have the potential to shape future research in material development for various advanced applications of polymeric materials.

References

1. O.J. Botlhoko, J. Ramontja, S.S. Ray, A new insight into morphological, thermal, and mechanical properties of melt-processed polylactide/poly(ε-caprolactone) blends. Polym. Degrad. Stab. **154**, 84–95 (2018)
2. L.A. Utracki, Economics of polymer blends. Polym. Eng. Sci. **22**, 1166–1175 (1982)
3. L.M. Robenson, *Polymer Blends: A Comprehensive Review* (Hanser, Munich, 2007)
4. R. Salehiyan, S.S. Ray, J.F. Stadler, V. Ojijo, Rheology-microstructure relationship in melt-processed polylactide/poly(vinylidene fluoride) blends. Materials **11**, 2450 (2018). https://doi.org/10.3390/ma11122450
5. R. Salehiyan, S.S. Ray, V. Ojijo, Processing-driven morphology development and crystallization behaviour of immiscible polylactide/poly(vinylidene fluoride) blends. Macromol. Mater. Eng. **303**, 1800349 (2018)
6. B.J. Briscoe, C.J. Lawrence, W.G.P. Mietus, A review of immiscible fluid mixing. Adv. Colloid Interface Sci. **81**, 1–17 (1999)
7. G.I. Taylor, The formation of emulsions in definable fields of flow. Proc. R. Soc. Lond. Ser. A **134**, 146, 501–523 (1934)
8. C. Koning, M. Van Duin, C. Pagnoulle, R. Jerome, Strategies for compatibilization of polymer blends. Prog. Polym. Sci. **23**, 707–757 (1998)
9. V. Ojijo, S.S. Ray, R. Sadiku, Role of specific interfacial area in controlling properties of immiscible blends of biodegradable polylactide and poly[(butylene succinate)-co-adipate]. ACS Appl. Mater. Interfaces **4**, 6690–6701 (2012)
10. A. Ajji, in *Polymer Blends Handbook*, ed. by L.A. Utracki, Chapter 4 (Kluwer Academic Press, Dordrecht, 2002)
11. S.B. Brown, in *Polymer Blends Handbook*, ed. by L.A. Utracki, Chapter 5 (Kluwer Academic Press, Dordrecht, 2002)
12. J.M. Machado, C.S. Lee, Compatibilization of immiscible blends with a mutually miscible third polymer. Polym. Eng. Sci. **34**, 59–68 (1994)
13. T.K. Kwei, H.C. Frisch, W. Radigan, S. Vogel, Ternary polymer mixtures. Macromolecules **10**, 157–160 (1977)
14. C.E. Locke, D.R. Paul, Graft copolymer modification of polyethylene–polystyrene blends. II. Properties of modified blends. J. Appl. Polym. Sci. **17**, 2597–2617 (1973)
15. R. Fayt, R. Jerome, Ph. Teyssie, Molecular design of multicomponent polymer systems. II. Emulsifying effect of a poly(hydrogenated butadiene-*b*-styrene) copolymer in high-density polyethylene/polystyrene blends. J. Polym. Sci. Part B: Polym. Phys. **19**, 1269–1272 (1981)
16. H. Gleiter, Nanostructured materials: basic concepts and microstructure. Acta Mater. **48**, 1–29 (2000)

17. H. Gleiter, Nanostructured materials: state of the art and perspectives. Nanostruct. Mater. **6**, 3–14 (1995)
18. N. Nasirpouri, *Electrodeposition of Nanostructured Materials.* Springer Series in Surface Sciences, vol. 62. https://doi.org/10.1007/978-3-319-44920-3_1
19. Y.S. Lipatov, *Polymer Reinforcement* (ChemTec Publishing Toronto, Canada, 1995)
20. S.S. Ray, S. Pouliot, M. Bousmina, L.A. Utracki, Role of organically modified layered silicate as an active interfacial modifier in immiscible polystyrene/polypropylene blends. Polymer **45**, 8403–8413 (2004)
21. S.S. Ray, J. Bandyopadhyay, M. Bousmina, Influence of degree of intercalation on the crystal growth kinetics of poly[(butylene succinate)-co-adipate] nanocomposites. Eur. Polym. J. **44**, 3133–3145 (2008)
22. S.S. Ray, M. Bousmina, Compatibilization efficiency of organoclay in an immiscible polycarbonate/poly(methyl methacrylate) blend. Macromol. Rapid Commun. **26**, 450−455 (2004)
23. S.S. Ray, M. Bousmina, Biodegradable polymers and their layered silicate nanocomposites: in greening the 21st century materials world. Prog. Mater. Sci. **50**, 962–1079 (2005)
24. S.S. Ray, M. Bousmina, Effect of organic modification on the compatibilization efficiency of clay in an immiscible polymer blend. Macromol. Rapid Commun. **26**, 1639–1646 (2005)
25. S.S. Ray, M. Bousmina, Poly(butylene sucinate-co-adipate)/montmorillonite nanocomposites: effect of organic modifier miscibility on structure, properties, and viscoelasticity. Polymer **46**, 12430–12439 (2005)
26. S.S. Ray, M. Bousmina, K. Okamoto, Structure and properties of nanocomposites based on poly(butylene succinate-co-adipate) and organically modified montmorillonite. Macromol. Mater. Eng. **290**, 759–768 (2005)
27. V. Ojijo, S.S. Ray, R. Sadiku, Effect of nanoclay loading on the thermal and mechanical properties of biodegradable polylactide/poly[(butylene succinate)-co-adipate] blend composites. ACS Appl. Mater. Interfaces **4**, 2395–2405 (2012)
28. V. Ojijo, H. Cele, S.S. Ray, Morphology and properties of polymer composites based on biodegradable polylactide/poly[(butylene succinate)-co-adipate] blend and nanoclay. Macromol. Mater. Eng. **296**, 865–877 (2011)
29. R. Salehiyan, S.S. Ray, Influence of nanoclay localization on structure-property relationship of polylactide-based biodegradable blend nanocomposites. Macromol. Mater. Eng. **303**, 1800134 (2018)
30. R. Salehiyan, S.S. Ray, J. Bandyopadhyay, V. Ojijo, The distribution of nanoclay particles at the interface and their influence on the microstructure development and rheological properties of reactively processed biodegradable polylactide/poly[(butylene succinate)-co-adipate] nanocomposites. Polymers **9**, 350 (2017). https://doi.org/10.3390/polym9080350
31. R. Salehiyan, S.S. Ray, Tuning the conductivity of nanocomposites through nanoparticle migration and interface crossing in immiscible polymer blends: a review on fundamental understanding. Macromol. Mater. Eng. **303**, 1800134 (2018)
32. O.J. Botlhoko, J. Ramontja, S.S. Ray, Thermal, mechanical, and rheological properties of graphite- and graphene oxide-filled biodegradable polylactide/poly(ε-caprolactone) blend composites. J. Appl. Polym. Sci. **134**, 45373 (2017)
33. O.J. Botlhoko, S.S. Ray, J. Ramontja, Influence of functionalized exfoliated reduced graphene oxide nanoparticle localization on mechanical, thermal and electronic properties of nanobiocomposites. Eur. Polym. J. **102**, 130–140 (2018)
34. O.J. Botlhoko, J. Ramontja, S.S. Ray, Thermally shocked graphene oxide-containing biocomposite for thermal management applications. RSC Adv. **7**, 33751–33756 (2017)
35. O.J. Botlhoko, J. Ramontja, S.S. Ray, Morphological development and enhancement of thermal, mechanical, and electrical properties of thermally exfoliated graphene oxide-filled biodegradable polylactide/poly(ε-caprolactone) blend composites. Polymer **139**, 188–200 (2018)
36. E.S. Ogunniran, R. Sadiku, S.S. Ray, N. Luruli, Morphology and thermal properties of compatibilized PA12/PP blends with boehmite alumina nanofiller inclusions. Macromol. Mater. Eng. **297**, 627–638 (2012)

37. E.S. Ogunniran, R. Sadiku, S.S. Ray, N. Luruli, Effect of boehmite alumina nanofiller incorporation on the morphology and thermal properties of functionalized polypropylene/polyamide 12 blends. Macromol. Mater. Eng. **297**, 237–248 (2012)
38. T. Malwela, V.M. Khumalo, R. Salehiyan, S.S. Ray, Characterization of polypropylene/polystyrene boehmite alumina nanocomposites: impact of filler surface modification on the mechanical, thermal, and rheological properties. J. Appl. Polym. Sci. **135**, 46376 (2018)
39. G. Fleer, M.A. Cohen-Stuart, J.M.H.M. Scheutjens, B. Cosgrove Vincent, *Polymers at Interfaces* (Chapman and Hall, London, 1993)
40. T. Cosgrove, T.G. Heath, J.S. Phipps, R.M. Richardson, Neutron reflectivity studies of polymers adsorbed on mica from solution. Macromolecules **24**, 94–98 (1991)
41. J. Klein, Y. Kamiyama, H. Yoshizawa, J.N. Israelachvili, G.N. Fredrickson, P. Pincus, L.J. Fetters, Lubrication forces between surfaces bearing polymer brushes. Macromolecules **26**, 5552–5560 (1993)
42. L.A. Utracki, *Clay-Containing Polymeric Nanocomposites* (RAPRA, Shawbury, Shewsbury, 2004)

Chapter 2
Fundamentals of Polymer Blends and Polymer Blend Nanocomposites

Abstract This chapter starts by introducing the fundamentals of polymer blends and offers an overview of polymer blend nanocomposites. It subsequently examines nanoparticle migration, crossing interfaces, particle localization, and the diverse factors that affect particle migration and localization. This knowledge is essential for creating blend nanocomposites with enhanced properties.

2.1 Introduction

Polymer blends constitute a significant category of polymeric materials. A neat polymer frequently fails to satisfy the property requirements of many applications. Blending the polymer with another polymer enhances its properties and achieves a synergy of characteristics, thus broadening the range of applications. Furthermore, blending often allows for property optimisation at a lower cost than a single polymer. The properties of a polymer blend are determined by the miscibility of the blend components and their volume fractions. In the case of immiscible polymer blends, the properties are significantly influenced not only by the blend components but also by blend morphology, which is governed by several factors, such as the ratio of the components, their viscosity ratio, and the processing conditions utilised.

Often, a polymer blend fails to meet all the property requirements of a product. In such cases, the addition of a small amount of a nanofiller may achieve the objective. The nanoparticles not only impart unique properties, such as electrical conductivity in the case of carbon nanotubes and graphene, but also enhance the mechanical and other properties of the blend. The properties of blend nanocomposites depend not only on the blend components and their volume fractions, blend morphology, the nanofiller used, and its loading, but also on the quality of dispersion of the nanoparticles and their location within the blend. When dispersed in a blend of two immiscible polymers, the nanoparticles may be present in both phases, preferentially localised in one of the phases, or even reside at the interface. The properties of the blend nanocomposite differ significantly in all three scenarios. Furthermore, the

S. S. Ray and R. Banerjee, *Polymer Blend Nanocomposites*, Springer Series in Materials Science 354, https://doi.org/10.1007/978-3-032-00309-6_2

localisation of the nanoparticles also profoundly influences the blend morphology, which impacts the properties.

The localisation of nanoparticles is governed by thermodynamics and kinetics [1]. By controlling the viscosity ratio of the blend components, correctly selecting the characteristics of the particles and the surface functionality of the nanofiller, and optimising processing parameters—including the mixing protocol (time and sequence of mixing)—one can control the migration of nanoparticles to achieve their desirable localisation in blend nanocomposites.

This chapter begins with the fundamentals of polymer blends and provides an overview of the essentials of polymer blend nanocomposites. It then explores nanoparticle migration, interface crossing, particle localization, and the various factors influencing particle migration and localization, offering insights that are vital for developing blend nanocomposites with improved properties.

2.2 Fundamentals of Polymer Blends

A polymer blend is a mixture of at least two macromolecular substances, including polymers or copolymers, in which the ingredient content exceeds 2% [2]. The benefits of polymer blends are numerous and have evolved over time. In the 1960s, blending was conducted to enhance specific polymer properties, primarily impact strength. During the following decade, blending aimed to reduce costs by diluting engineering polymers with commodity polymers. In the 1980s, blending was used to improve the processability of high-temperature specialty resins. In the present day, blending is performed to achieve targeted properties for specific applications. For example, the material used for automotive body panels should be easy to mold into precise dimensions, maintain its shape up to 85 °C, be impact-resistant at temperatures as low as −40 °C, and exhibit resistance to gasoline, motor oil, and soap solutions. Additionally, it should be paintable, cost-effective, recyclable, and more. It is not feasible to meet all of these requirements with a single polymer. The solution lies in combining the characteristics of several polymers by preparing a polymer blend [3].

Polymer blends fall into two broad categories: miscible and immiscible. A miscible polymer blend is homogeneous at the molecular level and is associated with a negative value of free energy of mixing ($\Delta G_m \approx \Delta H_m \leq 0$) and a positive value of the second derivative ($\partial^2 G_m / \partial \emptyset^2 > 0$). An immiscible polymer blend has a positive free energy of mixing ($\Delta G_m \approx \Delta H_m > 0$) [2].

Miscible blends typically exhibit properties that average those of the blend components. The miscible blend of polyphenylene oxide (PPO) with polystyrene (PS) demonstrates a lower processing temperature and cost than PPO, as well as a higher heat distortion temperature than PS. In contrast, an immiscible blend can display superior properties compared to either component. A common example is the enhancement of polymer toughness achieved by blending with 10–20% of an elastomer [4].

Mixing two polymers typically results in an immiscible system characterized by a coarse, unstable morphology and poor adhesion between the phases. Such blends contain large-sized domains of the dispersed phase and exhibit inadequate adhesion, leading to inferior and unreproducible performance, with the unreproducibility arising from morphological instability. The performance of immiscible blends can improve through compatibilization. Compatibilization refers to the modification of interfacial properties in immiscible polymer blends, resulting in a reduction of the interfacial tension coefficient and stabilization of the desired morphology, thus creating a polymer alloy. A polymer alloy is an immiscible, compatibilized polymer blend with a modified interface and morphology. The compatibilization process has three objectives: (i) adjusting the interfacial tension to achieve the desired degree of dispersion; (ii) ensuring that the morphology generated during the alloying stage will yield the optimum structure during the forming stage; and (iii) enhancing the adhesion between the phases in the solid state, thereby facilitating stress transfer and improving performance [5].

Compatibilization is typically achieved by (i) adding a small amount of a co-solvent that is miscible in both phases; (ii) adding a copolymer, with one part miscible in one phase and another part in the other phase; (iii) reactive compounding to modify at least one polymer to develop regions of local miscibility; and (iv) adding a large amount of a core–shell copolymer that serves as both an impact modifier and a compatibilizer [2]. These strategies result in diverse alloys with varying properties. For instance, the addition of a small quantity of block copolymer predominantly influences the interfacial tension coefficient and, consequently, the size of dispersion, though it typically has minimal effect on the shear sensitivity of the blend morphology. Research on compatibilization has primarily focused on di-block or tri-block copolymers. Adding a substantial amount of a core–shell copolymer has proven particularly effective in blends of two brittle, immiscible polymers that also require impact modification. The third strategy, reactive compatibilization, has been shown to generate a thick interphase, demonstrating excellent stability under high stress and strain, such as those produced during injection molding [6].

The properties of an immiscible polymer blend are determined by its morphology. A blend with droplet/matrix morphology will significantly improve impact properties compared to the matrix polymer. In contrast, a blend with fibrillar morphology will demonstrate superior unidirectional tensile properties, while one with sheet-like inclusions (lamellar morphology) will possess exceptional barrier properties [7–9]. In a co-continuous morphology, both components can fully contribute to the properties of the blend [9]. The type of morphology that develops depends on the nature of the polymers (interfacial tension, viscosities, and viscosity ratios), their volume fraction, and the processing conditions employed. The volume fraction of the minor phase corresponding to the onset of continuity of one phase within the other (percolation threshold) depends on the dispersed particles' shape [10]. This can be observed above 16 vol% of the minor component in droplet/matrix morphology. In the case of fibrillar morphology, this value may be much lower. However, the percolation threshold marks the onset of continuity of the minor phase. Not all material of the minor phase is co-continuous, then. As the volume fraction of the minor phase increases, more material of this component becomes part of the percolating structure, i.e., co-continuous, until a stage is reached when a fully co-continuous structure is formed. Willemse et al. [7] have shown that the volume fraction that marks the formation of a fully co-continuous structure is a function of matrix viscosity, interfacial tension, and the shear rate during blending. Most blends possess a droplet/matrix structure [11].

2.2.1 *Interphase and Compatibilization*

2.2.1.1 Interphase

The interface/interphase is of paramount importance in polymer blend technology. Compatibilization may be considered as the modification of the interface. Helf and Tagami [12, 13] reported that the density profile across the interface exhibits an exponential decay curve (Fig. 2.1). The intercepts of its steepest tangential with the horizontal lines (marking the volume fraction of either of the two polymeric ingredients, $f = 0$ and 1) represent the thickness of the interphase, Δl. Experimentally determined Δl values are presented in Table 2.1 and have been found to lie in the range of 2–60 nm; the lower limit has been observed for 'antagonistically' immiscible polymer pairs, while the upper limit is noted for reactively compatibilized polymer alloys. The addition of block copolymers has been observed to slightly increase the interfacial thickness to 6–8 nm. [14, 15]. The volume fraction of the interface will be larger than that of the dispersed phase when the interface thickness $\Delta l > 0.13D$ (D is the droplet diameter).

Helf and Tagami lattice theory is based on the mean-field approach: (i) the two homopolymers are assumed to have the same degree of polymerization; (ii) the complex set of equations obtained for the segmental density profile ρ_i, where $i = A$ or B, is solved for entangled macromolecules, $M_w \rightarrow \infty$; (iii) the isothermal compressibility is assumed to be negligible; and (iv) there is no change in volume due to blending. The predictions of the theory are: (i) the product, $\Delta l_\infty \nu_\infty$ is independent of χ_{ij}, where Δl_∞ and ν_∞ are the values of interfacial thickness and the interfacial coefficient, respectively, for infinite molecular weight blends and χ_{ij} is the thermodynamic binary interaction parameter; (ii) the surface free energy is proportional to $\chi_{ij}^{1/2}$; (iii) the chain-ends of both polymers are concentrated at the interface; and (iv) low molecular weight third components are repelled to the interface. Theoretically, the interfacial tension coefficient, ν_{ij}, can be calculated from the molecular structure of two polymers using Hansen's solubility parameters [16–18].

According to Leibler [19], for the most efficient lowering of the interfacial tension coefficient between two immiscible homopolymers, A and B, the two blocks of an A-B block copolymer must be of equal length. When the interface becomes saturated with the copolymer, the interfacial tension coefficient ν_{ij} reaches its lower plateau, $\nu_{ij} = \nu_{CMC}$, and the copolymer macromolecules start forming micelles, i.e., the critical micelle concentration, CMC, is reached, $\varphi = \varphi_{CMC}$.

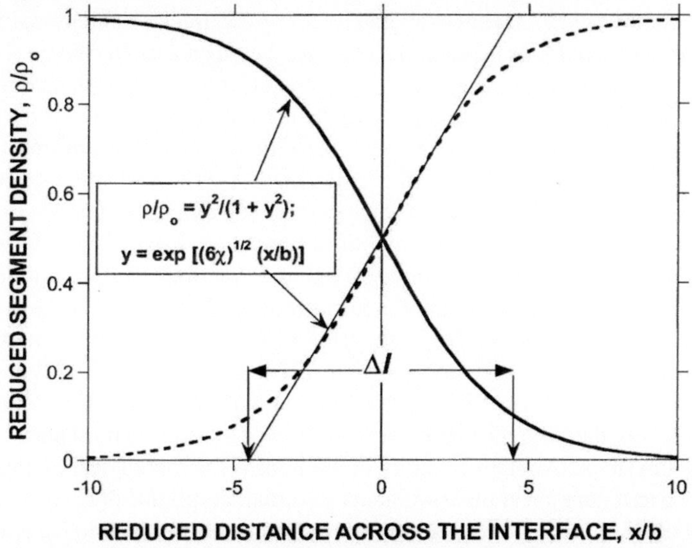

Fig. 2.1 Density profile across the interface; definition of the interphase thickness. Reproduced with permission [5]. Copyright 2002, Wiley Periodicals LLC

Table 2.1 Interphase thickness

Type of blend	Thickness (nm)
Immiscible	2
Block copolymer	4–6
Polymer/copolymer	30
Reactive compatibilization	30–60
Radius of gyration, $\langle R_g^2 \rangle^{1/2}$	5–30

Reproduced with permission [5]. Copyright 2002, Wiley Periodicals LLC

Assuming that the compatibilizer is only present at the interface (i.e., there is no compatibilizer dissolution or micellization), the amount of compatibilizer needed for saturation (w_{cr}) is given by equation (2.1):

$$w_{cr} = \frac{K\varphi}{D} \tag{2.1}$$

Factor K depends on whether the compatibilizer molecules only intersect the interface or coil on the surface of the dispersed phase once. The first case produces proportionality between compatibilizer molecular weight (M_W) and w_{cr}, whereas the second eliminates such proportionality. The reality lies somewhere in between these two ideal cases.

Various methods [20, 21] are available to determine interfacial tension. The principle behind these methods is based on a balance between a driving force (Brownian forces, gravitational and inertia forces, or viscous forces) and resistance because of the interfacial force that tends to minimize the area of contact between the phases.

The various methods can broadly be classified into equilibrium and transient. The equilibrium methods include the "pendant drop" [22–24], the "sessile drop" [25], and the "spinning drop" [26, 27] techniques. The dynamic methods include the "breaking thread" [28–30], the "retraction of deformed drop" [17], the "imbedded fiber retraction" [31, 32] and the "imbedded disk retraction" [33] techniques.

In equilibrium methods, the interfacial tension is determined in the final state, corresponding to the equilibrium state. In contrast, in the dynamic methods, the liquid–liquid interfacial shape varies with time. The interfacial tension is determined from the time evolution of the interfacial profile with the help of a model for the thread distortion, the fibre retraction, or the return of a distorted fluid particle toward its equilibrium shape. Another category of methods is the rheological method, which is based on oscillatory shear measurements at a small amplitude of deformation [34].

Xing et al. [34] conducted studies on polystyrene (PS)/polyamide (PA-6) blends and presented a critical comparison of five techniques: (i) breaking thread; (ii) embedded fibre retraction; (iii) retraction of deformed drop; (iv) pendant drop; and (v) rheological method. All experiments were conducted at the same temperature. Systems with compatibilizers were not studied because of the complexity of such systems and the difficulty in identifying the location of the compatibilizer. For each method, the principle, the basic equations, and the experimental results were critically evaluated in terms of advantages, limitations, and error sources, as presented in Table 2.2.

The experimental techniques used for the determination of the interface thickness Δl include ellipsometry [15, 35], elastic recoil detection [36, 37], dynamic secondary ion mass spectrometry [38, 39], electron microscopy [15, 40], nuclear reaction analysis [41, 42], small-angle X-ray scattering [43, 44], neutron reflectometry [35, 45, 46], small-angle neutron scattering [47, 48], and X-ray reflectometry [46, 49]. Since the interface thickness for immiscible polymers is typically lower than 50 nm, the resolution of techniques should be suitable for determining interface thickness and profile [39, 50]. In addition to resolution requirements, adequate contrast must be present between the blend components to study the interface. This contrast can be created in neutron reflectometry (NR) by deuterating one component. Hence, compared to X-ray reflectometry, whose use is restricted because of the small contrast between most polymers, NR has a much broader scope of applications because of deuteration and can be used in multicomponent polymer blends, in which one polymer and its interfaces with the other polymers can be made visible for neutrons. The range of applications may be limited by the sample geometry required [46, 51]. Deuteration can also generate contrast in elastic recoil detection (ERD). However, ERD offers a resolution no better than 10 nm. The narrow interface thicknesses between immiscible polymers are challenging to resolve accurately using ERD. NR is possibly the most versatile technique for studies on very thin interfaces between blend components and gives the most accurate results [51].

Table 2.2 Summary of the different experimental methods

Method	Advantage	Limitation	Difficulty
Breaking thread method (BTM)	Short measurement time	Polymer matrix must be transparent	Straight thread with a uniform diameter and well-immersed in matrix melt.
	Does not require the density values	Viscosity $<10^5$ Pa s for both polymers	
	Requires simple tools	Does not take into account viscoelasticity	Requires clean samples without impurities or bubbles
		Not valid for polymers with close softening points	
Imbedded fiber retraction method (IFRM)	Short test time	Polymer matrix must be transparent	Straight thread with a uniform diameter and well immersed in matrix melt.
	Valid for high-viscosity systems	Empirical equation	
	Does not require the density values	Viscosity ratio must be between 10^{-3} and 10^3	Requires clean samples without impurities nor bubbles
	Requires simple tools	Not valid for polymers with close softening points	
Retraction of deformed drop method (DDRM)	Short test time	Polymer matrix must be transparent	Well-deformed drop with symmetrical section
	Repeated measurements on the same sample	Viscosity ratio should be lower than 4	Requires clean samples without impurities nor bubles
	Easy to prepare samples	Does not take into account viscoelasticity	
	Requires simple tools		
Pendant drop method (PDM)	Accurate interfacial tension value	Long experimental time	Difficult to obtain accurate density values of both polymer melts
	Valid for Newtonian and viscoelastic fluids	Polymer matrix must be transparent and possess the lowest density	Possible thermal degradation for highly viscous polymers

(continued)

Table 2.2 (continued)

Method	Advantage	Limitation	Difficulty
Rheological method (RM)	Valid for viscoelastic system	Limited by the torque and frequency accessible with the rheometer	Difficult to obtain accurate diameter of the dispersed drops
	Valid for high-viscosity system	Limited to small deformation flow (linear viscoelasticity)	Many step experiments are required
	Results are obtained on several droplets rather than a unique isolated droplet	Limited to the spherical shape of the dispersed particles	

2.2.1.2 Compatibilization by Addition

In the past, the most widely used compatibilization technique was by adding a third component, though nowadays this method is not of much commercial importance. Usually, the third component was either a block or graft copolymer. Since the primary requirement is miscibility, the copolymer need not have identical chain segments as those of the blend components. It is sufficient if the copolymer has segments having specific interactions with the blend components, such as hydrogen bonding, dipole–dipole, dipole-ionic, and Lewis acid–base. According to the results of theoretical calculations, an increase in the molecular weight of a compatibilizer increases its efficiency. However, thermodynamics requires that the added compatibilizer concentrates at the interface and is also soluble in both phases, where it may form micelles. It is necessary to design the compatibilizer such that it migrates to the interface, thereby widening the segmental concentration profile. The addition of a block or graft copolymer lowers the interfacial tension and modifies the molecular structure at the interface. Hence, compatibilization by addition alters not only the interfacial properties but may also influence the flow behavior and performance.

One of the limitations of the addition method is the tendency of the compatibilizer to form micelles. Micelle formation lowers the efficiency of the compatibilizer, enhances the blend viscosity, and may lower the mechanical properties of the blend. Due to these reasons, the copolymer molecules should be designed in such a way to: (i) maximize the miscibility of the appropriate segments of the molecules with the specific blend component; (ii) minimize its molecular weight to nearly about the entanglement molecular weight for every interacting block; and (iii) minimize the concentration of the copolymer in the blend. The addition of 0.5–2 wt% of a well-designed, tapered block copolymer has been found to satisfy the requirements [52].

While lowering the interfacial tension is relatively easy, the other two objectives of compatibilization (stabilization of blend morphology and enhancement of interfacial adhesion in the solid state) are seldom simultaneously accomplished. Often, a combination of compatibilizers or various other strategies, such as crosslinking of one of the three phases, may yield the desired results [11]. Morphology stabilization may also be achieved by the addition of a small amount of a third polymer to a binary blend, such that the added polymer has limited miscibility with either component of the binary blend. The third objective of compatibilization, the enhancement of adhesion in the solid state, can be achieved either by ensuring that an adequate concentration of covalent bonds crosses the interface by introducing a compatibilizer that can act as an adhesive between two polymers, and/or by controlling the blend morphology, especially by achieving phase co-continuity [5].

2.2.1.3 Reactive Compatibilization

Reactive compatibilization is the most widely used method of compatibilization at present. It is based on a chemical reaction between two polymers during mechanical blending at a high temperature. The chemical reaction takes place at the interface, and thus, the interfacial agent is produced in situ. Reactive processing usually results in the formation of block or graft copolymers. Since the compatibilizing copolymer is at the interface, where it should stay, the highest molecular weight copolymer is the most desirable, in contrast with copolymerization by addition. Reactive compatibilization not only facilitates the production of superior performance but is also economically viable. The process creates a thick interphase that stabilizes the morphology during subsequent processing, such as injection molding, though it increases the melt viscosity.

Reactive processing is more complex than compatibilization by addition and is based on the principles of polymer chemistry and polymer processing (melting, compounding, extruding, and forming). To successfully design a commercial process involving reactive compatibilization, one must know chemical kinetics and the reactants' and products' rheological and thermal properties. The criteria for successful reactive processing include: (i) sufficient dispersive and distributive mixing for achieving the necessary renewal of the interface; (ii) presence of a reactive functionality, capable of reaction across the interphase; (iii) adequate reaction rate for the production of enough quantity of the compatibilizer within the residence time in the processing unit; (iv) stability of the produced chemical structures; and (v) stability of the morphology.

The initial step of reactive processing is often modifying an addition polymer, such as grafting it with either maleic anhydride or glycidyl methacrylate. The next step is compounding the product with a condensation polymer, such as a polyamide (PA) or a polyester (PEST). Finally, the resulting copolymer may be compounded into a mixture of the main blend components.

A summary of the five principal chemical processes by which interchain copolymer formation has been accomplished in an extruder, as classified by Brown [53] is presented in Table 2.3. Interchain block copolymer formation by reactive extrusion is of great use. Block copolymers can be preferentially formed in reactions between the end groups of two polymers. Because of the low probability of two end groups reacting within the typical residence time in an extruder, highly reactive functionalities, a catalyst, and/or high concentration of end groups (i.e., low molecular weight polymers) may have to be used. Most polymers modified by reactive processing have nucleophilic end-groups, which readily form covalent bonds with suitable electrophilic functionalities, producing the desired copolymer.

Trans-reaction is a slow process involving ester, amide, or amide-ester groups. It may lead to the destruction of crystallinity and/or the lowering of glass transition temperature (T_g), as observed for polyethylene terephthalate (PET)/polycarbonate (PC) or PET/PA-6.

A reactively compatibilized blend has a thicker interface than a blend compatibilized with an added copolymer. Yukioka and Inoue [14] conducted research on the determination of the interfacial thickness of reactively compatibilized blends by time-resolved ellipsometric method during annealing. The researchers reported that the interfacial thickness increased with time to a plateau, the value of which was found to depend on the temperature and the net concentration of the reactive sites.

Table 2.3 Classification of reactive compatibilization strategies

1	Redistribution or trans-reactions ⇒ Block and random copolymers
1a	Reactive end-groups of polymer-1 (P-1) attack main chain ofpolymer-2 (P-2)
1b	Chain cleavage/recombination involving all polymers
2	Graft copolymer formation ⇒ Graft copolymers
2a	Direct reaction of end-group of P-1 with pendent groups of P-2
2b	Reaction of the end-group of P-1 with the pendent group of P-2 in the presence of a condensing agent
2c	Reaction of the end-group of P-1 with a pendent group of P-2 in the presence of a coupling agent (c)
2d	Reaction of a pendent group of P-1 with main chain of P-2 in a degradative process
3	Block copolymer formation ⇒ Block copolymers
3a	Direct reaction of end-group of P-1 with end-group of P-2
3b	Reaction of the end-group of P-1 with the end-group of P-2 in the presence of a condensing agent
3c	Reaction of end-group of P-1 with end-group of P-2 in the presence of a coupling agent (c)
3d	Reaction of the end-group of P-1 with the main chain of the P-2 in a degradative process
4	Crosslinked copolymer formation ⇒ Crosslinked structures
4a	Direct reaction of pendent functionality of P-1 with pendent functionality of P-2
4b	Reaction between pendent functionalities of P-1 and P-2 in the presence of a condensing agent
4c	Main chain of P-1 reacts with main chain of P-2 in the presence of a radical initiator
4d	Reaction between pendent functionalities of P-1 and P-2 in presence of a coupling agent (c)
5	Ionic bond formation ⇒ Block, graft, or crosslinked structures
5a	Ion-ion association mediated by metal cations as linking agents (c)
5b	Ion-neutral donor group association mediated by metal cations
5c	Ion-ion association mediated by interchain protonation of a basic polymer by an acidic polymer

Interfacial thickness was high ($\Delta l = 10$–60 nm), more than the radius of gyration of the copolymer.

Reactive compatibilization is usually carried out in a twin-screw extruder (TSE). The process can widely vary depending on the specifications of the processing equipment (such as the length of the barrel and the screw configuration), the processing conditions (such as the temperature and the screw speed, the reactants, and the blend components. Reactive compatibilization requires a proper selection of ingredients and appropriate processing conditions. Also important is the sequence of the addition of the ingredients [5].

2.2.2 *Blend Morphology Development*

2.2.2.1 Fundamentals of Morphology Development

Most immiscible polymer blends are manufactured by melt processing in extruders. The melting and mixing process can result in various morphologies, from dispersed droplets to fibres, lamellae, and co-continuous structures, as presented in Fig. 2.2. Controlling these morphologies is crucial for achieving the desired blend performance.

During melt processing, high stresses are exerted on the softening polymer pellets or powder in batch mixers and the kneading elements of the screws in twin screw extruders. These stresses shear off the polymer from the pellet surface, forming sheets of the dispersed phase. Scott and Macosko [54, 55] demonstrated the formation of sheets by quenching samples from a batch mixer and dissolving the matrix polymer selectively with a suitable solvent. According to the researchers, the sheets or ribbons of the dispersed phase become unstable because of the influence of flow and interfacial tension. Holes are formed in the ribbons, which exhibit growth in size and concentration until a fragile lace structure is formed. The lace structure breaks into irregularly shaped particles, subsequently broken into nearly spherical particles. This mechanism rapidly forms small, dispersed phase particles, almost the same size as those obtained at long mixing times. Prolonged mixing primarily lowers the size of the largest particles in the size distribution.

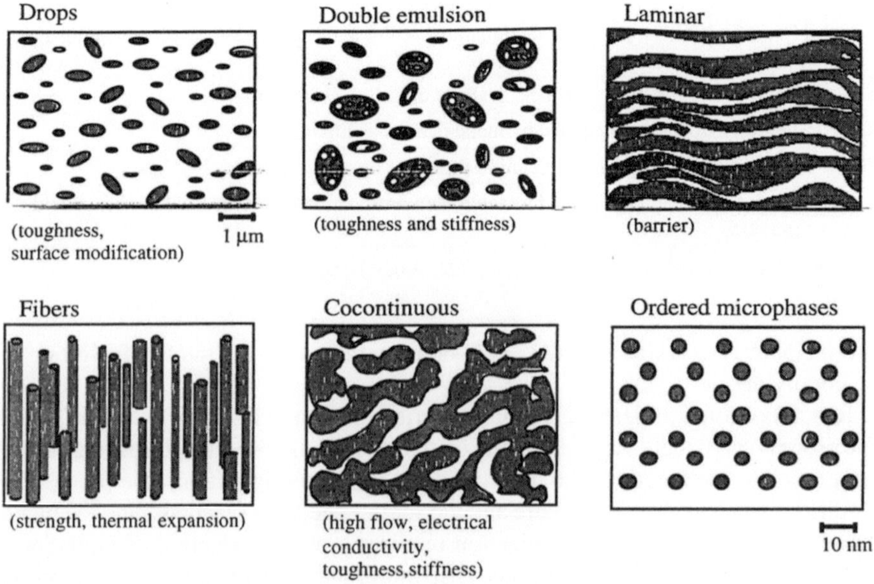

Fig. 2.2 Schematic of useful morphologies that can be produced by polymer–polymer melt blending. Reproduced with permission [4]. Copyright 2000, Wiley Periodicals LLC

Lindt and Ghosh [56] reported that the sheeting mechanism occurs during melt blending in a single screw extruder. Sundararaj et al. [57] reported that the exact mechanism is followed in twin screw extruders, as well as in small cup and rotor mixers. According to the researchers, the dispersed phase is stretched into sheets and ribbons. The sheets and ribbons break into cylinders, from which droplets are subsequently created via Rayleigh-type instabilities. Droplet size distributions can be accurately predicted if the size of cylinders formed in the high shear fields of the mixer is known. Quenching time has a profound influence on blend morphology. Hence, blends should be quenched extremely rapidly (well within a minute) for the proper evaluation of mixing. Otherwise, the morphology development during the quenching time should be considered, which may be irrelevant to the mixing process. The shear stress in miniature mixers is uniform, unlike other mixers, which generate varying stress levels. A high-stress level followed by a lower stress level is essential for efficient mixing in polymer blending. The dispersed phase is stretched into extended shapes at high stress levels, which experience instabilities and break up upon entering the low-stress level. In the miniature mixer, the dispersed phase experiences only one stress level; thus, highly extended shapes remain at the end of mixing. Almost identical dispersions are achieved in twin screw extruders and internal mixers at matched maximum shear rates. For comparable shear rates in the miniature mixer, the final dispersion of reactive blends is like that achieved in other mixers. However, the miniature mixer performs poorly in dispersing high viscosity uncompatibilized blends, and the mixing conditions must be modified to achieve efficient mixing. Sundararaj et al. [58, 59] further showed that sheets are formed when large molten droplets are sheared at sufficiently high shear rates. The researchers reported the creation of micrometer-thick sheets from millimeter-sized pellets by shearing in the parallel disc geometry. Extensional flow was found not to be required for the creation of sheets. Pellets were observed to break in three ways: (i) by stretching into cylinders with droplets streaming off the end; (ii) by extending sheets that formed fingers at the edges; and (iii) by stretching into thin sheets that broke by the formation of holes. A map of the different regions of breakup was proposed using the Deborah number and the ratio of the first normal stress difference of the matrix to the restoring stress of the pellet (droplet). The droplet restoring stress was the sum of the surface stress resulting from interfacial tension and the first normal stress difference of the drop. A combination of stress levels was required for the breakup, such that the dispersed phase domains were affinely stretched in a high-stress region and then allowed to relax and break up in a lower-stress region. Interfacial reaction stabilized the dispersed phase domains, and when the reaction formed a viscous, elastic shell around large drops, they were difficult to break up. The researchers, thus, also proposed that the sheets were stable and did not form holes. Figure 2.3 schematically represents the morphology development from sheets to fibres to droplets during the melt processing of polymer blends.

Prediction of morphology and dispersed sizes obtained by blending immiscible polymers is challenging due to the complicated nature of the blending process. The classical approach for predicting the dispersed phase size in immiscible polymer

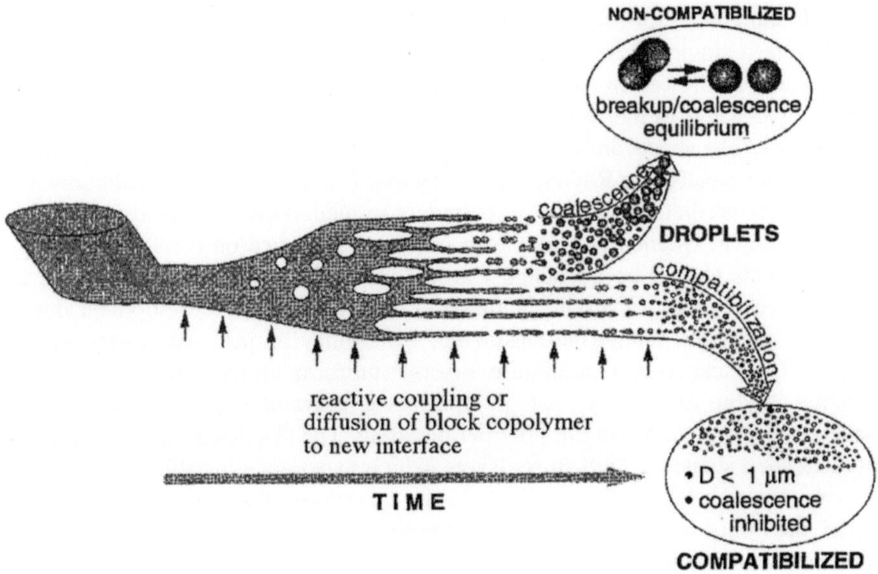

Fig. 2.3 Schematic of morphology development during melt blending. Reproduced with permission [4]. Copyright 2000, Wiley Periodicals LLC

blends is based on the elementary process of break-up of isolated droplets in a flow field of the matrix polymer. This process is governed by the Capillary number (Ca), equation (2.2):

$$Ca = \frac{\tau R}{\sigma} = \frac{\eta_m \dot{\gamma} R}{\sigma} \tag{2.2}$$

where τ is the shear stress, η_m is the viscosity of the matrix phase, $\dot{\gamma}$ is the shear rate, σ is the interfacial tension, and R is the radius of the droplet. Above a critical value, Ca_{crit}, the shear stress predominates over the interfacial stress, and the droplet is stretched till it breaks. The resulting droplet diameter $B_{\min}$ is given by equation (2.3):

$$B_{\min} = \frac{2Ca_{crit}\sigma}{\eta_m \dot{\gamma}} \tag{2.3}$$

Grace [60] demonstrated that for the break-up of Newtonian droplets in stationary flow, Ca_{crit} depends on the flow type and the viscosity ratio λ (given by equation (2.4)).

$$\lambda = \frac{\eta_d}{\eta_m} \tag{2.4}$$

where η_d is the viscosity of the dispersed phase.

Grace's observations are applicable only to isolated droplets. But during the melt processing of polymer blends, a large number of droplets are in close proximity to one another. This can cause their coalescence, leading to an increase in the dispersed phase sizes. Furthermore, Grace's values for Ca_{crit} provide us only with the conditions for the stability of a droplet in a stationary flow. However, in polymer processing, the droplets are subjected to non-stationary deformation. At adequately high values of Ca, the droplets will be elongated into long threads, which will ultimately break up into a line of droplets [61, 62]. Due to elongation, the diameter of the stretching droplet decreases and, hence, the local capillary number decreases until the critical value for break-up, $Ca_{crit} = 1$, of the extended thread is reached [62]. The extended thread will then break up according to the Raleigh mechanism, forming a line of droplets. The diameter of these droplets will be approximately two times the thickness of the rupturing thread [63].

The size of the dispersed phase is thus dependent on the total deformation of the original droplet or pellet. The diameter of the thread, B_{thread}, depends on the total amount of imposed strain ($\gamma = \dot{\gamma}t$) and is given by equation (2.5).

$$\frac{B_{thread}}{2R_0} = \left(1 + \gamma^2\right)^{-1/4} \tag{2.5}$$

where R_0 is the radius of the original droplet, assuming affine deformation. This affine deformation is possible in shear only if $Ca/Ca_{crit} > 2$ [64]. In elongational flow, affine deformation occurs only if $Ca/Ca_{crit} > 5$ [62]. However, in most mixing equipment, shear flow predominates because elongation is challenging to sustain for long periods.

Break-up of threads according to the Raleigh mechanism will take place only if the residence time of the thread is greater than the time required for break-up. The break-up time t_b can be determined using Tomotika's theory [25]:

$$t_b = \frac{\eta_m B_{thread}}{\sigma \Omega_m} \ln\left(\frac{0.81 B_{thread}}{2\alpha_0}\right) \tag{2.6}$$

where α_0 is the amplitude of the initial disturbance, usually found to lie in the range of 10^{-7} and 10^{-9} m [65, 66] and Ω_m is a function of the viscosity ratio [67]. However, (equation 2.6) is valid only for Newtonian fluids in quiescent conditions, which is not the case in the processing of polymer blends. Stretching has a stabilizing effect against the break-up of the liquid thread, leading to thinner threads [66]. The prediction of the dispersed phase size using (2.3) and (2.5), if applicable, will become invalid if droplet coalescence occurs, increasing the average size of the dispersed phase. Coalescence occurs only when the concentration of droplets is adequately high and when the droplet diameter does not exceed a critical diameter [62].

The theories discussed above apply only to dispersed droplets and threads and are not valid for the mechanism of their initial formation. A quick morphology change occurs in the initial stages of polymer blend processing. This change involves the formation of striated structures [55, 56, 59, 68–70]. In this direction, Plochocki

[68] proposed an 'abrasion' mechanism for this change in morphology. According to Macosko et al. [55, 59, 70], the proposed 'sheeting' mechanism involves stretching the pellets into ribbons or 'sheets' upon melting. Sundararaj et al. [20] mapped the conditions under which 'sheets' or threads may be created. The researchers used the ratio of normal stresses and the Deborah number to influence the formation of sheets and threads. They demonstrated that sheets can easily be created in shear flow and that threads are created only at low shear rates (<5 s^{-1}). Sheets are not converted into threads because the interfacial forces are far too weak under most conditions to alter the shape of the cross-section of the sheets significantly [59, 61].

In the presence of shear flow, the thickness of the sheets decreases till the initiation of sheet break–up by interfacial disturbances. The sheets' thickness decrease can be calculated as a striation thickness [71]. A sphere with an initial radius of R_0 will be transformed into a ribbon ('sheet') in shear flow. The width of the ribbon will be equal to the radius of the original sphere, but the thickness B_{sheet} will be much smaller, though the volume of the ribbon will be equal to that of the original sphere. B_{sheet} is given by equation (2.7) when $\gamma >> 1$ [63].

$$\frac{B_{sheet}}{R_0} = \frac{2}{3}\left(1 + \gamma^2\right)^{-1/2} \tag{2.7}$$

Figure 2.4 presents the variation of sheet thickness B_{sheet} with total deformation. It can be observed that the lowering of sheet thickness with deformation is much more rapid than that of thread diameter.

This figure shows that the sheet thickness decreases much more rapidly with the deformation than the thread diameter does. At a specific thickness, the break-up of the sheet begins with the formation of holes. This breakup may be caused by the interfacial instabilities arising in the laminar structures during flow [72] or by the impurities in the blend [70]. The holes grow larger due to the interfacial forces, and finally, they coalesce, resulting in a network of filaments. Threads and droplets

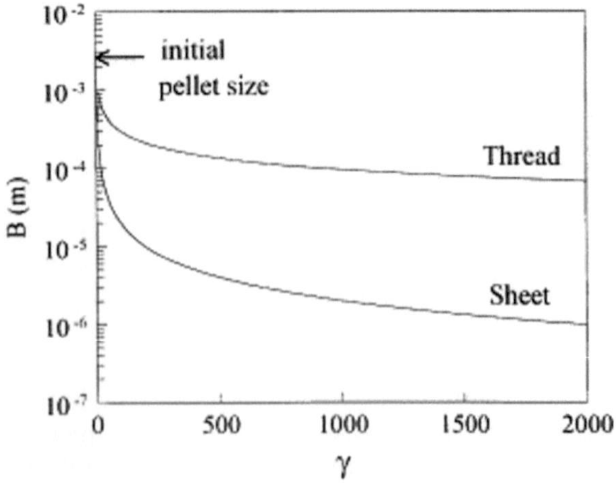

Fig. 2.4 The diameter of the thread and the thickness of the sheet in the case of affine deformation of a pellet of 3 mm thickness into a thread equation (2.5) or a sheet equation (2.7), respectively. Reproduced with permission [63]. Copyright 1999, Elsevier Science Ltd.

can be formed by the break-up of this network [56], depending on the processing conditions. The dimensions of the final dispersed phase structures are of the same order of magnitude as the final thickness of the sheets. Equation (2.7) can thus be used to predict the final size of the dispersed phase particles. Figure 2.4 shows that the typical phase sizes of 0.1–1 mm found after blend processing are indeed predicted by equation (2.7). Thus, a 'sheeting' mechanism can explain the rapid change in phase dimensions found in various polymer blend processing techniques [55, 56, 59, 68–70]. Shear flow is therefore an effective way of achieving dispersion, because striation thickness is lowered by breaking the sheets into dispersed structures. Reduction in phase size by the deformation of threads, followed by their breakup into droplets, has much lower effectiveness.

Willemse et al. [63] in their studies on processing of polyethylene(PE)/PS blends, observed the occurrence of the "sheeting mechanism", in single screw extrusion as well as in static mixing. A rapid reduction in phase dimensions was observed, as predicted theoretically. After a breakup, no further significant lowering of phase dimensions was found to occur. The type of morphology was found to depend on the stability of the fibres formed after the break-up of sheets. Fibrillar morphology was observed when the capillary number was higher than one. At lower values of the capillary number, droplet matrix morphology was formed. The final phase sizes were found to be independent of the capillary number.

When the concentration of the dispersed phase is greater than 1%, collision between droplets occurs, and droplet size increases due to coalescence. An equilibrium between the coalescence of droplets and the breakup of fibres into droplets determines the final droplet size [57]. Compatibilizers are added to prevent coalescence. As discussed in Sect. 2.2, reactive compatibilization is known to be much more effective than compatibilization by addition. When a pre-made block copolymer is added, a large proportion of it remains in micelles [73, 74]. Relatively high molecular weight blocks will be swollen by the matrix and form a wet brush, providing good adhesion. However, removing the high molecular weight block copolymer from the micelles is difficult. Random copolymers don't form micelles and can readily go to the interface, but are not anchored. They are unable to prevent coalescence, though they may lower the rate of coalescence [75–77]. ABC triblock copolymers do not tend to form micelles and may get anchored at the interface [78, 79].

The limitations associated with compatibilization using pre-made block copolymers are overcome in reactive compatibilization. Coupling reactions between two immiscible polymers can only occur at the interface where the compatibilizer is required. Suppose the reaction is fast, as observed in reactive compatibilization of several blends. In that case, enough interfaces will be covered with the copolymer, as soon as the copolymer is produced, resulting in very effective compatibilization [80–84]. There are coupling reactions, which, despite being slower, are known to be effective [57, 74, 85, 86]. Whatever the type of compatibilizer (pre-made copolymer or reactively formed copolymer), the interface should be rich in the copolymer. Macosko et al. [73], in their studies on compatibilization with pre-made block copolymer, they observed that if the time for the compatibilizer to diffuse to the interface is less than the time required for the formation of a new interface, the interface will be rich in the compatibilizer.

Macosko et al. determined [73] the fractional coverage required to stabilize the blends against static and dynamic coalescence. Studies were carried out on PS/PMMA blends with 30 wt% PMMA content, which was compatible with pre-made symmetric block copolymer poly(s-b-MMA). Blends were compatible with different amounts of the compatibilizer. Studies were also conducted with copolymers having different proportions of the comonomers. The researchers assumed all the copolymer to be located at the interface and calculated the number of block copolymer chains per unit area (Σ) using equation (2.8):

$$\Sigma = \frac{Chains/Volume}{Interface/Volume} = \frac{N_A \rho \varphi}{MS_{sp}} \tag{2.8}$$

where ρ, φ, and M are the density, volume fraction, and molecular weight of the copolymer, respectively, and N_A is Avogadro's number. S_{sp} is the interfacial area per unit volume and is calculated from the mean diameter of the dispersed phase (D), using equation (2.9):

$$D = \frac{6\varphi_p}{S_{sp}} \tag{2.9}$$

where φ_p is the volume fraction of the dispersed phase.

The maximum surface concentration of block copolymer, Σ_0, was determined by assuming a dense monolayer at the interface. Considering symmetric diblock copolymers and ignoring the radius of curvature of the particles, the thickness of the monolayer was taken as Λ/2, Λ being the lamellar spacing. Σ_0 was calculated using equation (2.10):

$$\Sigma_0 = \frac{Thickness\ of\ copolymer\ monolayer}{Volume\ of\ one\ chain} = \frac{\Lambda/2}{M/\rho N_A} \tag{2.10}$$

Using Λ = 39.8 nm, as obtained from literature, Σ_0 values were calculated, and the Σ_0 values were used to determine Σ/Σ_0. The blends were stable after melt annealing, i.e., stable against static coalescence, when Σ/Σ_0 was in the range 0.15–0.20. However, particle size reduction, i.e., preventing dynamic coalescence of the droplets, was observed at much lower fractional coverage of the interface (~0.01). Similar values of Σ/Σ_0 required for prevention of static coalescence were also obtained in blends compatibilized with reactively formed block copolymers [85, 87]. However, in contrast to the behaviour observed in blends compatible with pre-made block copolymers, Majumdar et al. [88] observed that in reactive compatibilization, a higher concentration of copolymer chains per unit area may be needed to stabilise dispersed droplets against dynamic coalescence compared with static coalescence.to quiescent conditions.

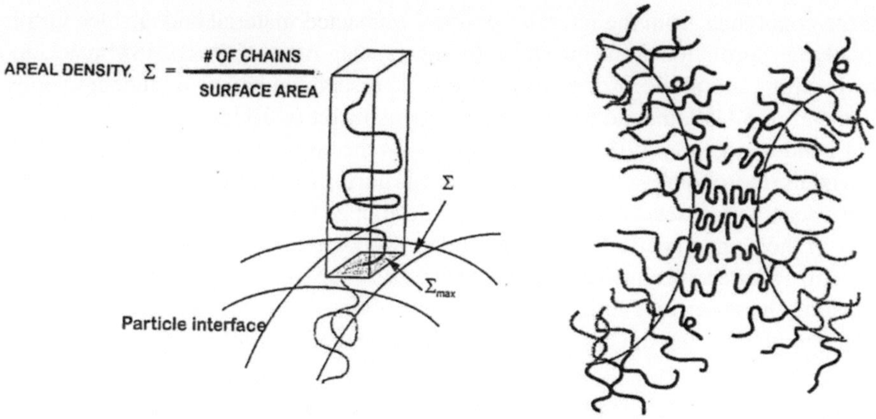

Fig. 2.5 Definition of chain area density, Σ (left), and schematic of two approaching drops. Block copolymer in the interface prevents van der Waals attraction from coalescing the drops. The fraction of surface covered is Σ/Σ_{max}. A minimum coverage is required to avoid the chains from sliding out of the contact region. Reproduced with permission [4]. Copyright 2000, Wiley Periodicals LLC

Figure 2.5 presents the mechanism for preventing coalescence by block copolymers. As the droplets approach each other, the blocks serve as entropy springs, compressing to resist the van der Waals forces of attraction. A minimum interfacial coverage, Σ/Σ_0, is required to prevent the block copolymers from moving out of the contact region.

Bhadane et al. [89] compared morphology development in reactively compatibilised blends with classical interfacially modified blends. Studies were conducted on polyamide (PA)/brominated poly(isobutylene-*co-p*-methylstyrene) (BIMSM) blends. The presence of pendent benzylic bromine in BIMSM facilitated the rapid in situ production of BIMSM-PA graft copolymer. PA/poly(isobutylene-*co-p*-methylstyrene) (IMSM) blends were also compared. The reactively compatibilized blend exhibited several features, which contrasted with those exhibited by classical interfacially modified blends. The reactive system exhibited a 37-fold reduction in volume average diameter compared to the non-reactive one. A high phase size distribution was observed at all compositions, with the fine droplets in sizes 50–80 nm. 46% of the total blend material reacted over a short mixing time. A linear drop in particle size was observed, and a much higher concentration of copolymer was required to reach a plateau value in the case of the reactively compatibilized system. This was much higher than the copolymer needed to saturate the interface. The researchers proposed a mechanism of reactive morphology development, which they named "interfacial erosion". The mechanism considers the production of a very high viscosity graft copolymer at the interface during dynamic mixing, due to the mutual contact of the BIMSM and the PA molecules. The viscosity mismatch between the formed graft copolymer and the other blend components results in the subsequent erosion of the graft copolymer from the interface, during dynamic mixing, resulting in the formation of fine, nanometer-sized micelles in the bulk. The removal of the copolymer from the interface exposes unreacted material and enables further copolymer formation at the interface. In this manner, most of the BIMSM react, and the resulting morphology comprises nanoscale dispersion having a number-average diameter of 50–80 nm and a volume average diameter of 300 nm.

Li and Sundararaj [90] conducted studies on the effect of reactive compatibilization on the morphology development of blends during melt processing in an extruder. The researchers observed that morphology developed faster in compatibilized blends than in their uncompatibilized counterparts. Rheological characterisation revealed that slip occurred in compatibilized blends and was suppressed by reactive compatibilization. The viscosities (η_{blend}) of both compatibilized and uncompatibilized blends were predicted using the log-additive mixing rule, equation (2.11):

$$\log\eta_{blend} = \varphi_1 \log\eta_1 + \varphi_2 \log\eta_2 \tag{2.11}$$

where φ_i and η_i are the volume fraction and viscosity of the ith component, respectively. The viscosities of the uncompatibilized blends showed significant negative deviation from the predicted values. This was attributed to interfacial slip. In contrast, the viscosities of the compatibilized blends were very close to the predicted values, indicating the suppression of interfacial slip by reactive compatibilization. The researchers concluded that compatibilization facilitates and speeds up the break-up process in the initial stages of melt blending and prevents droplet coalescence in the later stages. The facilitation and increased speed of morphology development are due to the reduction of interfacial slip. Chain segments of the block copolymer, formed by in situ reaction, penetrate both the dispersed phase and the matrix, entangled with the bulk polymer chains. This lowers interfacial slip and enhances adhesion between the matrix and the dispersed phase. This facilitates more effective transfer of shear stress from the matrix to the dispersed phase.

Kim et al. [91] conducted studies on the effect of the molecular architecture of in situ reactive compatibilizer on the efficacy of compatibilization. Two types of compatibilizers, poly(ethylene-ran-acrylic acid) (PE-r-AA) and poly(propylene-graft-acrylic acid) (PP-g-AA) were added to polystyrene (PS)/poly(styrene-ran-glycidyl methacrylate) (PS-GMA) system. Melt processing was carried out in an internal mixer. It is well-known that the acrylic groups in both PE-r-AA and (PP-g-AA) react readily with the epoxy groups of PS-GMA, forming PE (or PP)-graft-PS copolymer. The researchers observed that PE-r-AA was a very effective compatibilizer in reducing the domain size of the blend. However, PP-g-AA was found to be ineffective in lowering domain size. This was attributed to an inhomogeneous distribution of AA functional units in PP-g-AA and poly(acrylic acid) (PAA). The superior efficacy of PE-r-AA in lowering the domain size was attributed to the well-distributed functional units of AA across the copolymer chains and the absence of PAA. The presence of PAA in PP-g-AA was confirmed by IR spectroscopy and by the determination of T_g using the loss modulus peak in DMA studies. The inhomogeneous distribution of AA functional groups in PP-g-AA was demonstrated by the phase-separated domains of size 0.04–0.35 μm observed in scanning electron microscopy (SEM) and transmission electron microscopy (TEM) studies. The researchers concluded that unless the functional groups are homogeneously distributed across the entire chain, the better interfacial activity expected by a reduction in domain size due to a graft polymer formed during melt processing will not occur even if an adequate number of functional groups are present.

2.2.2.2 Co-continuous Morphology Development

There are two essential approaches used in defining co-continuous structures. The first approach employs a classical method of describing an ideally co-continuous structure. Co-continuity has been defined as "the co-existence of at least two continuous structures within the same volume" [92]. In a polymer blend, this implies 3D spatial continuity of both polymers on some finite scale of blending, each polymer possessing its internal structure from which its properties arise [92, 93]. Other terms used in literature for describing co-continuous polymeric structures are interpenetrating polymer networks (IPNs) [94], interpenetrating phase structures (IPSs) [95], interpenetrating polymer blends [96], structures with dual-phase continuity [97], and co-phase continuity [98]. In a binary co-continuous blend, the surfaces of the two phases are exact topological replicas. Co-continuous structures can be produced over a specific range of volume fractions. This range can be very narrow or broad and is centred around the phase inversion composition (φ_{PI}), also known as the phase inversion point.

The second approach is based on the percolation threshold theory. According to this approach, melt blending of polymers results in the coexistence of different types of structures and does not create an ideal network. Lyngaae-Jørgensen and Utracki [10, 95, 99, 100] have defined a co-continuous structure as a structure in which at least a portion of each phase forms a coherent continuous structure permeating the whole volume. This definition also permits the existence of some components as discrete domains that are not part of the network structure. The volume fraction corresponding to the change of the system from one containing only discrete domains of one phase to one containing an infinite structure (and discrete domains) is the transition point and is known as the critical volume fraction, φ_{cr}, or percolation threshold, as shown in Fig. 2.6. Significant changes in the blend properties are observed at this point. Further increase in the volume fraction increases the proportion of the minor component becoming a part of the percolation structure. At a certain volume fraction, a fully co-continuous structure is achieved. The continuity index φ_I [95], also known as the degree of co-continuity [10, 99, 100] interpenetration index [10, 99] or continuity fraction [101], is the portion of a component that forms a part of the percolating structure [92].

The first definition can be regarded as a special case of the second one in the situation when the value of the continuity index is 1 for both phases. Most researchers use the term "co-continuous structure," which is generally used more in the context of the first definition, i.e., a complete network structure of both polymers. Sometimes the term "fully co-continuous structure" distinguishes this case from the second definition. The disadvantage of the second definition is that the "continuity index"

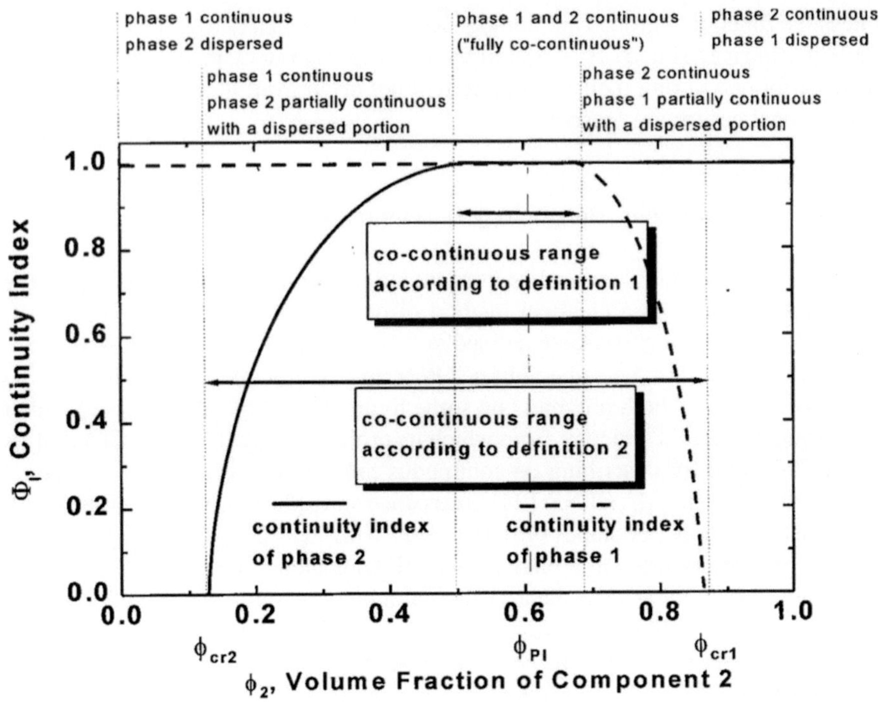

Fig. 2.6 Co-continuous structure composition range using definition 1 and definition 2 using the critical volume fraction φ_{cr}, the continuity index φ_I, and the phase inversion composition φ_{PI}. Reproduced with permission [92]. Copyright 2003, Taylor and Francis

is needed to describe the structure and its influence on properties. However, the advantage of this definition is the connection of the continuity index with properties.

In the presence of shear, the most effective mixing in developing a co-continuous structure can be achieved when the viscosities and the volume fractions of the two blend components are equal. Equal volume fractions maximize the opportunity for connectivity since neither element is present in a minor quantity. However, in most systems, the viscosities of the components are unequal. The phase with lower viscosity tends to be continuous, which results in minimum energy dissipation in the flow field. This tendency can be compensated for if the volume fraction of the component having higher viscosity is increased to the same extent as the difference in viscosities of the components. In this way, connectivity between phases with different viscosities can be maintained. This relation is shown qualitatively in Fig. 2.7. These considerations form the basis for some of the equations proposed for calculating the phase inversion composition.

Various equations have been proposed to determine the phase inversion composition. Equation (2.12) is a simple relationship that describes the phase inversion composition in terms of volume fractions and viscosities [96, 102]:

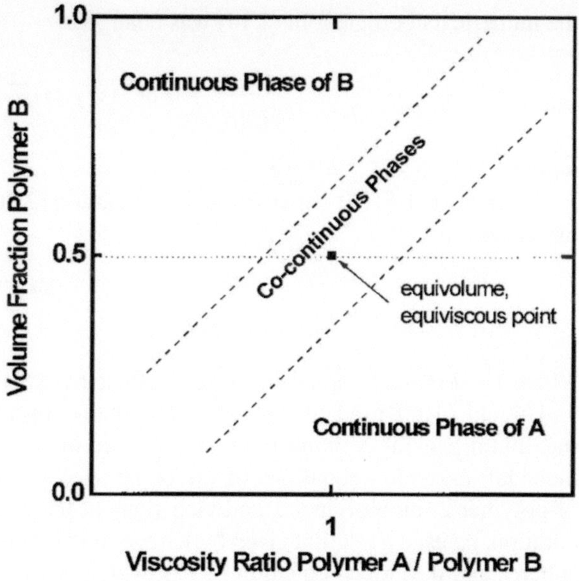

Fig. 2.7 Relation between the volume range of co-continuous structures and the viscosity ratio. Reproduced with permission [92]. Copyright 2003, Taylor and Francis

$$\frac{\varphi_{1,PI}}{\varphi_{2,PI}} \times \frac{\eta_2}{\eta_1} \cong 1 \tag{2.12}$$

where η_i and $\varphi_{i,PI}$ are the viscosity and the volume fraction, respectively, of component i at phase inversion. Since $\lambda = \eta_1/\eta_2$ and $\varphi_{1,PI} = 1 - \varphi_{2,PI}$, equation (2.12) can be written as:

$$\varphi_{2,PI} = \frac{1}{1+\lambda} \tag{2.13}$$

The viscosity ratio should be determined at the shear stress used for the preparation of the blends [100, 103, 104]. Since polymer blends are often processed in a torque rheometer or similar processing equipment, Avgeropoulos et al. [105, 106] used the torque ratio (τ_1/τ_2) instead of the viscosity ratio for convenience.

The basic equation was found to be valid for blend systems having viscosity ratios close to unity [107–110]. However, substantial deviations were observed when the viscosity ratio was far from unity [100, 111, 112]. Thus, many researchers modified the relation to obtain better fits with their experimental results. For example, Ho et al. [113] conducted studies on PP/ethylene propylene rubber (EPR) and PS/styrene-butadiene copolymer rubber (SBR) blends processed in an internal mixer and introduced an exponent and a prefactor, resulting in equation (2.14):

$$\frac{\varphi_{1,PI}}{\varphi_{2,PI}} = 1.22(\tau_1/\tau_2)^{0.29} \tag{2.14}$$

Kitayama et al. [114] reported that a good representation of their results for uncompatibilized PA6/styrene-acrylonitrile copolymer (SAN) blends were obtained when the numerical coefficient in (2.14) was changed:

$$\frac{\varphi_{1,PI}}{\varphi_{2,PI}} = 0.887(\tau_1/\tau_2)^{0.29} \tag{2.15}$$

where 1 = PA6 and 2 = SAN.

Everaert et al. [115] observed that the equation (2.16) described the PP/(PS/PPO) blend system:

$$\frac{\varphi_{1,PI}}{\varphi_{2,PI}} = (\eta_1/\eta_2)^{0.3} \tag{2.16}$$

where 1 = PP and 2 represents different miscible PS/PPO mixture compositions.

Utracki [100] used an approach based on "intrinsic viscosities" [η] and the maximum packing volume fraction φ_{max}. He included the viscosity concentration behavior, assuming equalities of the blend viscosities resulting from the addition of polymer 2 into polymer 1 and vice versa at the phase inversion composition. In addition, he used a relation given by Krieger and Dougherty for monodispersed hard spheres and developed equation (2.17), which can be simplified to equation (2.18) for $0.1 < \lambda < 10$.

$$\lambda = \left[\frac{(\varphi_{max} - \varphi_{2,PI})}{(\varphi_{max} - \varphi_{1,PI})}\right]^{[\eta]\varphi_{max}} \tag{2.17}$$

$$\varphi_{2,PI} = \frac{(1 - \log\lambda/[\eta])}{2} \tag{2.18}$$

where [η] was assumed to be 1.9 for spherical domains and $\varphi_{max} = 0.84$. By using other equations for the variable [η], interfacial viscosity or interfacial tension effects can also be taken into consideration [100]. The validity of the equation was compared with literature data for different mixing conditions and viscosity ratio calculation methods. The results agreed with the data much better than (5.14), especially when the viscosity ratios were far from unity [100, 107, 112].

Metelkin and Blekht [116] developed a model using the theory of Tomotika [67] according to which, a liquid thread surrounded by another liquid will break up because of interfacial tension-driven instabilities. The researchers proposed a relation between phase composition, viscosity ratio, and a function of viscosity ratio [F(λ)], given by equation (2.19):

$$\varphi_{2,PI} = [1 + F(\lambda) \times \lambda]^{-1} \tag{2.19}$$

Utracki [103] demonstrated that the experimental F(λ) is given by equation (2.20):

$$F(\lambda) = 1 + 2.25\log(\lambda) + 1.81[\log(\lambda)]^2 \tag{2.20}$$

Using (2.20) for F(λ), researchers [107, 112] observed that the model proposed by Metelkin and Blekht overestimates the phase inversion composition, especially when the viscosity ratio is higher than unity.

Steinmann et al. [117] developed a model based on the assumption that the shape relaxation times of domains of components 1 and 2 meet in a maximum at the phase inversion composition. The researchers proposede equation (2.21) to predict the phase inversion composition:

$$\varphi_{2,PI} = \frac{1}{\left(\lambda^{1/z} + 1\right)} \tag{2.21}$$

where the parameter Z depends on the blend system and is a characteristic property of both the specific interfacial area and the relaxation rate of the structure. Z must be determined from experimentally determined phase inversion composition and viscosity ratio data for each system under investigation. This model cannot be used for predictions as the value of Z has to be determined experimentally.

The models discussed so far do not include the effects of melt elasticity or interfacial tension and often fail to predict the phase inversion composition accurately. Some researchers [101, 107, 118] have attributed the inaccuracy of the prediction to the effects of melt elasticity. According to Bourry and Favis [119] elasticity is an essential factor influencing phase inversion. They expressed phase inversion composition in terms of the elasticity ratio of the blend components, using storage modulus G′(ω) or tan (δ) to represent the elasticity of the components:

$$\frac{\varphi_{1,PI}}{\varphi_{2,PI}} = \frac{G_2'(\omega)}{G_1'(\omega)} \tag{2.22}$$

$$\frac{\varphi_{1,PI}}{\varphi_{2,PI}} = \frac{\tan\delta_2}{\tan\delta_1} \tag{2.23}$$

G′(ω) and G″(ω) were measured at the frequency corresponding to the shear rate of the mixing process, based on the assumption that the Cox-Merz rule is applicable. Using equations (2.22) and (2.23), the researchers achieved good agreement with experimental results for HDPE/PS blends melt processed in a twin-screw extruder. An interpretation of experimental results based on the viscosity ratio alone failed.

Interfacial tension is the driving force behind the change of a co-continuous structure into a dispersed one. Low interfacial tension facilitates the formation of co-continuous structures and lowers the rate of breakup. Willemse et al. developed a model that includes interfacial tension [7]. The model is based on material properties and processing conditions. It combines co-continuity formation requirements related to the shape of the minor phase and micro-rheological conditions for the stability of extended structures. This model provides a range of compositions and describes an equilibrium structure during flow. The researchers assumed that morphology development begins with extended structures produced during the breakup of sheets, according to the mechanism proposed by Scott and Macosko [54, 55]. To simplify the mathematical description, the minor phase is considered a dense packing of randomly aligned rod-like particles touching each other. Continuity is achieved when the packing density is maximum. The maximum packing density P_{max} is dependent on the aspect ratio L/B, where L is the length and B is the diameter of the thread, and is given by equation (2.24):

$$\frac{1}{P_{max}} = 1.38 + 0.0376(L/B)^{1.4} \tag{2.24}$$

The material characteristics and processing conditions influence L/B, including the effects of interfacial tension, viscosities, and shear rate. Equation (2.25) is applicable to a droplet having radius R_0 that has been deformed into a cylinder L and diameter B.

$$B = 2R_0(2/3)^{1/3} + (L/B)^{-1/3} \tag{2.25}$$

Such elongated structures exhibit stability when $C_a > 1$, where C_a is given by equation (2.26).

$$Ca = \frac{\eta_m \dot{\gamma} R_0}{\sigma} \tag{2.26}$$

where σ = interfacial tension, η_m = matrix viscosity and $\dot{\gamma}$ = shear rate.

Thus, $C_a = 1$ has been used as the criterion for stability. P_{max} is the minimum volume fraction $\varphi_{d,cc}$ of the minor component at which co-continuity exists for a given value of L/B. Equation (2.27) is obtained by combining equation (2.24)–(2.26) for $C_a = 1$:

$$\frac{1}{\varphi_{d,cc}} = 1.38 + 0.0213\left(\frac{\eta_m \dot{\gamma} R_o}{\sigma}\right)^{4.2} \tag{2.27}$$

Equation (2.27) gives the lower limit of co-continuity $\varphi_{d,cc}$ in terms of the interfacial tension, the major phase viscosity η_m, the shear rate $\dot{\gamma}$ and the equivalent sphere diameter R_0. When the volume fraction of the minor phase is higher than $\varphi_{d,cc}$, the minor phase becomes continuous. The upper limit is obtained using the same equation, but the values of η_m and R_0 are usually different for the other component.

The model was valid for various blends (PE/PS, PE/PA6, PE/PP) produced in a laboratory extruder at an average shear rate of 22 s^{-1}. For each system, R_0 was determined after blend processing and was found to be in the range of 0.25–0.7 mm [7, 120]. However, a limitation of the model is that R_0 has to be determined afterwards. However, the model provides a better explanation of the effect of processing conditions on structure development than that provided by the dependence of the viscosity ratio on shear stress [92].

Lyngaae-Jørgensen and Utracki [10, 95] described the co-continuous composition range using equations based on the percolation theory. According to the researchers, the continuity index φ_I is dependent on the volume fraction at the onset of co-continuity φ_{cr}, and an exponent of 0.45 reflects its rise beyond the onset composition, as depicted in equation (2.28):

$$\varphi_I = k(\varphi - \varphi_{cr})^{0.45} \tag{2.28}$$

where k is an empirical coefficient. This equation is analogous to the more general relationship from percolation theory, describing any property of the material, and having the form:

$$Property \propto (\varphi - \varphi_{cr})^x \tag{2.29}$$

where x is a universal constant that depends on the property. This relationship is valid for several properties of various polymer blends [121–125].

Lyngaae-Jørgensen and Utracki [95] assumed that the φ_{cr} for binary polymer blends can be calculated from the shape of the dispersed phase particles [27]. Furthermore, the model assumes complete immiscibility of the blend components and the ability of the phases to coalesce. For completely immiscible blends with monodispersed spherical particles, the theoretical value of φ_{cr} is 15.6 vol%, which agreed with the experimental results [10]. An increase in φ_{cr} was predicted for spherical domains with a broad domain size distribution, which agreed with experimental results. The researchers also considered φ_{cr} values for non-spherical domains, assuming simplified typical polymer blend morphologies. The prediction of the model was found to be in reasonable agreement with experimental results for PS/poly(methyl methacrylate) (PMMA) blends having low contents of PS, which exhibited coexistence of ellipsoids and droplets at higher shear stresses encountered during processing [92].

The stability of co-continuous structures is influenced by some of the same factors that govern their formation, such as interfacial tension, viscosities of the phases, and thread dimensions. Studies by various researchers [97, 126, 127] indicate that there is a critical volume concentration $\varphi_{cc,stable}$ above which the co-continuous structures are stable during melt annealing. Below $\varphi_{cc,stable}$, co-continuous structures are converted to dispersed structures during annealing. Willemse et al. [127] established a relationship between $\varphi_{cc,stable}$, and the length of the threads (ligaments) between their crossover points. If the length of a thread is shorter than the dominant wavelength of a sinusoidal disturbance, then the thread will not break, and the co-continuous structure will be inherently stable in the calm melt. Lower interfacial tension benefits the development and stability of co-continuity. Reduced coalescence hinders the formation of co-continuity structures. However, once formed, it enhances the structure's stability. The increase in interfacial viscosity prevents drop or thread deformation and also prevents breakup [128].

In the co-continuous composition range, the stability of co-continuous structures can be significantly increased through the selective filling of polymer blends. This method allows for a notable retardation of the breakup and coalescence processes, and the increase in melt viscosity of the filled phase contributes to this [101, 129].

Willemse et al. [127] reported the stability of co-continuous morphology after injection molding, even at compositions below the critical fraction. For the same composition, co-continuity was lost after compression molding. However, this effect was found to depend on the local capillary number, which should be more critical than the critical capillary number [130]. During injection molding, it was found that the residence time in the reservoir was the primary factor influencing undesirable coarsening.

Compatibilizers lower interfacial tension and coalescence at the interface. However, they also raise interfacial viscosity, applied shear stress, and interfacial adhesion. Moreover, the interaction of these effects differs between non-reactive and reactive compatibilizers. Additionally, when reactive compatibilizers preferentially react with one of the components, the melt viscosity of that phase increases, resulting in a shift in the interplay of viscosities. These factors work both cooperatively and counteractively, complicating the analysis of the effects of compatibilization on the development of co-continuous structures [92].

Blends of block copolymer-type thermoplastic elastomers (TPEs) with a thermoplastic or another TPE frequently demonstrate co-continuity. Figure 2.8 shows a TEM micrograph of a co-continuous PS/SEBS blend containing 50 wt% PS content. The TPEs consist of hard segments and soft segments, where the moldable soft segments contribute to the elastomeric properties, while the hard segments facilitate physical crosslinking effects. The presence of physical crosslinks in the melt of block copolymers leads to highly elastic properties and a notable yield stress in the microphase-separated state, which stabilizes elongated structures. Thread breakup and retraction are significantly minimized or even prevented in blends of block copolymer TPEs with other polymers. As a result, these blends demonstrate remarkable stability over a wide range of processing conditions [9, 93, 130, 132–135], in contrast to other co-continuous blends.

2.2.3 *Effect of Processing Conditions on Blend Morphology*

Since the development of polymer blend morphology relies on the shear rate, mixing time, and viscosity ratio of the components, processing conditions significantly affect these factors and therefore have a profound impact on blend morphology. Various research groups have investigated how processing conditions influence polymer blend morphology.

Xu et al. [136], in their studies on PP/ethylene-octene copolymer (EOC) blends prepared in a rubber mixer, reported the effect of rotor speed on blend morphology. Blends containing 20 wt% EOC were used for the studies. As indicated in Table 2.4, the viscosity ratio remained the same at the shear rates employed in the study. However, an increase in shear rate resulted in larger dispersed phase domains and a broader size distribution. This was attributed to the increased elasticity and elasticity ratio at higher shear rates, which stabilized the domains and prevented them from breaking [137]. The most prominent effect, as seen in Fig. 2.9, was the variation in

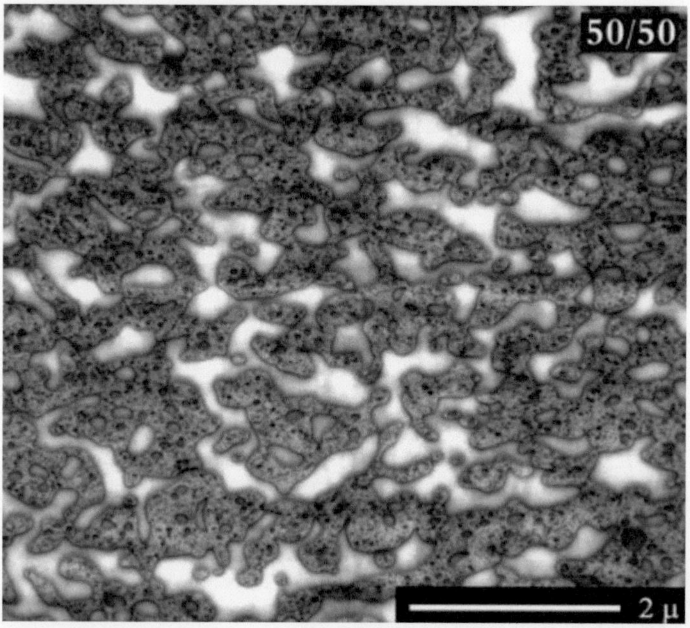

Fig. 2.8 TEM image of co-continuous 50/50 hPS/SEBS blend. Reproduced with permission [131]. Copyright 2012, Society of Plastics Engineers

the shape of the domains with rotor speed. The domains deformed from spherical droplets to elliptical droplets, ultimately forming fibrillar or sheet morphology. The change in domain shape was quantified by means of the shape factor, defined as the ratio of the length (major axis of ellipse) to the width (minor axis of ellipse) of the dispersed particles. A value of 1–1.5 was taken as indicative of spherical domains, while higher shape factor values indicated the evolution of ellipsoid, fibrillar, and lamellar morphologies. As shown in Fig. 2.9 and Table 2.4, an increase in rotor speed led to the transformation from spherical to fibrillar morphology, again attributed to the increase in shear rate and elasticity ratio.

Several research groups have reported on the effects of shear rate and mixing time on blend morphology. Shi and Utracki [138], in their studies on theoretical modeling of blend morphology development in a twin-screw extruder, predicted that at a given throughput, an increase in screw speed would lead to a finer morphology. This was attributed to the fact that increasing screw speed would provide greater energy to the material, enhancing the efficiency of deformation and breakup processes of the dispersed phase. The researchers also predicted that at a constant screw speed, an increase in throughput would result in a coarser morphology, with the influence of throughput being more pronounced than that of screw speed. Moreover, the coarseness from the increase in throughput can be mitigated by raising the screw speed. Willis et al. [139], in their studies on polystyrene-co-maleic anhydride/bromobutyl rubber blends, observed that an increase in residence time during compounding in

Table 2.4 The various parameters of different rotor speed in the PP/EOC (80/20) blends

Rotor speed (rpm)	Shear rate (s^{-1})	Viscosity ratio[a]	Elasticity ratio[b]	D_n (μm)	D_v (μm)	Size distribution[c]	Shape factor
32	51.2	0.27	0.68	0.33	0.46	1.39	1.50
40	64	0.26	0.76	0.42	0.66	1.57	1.45
50	80	0.27	0.83	0.34	0.64	1.88	2.82
60	96	0.27	0.94	0.40	0.80	2.00	3.32
70	112	0.28	–	Sheet	Sheet	–	4.45

Reproduced with permission [136]. Copyright 2007, Springer

[a] Viscosity ratio based on the ratio between η_{EOC} and η_{PP}

[b] Elasticity ratio based on the ratio between G'_{EOC} and G'_{PP}

[c] Size distribution based on the ratio between D_V and D_n

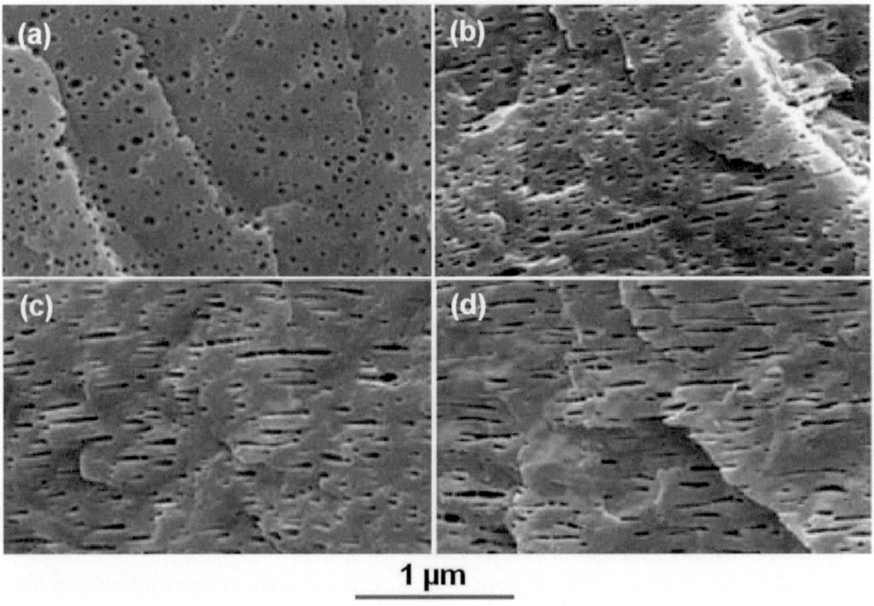

Fig. 2.9 SEM micrographs showing the morphology evolution of PP/EOC (80/20) blends as a function of rotor speed: **a** 32 rpm, **b** 50 rpm, **c** 60 rpm, **d** 70 rpm. Reproduced with permission [136]. Copyright 2007, Springer

a twin-screw extruder led to a finer morphology and a significant enhancement in impact strength. The significance of higher residence time (lower screw speeds) in achieving superior blend properties has been demonstrated by other research groups [140–142]. However, excessive residence time has also been found to result in a coarser morphology due to the coalescence of dispersed particles [142–144].

Lee and Han [145] conducted studies on five different blend systems to understand the role of processing temperature in the evolution of blend morphology in an internal mixer. They reported that, from a thermodynamic standpoint, an amorphous polymer may be considered a liquid above its Tg. However, from a rheological perspective, it cannot be regarded as a liquid until it reaches a temperature significantly higher than its T_g. In their analysis of morphology development in blends containing two amorphous polymers (PMMA/PS and PC/PS) or an amorphous polymer and a crystalline polymer (PS/HDPE and PS/PP blends), they utilized the concept of 'critical flow temperature' (T_{cf}), which was first introduced by Han [146] and is given by equation (2.30):

$$T_{cf} \cong T_g + 55\,^{\circ}\mathrm{C} \tag{2.30}$$

The researchers presented a schematic summary of morphology development (Fig. 2.10), which considered an immiscible blend comprised of two crystalline polymers. When analyzing an immiscible blend formed from two amorphous polymers or an amorphous polymer and a crystalline polymer, they replaced the T_m in Fig. 2.10 with T_{cf} for the amorphous polymer(s). The T_m (or T_{cf}) of the blend components was found to play a crucial role in the development of the blend morphology. When the processing temperature T was between the two T_ms or between the two T_{cf}s of the blend components (for instance, $T_{m,A} < T < T_{m,B}$ or $T_{cf,A} < T < T_{cf,B}$), component A initially formed the matrix phase, in which component B, still in a solid state, remained suspended. When $T > T_{m,A} > T_{m,B}$ or $T > T_{cf,A} > T_{cf,B}$, a dispersed two-phase liquid was formed, with droplets of component B dispersed in the matrix of component A. At this processing temperature, the lower viscosity of component B compared to that of component A, and/or component B being the major component, resulted in the possibility of phase inversion, leading to the formation of a blend with component B as the continuous phase and component A as the discrete phase. When phase inversion occurred, the two-phase liquid transitioned through a transitory morphological state, i.e., a co-continuous phase.

Elkovitch et al. [147] conducted studies on the effect of supercritical carbon dioxide ($scCO_2$) on the morphology development of PS/PMMA blends produced through melt processing. Compounding was carried out in a twin-screw extruder as well as in a high-pressure batch mixer. The viscosity reduction of PMMA and PS was measured using a slit die rheometer attached to the TSE. During melt processing, 0.5, 1.0, 2.0, and 3.0 wt% CO_2 was added, based on the polymer melt flow rate. The viscosity reduction due to the addition of carbon dioxide reached up to 80% for PMMA and up to 70% for polystyrene. A sharp decrease in the size of the dispersed phase was observed near the injection point of carbon dioxide in the TSE for blends with a viscosity ratio, η_{PMMA}/η_{PS}, of 7.3, at a shear rate of 100 s^{-1}. However, further processing resulted in the coalescence of the dispersed phase. TEM studies (Fig. 2.11) revealed that adding $scCO_2$ did not alter the path of morphology development; however, the final domain size was reduced. In both batch and continuous mixing, de-mixing occurred upon carbon dioxide venting. As shown in Fig. 2.12, the size reduction of the PMMA phase was lost after venting carbon dioxide, and

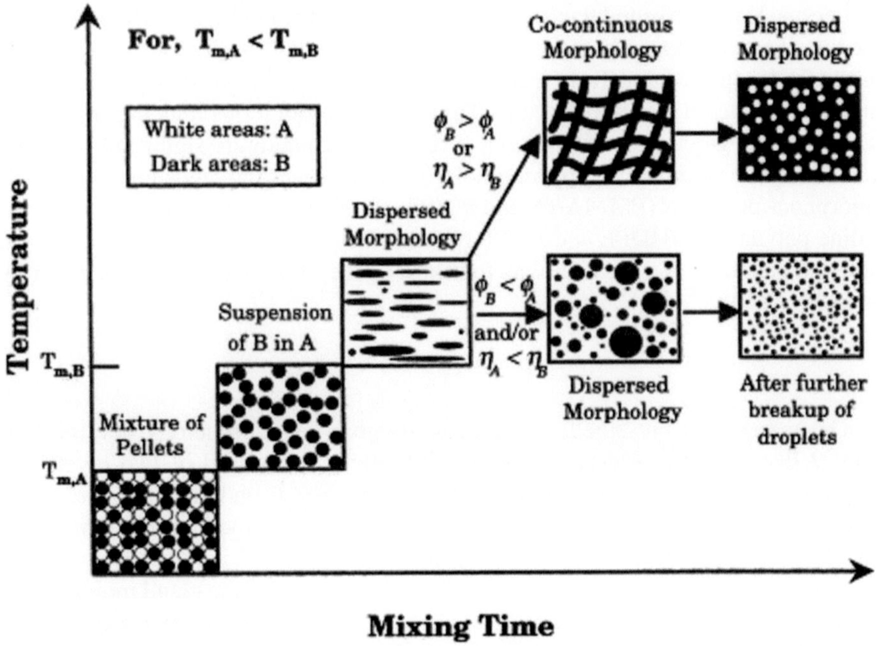

Fig. 2.10 Schematic describing the evolution of blend morphology during compounding of two immiscible polymers in an internal mixer, where the melting point of polymer A is assumed to be lower than that of polymer B. Reproduced with permission [145]. Copyright 1999, Elsevier Science Ltd.

the resulting morphology was similar to that without the addition of $scCO_2$. Adding small quantities of fillers (e.g., carbon black, calcium carbonate, or nano-clay particles) helped prevent the de-mixing of the polymer blend system when carbon dioxide was released, as evident in Fig. 2.13. Thus, the researchers demonstrated the efficacy of $scCO_2$ and small amounts of filler particles in achieving blends with superior morphology.

2.3 Fundamentals of Polymer Blend Nanocomposites

The morphological characteristics and, hence, the properties of blend nanocomposites are guided by the localization of nanoparticles, which, in turn, depend on thermodynamics and kinetics [1]. This section covers the fundamentals of nanoparticle migration, interface crossing, particle localization, and all factors influencing these processes.

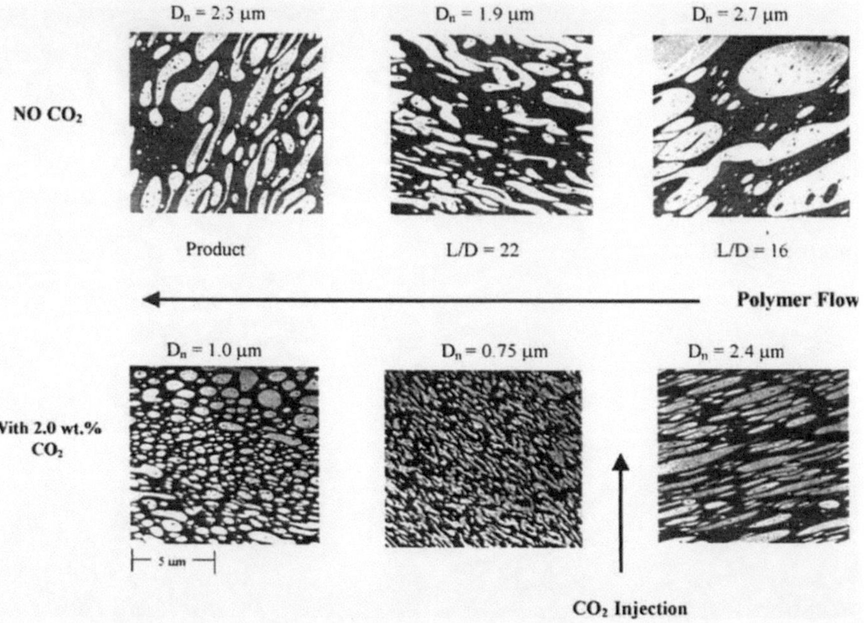

Fig. 2.11 Transmission electron microscopy images of 50/50 HMWPMMA/polystyrene: morphology development in the twin screw extruder. Reproduced with permission [147]. Copyright 2000, Society of Plastics Engineers

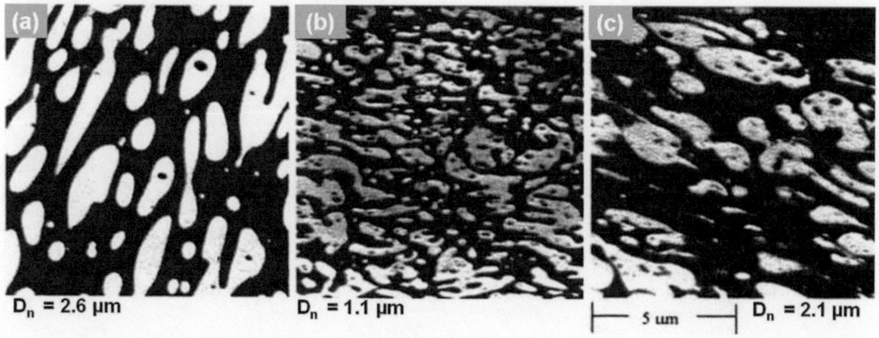

Fig. 2.12 Transmission electron microscopy images of 25/75 HMW-PMMA/polystyrene: **a** no CO_2, **b** with CO_2, no venting, **c** with CO_2, and venting. Reproduced with permission [147]. Copyright 2000, Society of Plastics Engineers

2.3.1 Particle Localization at the Equilibrium State

Particle localization in polymer blend nanocomposites can be predicted from the wetting parameter ω_{AB}, calculated using Young's equation (2.31) [20, 148]:

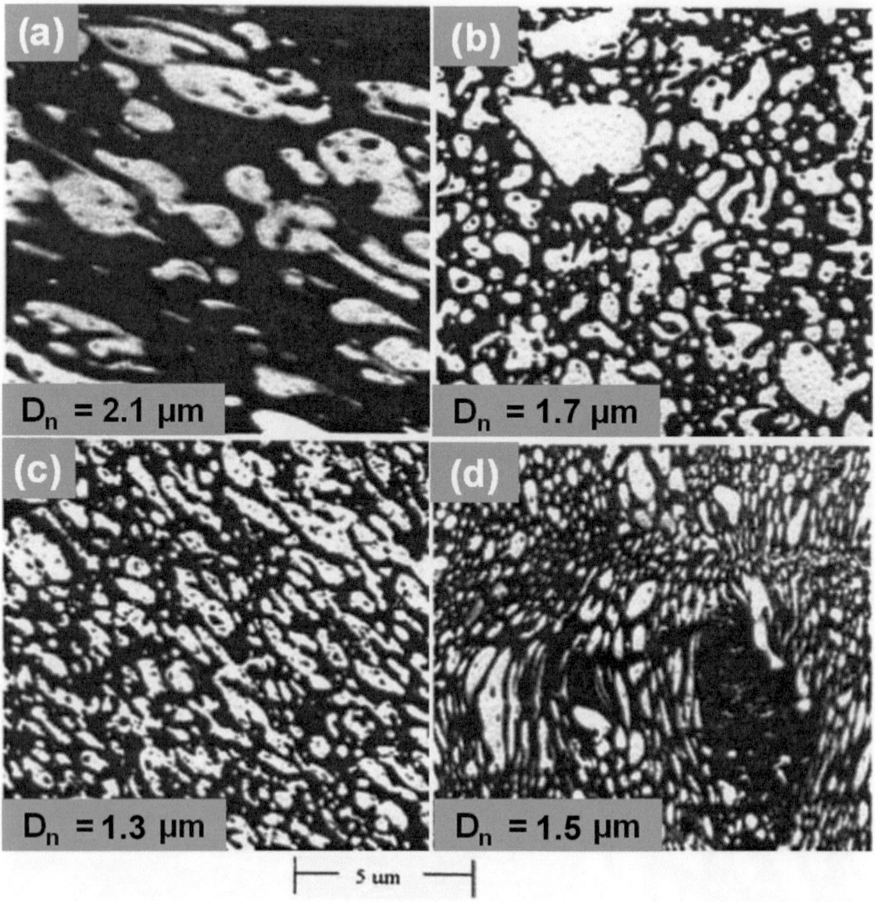

Fig. 2.13 TEM Micrographs of 50/50 HMW-PMMA/polystyrene: **a** CO_2 venting, **b** calcium carbonate, **c** nano-clay particles, and **d** carbon black. Reproduced with permission [147]. Copyright 2000, Society of Plastics Engineers

$$\omega_{AB} = \frac{\gamma_{PB} - \gamma_{PA}}{\gamma_{AB}} \tag{2.31}$$

where γ_{PB} and γ_{PA} are the interfacial tensions between the particle and polymer B and polymer A, respectively, and γ_{AB} is the interfacial tension between polymers A and B. If $\omega_{AB} > 1$, particles are expected to be localised in polymer A; if $\omega_{AB} < -1$, they are expected only in polymer B; and if $-1 < \omega_{AB} < 1$, they are expected to be localized at the interface between polymers A and B [149]. Interfacial tensions can be determined experimentally and theoretically using a variety of methods [150, 151]. The most popular theoretical methods are those proposed by Owens and Wendt [151] and Wu et al. [20], which employ geometric mean equation (2.32) and harmonic mean equation (2.33), respectively:

$$\gamma_{12} = \gamma_1 + \gamma_2 - 2\left(\sqrt{\gamma_1^d \gamma_2^d} - \sqrt{\gamma_1^p \gamma_2^p}\right) \tag{2.32}$$

$$\gamma_{12} = \gamma_1 + \gamma_2 - 4\left[\frac{\gamma_1^d \gamma_2^d}{\gamma_1^d + \gamma_2^d} + \frac{\gamma_1^p \gamma_2^p}{\gamma_1^p + \gamma_2^p}\right] \tag{2.33}$$

where d and p represent the dispersed and polar contributions to the surface tension values of components 1 and 2, respectively. The geometric mean and harmonic mean methods are reported to be more accurate for materials with high and low surface energy, respectively [151].

The prediction of nanoparticle localization at equilibrium using this method appears promising. However, discrepancies have been noted between experimental observations and theoretical calculations. This arises because theoretical predictions rely entirely on thermodynamic equilibrium wetting conditions. During the melt processing of polymer blend nanocomposites, kinetic and processing conditions can alter thermodynamic equilibrium, causing the migration of nanoparticles and profoundly influencing their localization [152–155]. Often, the processing time is insufficient for achieving the thermodynamically preferred localization of the filler particles, as theoretically predicted. This particularly occurs when the filler is initially dispersed in the less favorable phase. This can happen during pre-mixing or in a single-step mixing process when the less favorable phase melts first, causing the filler particles to be dispersed in it initially and then gradually migrate to the more favorable phase. If insufficient time is allowed, thermodynamic equilibrium may not be reached, resulting in deviations from theoretical predictions [156].

Additionally, the interfacial tension between a polymer and a nanofiller, and consequently, the localization of nanoparticles are affected by the particle characteristics of the filler [157, 158]. This fact is not taken into consideration in (2.32) and (2.33). Thus, thermodynamics, kinetics, processing conditions, and the shape and size of the filler particles must be considered when predicting the localization of filler particles in polymer blend nanocomposites [159].

2.3.2 Migration Mechanisms

Although there is significant literature on the various factors affecting particle localization in polymer blend nanocomposites and experimental observations of nanoparticle migration, there is limited literature on the mechanisms involved in the migration of these particles from one phase to another across the interface during processing [160].

In this context, Elias et al. [161] have provided a qualitative discussion on the three mechanisms involved in particle migration in polymer blend nanocomposites: The first mechanism involves the diffusion of filler particles through Brownian motion. The diffusion coefficient D_0 of a spherical particle with radius r, moving in a fluid with viscosity η at temperature T, is expressed by equation (2.34):

Table 2.5 Rotary diffusion coefficients of particles of various shapes

Aspect ratio (L/D)	Rotary diffusivity D_{ro} (s^{-1})
Nearly spherical shape (L/D $\approx$ 1)	$D_{ro} = \frac{K_B T}{\pi \eta_s D^3}$
Platelet-like particle (L/D < 1)	$D_{ro} = \frac{3K_B T}{4\eta_s D^3}$
Prolate spheroids (L/D > 1)	$D_{ro} = \frac{3K_B T[\ln 2(L/D)-0.5]}{\pi \eta_s L^3}$
Fibers, rods (L/D $\gg$ 1)	$D_{ro} = \frac{3K_B T[\ln(L/D)-0.8]}{\pi \eta_s L^3}$

$$D_o = \frac{K_B T}{6\pi \eta r} \tag{2.34}$$

where K_B is the Boltzmann constant. The time t_D taken by the particle to diffuse a distance equal to its radius r is given by equation (2.35)

$$t_D = \frac{r^2}{D_0} = \frac{6\pi \eta r^3}{k_B T} \tag{2.35}$$

Table 2.5 shows the rotary diffusion coefficients of particles of various shapes, i.e., aspect ratios [162, 163].

For spherical nanoparticles, assuming that r $\approx$ 100 nm, t_D = 7500 s for η = 2.5 $\times$ 10^3 Pa s (a typical viscosity value for a polymer melt) and T = 473 K. This approximate calculation indicates that the order of magnitude of the time required for the diffusion of nanoparticles is significantly greater than the processing time of a few minutes.

Overall, the high viscosity of a polymer melt is not favorable for particle migration by Brownian motion in the absence of shear. Lee et al. [164] reported that without shear, the diffusion time (t_D) for CNTs to diffuse 1 μm^2 in a polystyrene (PS)/poly(phenylene ether) blend was 2.57 $\times$ 10^4 s. Likewise, Bai et al. [165], in their studies on the diffusion of reduced graphene oxide (r-GO) in the high-viscosity polylactide (PLA) melt, stated that an annealing time of 1 h was necessary at 180 °C for r-GO to diffuse only 50 nm.

In contrast, Yoon et al. [166] proposed that temperature is the primary driving force; when there is sufficient annealing time, diffusion can take place through Brownian motion. The researchers stacked two sheets of polypropylene (PP)/CNT composite and neat polycarbonate (PC) on top of each other. The assembly was annealed at 200 and 300 °C for 5–20 min. Migration of certain CNTs to PC was observed to occur via diffusion. The extent of diffusion was greater at 300 °C and increased with prolonged annealing time.

Although the movement of nanoparticles through diffusion and Brownian motion across the interface is thermally induced, some level of compatibility is also necessary [159]. Yoon et al. [166] reported that no diffusion of CNTs was detected from the PC/CNT sheet to the PP sheet; however, CNTs did diffuse from the PP/CNT sheet to the PC sheet. This was attributed to the fact that the interfacial tension between PC and CNT is lower than that between PP and CNT. Doan et al. [167], in their studies on the migration of silica nanoparticles in laminated sheets of polybutadiene (BR) and poly(styrene-co-butadiene) (SBR) during annealing, observed that silica particles migrated from SBR/silica to BR, and the migration rate increased with rising temperature. However, there was no migration of silica from BR/silica to SBR. This was attributed to the affinity between silica nanoparticles and BR. The researchers concluded that the direction of transfer is determined by interfacial tension.

In another report, Wiwattananukul et al. [168] observed that CNTs could migrate from polyethylene (PE) to the PC/CNT phase during melt mixing. This contradicted the observation that CNTs migrated from PE/CNT to PC when studies were conducted on laminated sheets. The researchers attributed this to the oxygen on the CNT surface, which facilitated interactions in the blend, leading to the migration of CNTs to PE during melt mixing.

It can be summarized that the mechanism of nanoparticle migration in polymer blends is quite different. Many factors are involved, and interphase diffusion solely from annealing is quite unlikely. The second mechanism that has been proposed is shear-induced migration. The motion of nanoparticles is more translational than Brownian and is influenced by the Peclet number $\left(\frac{\dot{\gamma}}{D_0} \propto \frac{\dot{\gamma}}{t_D}\right)$, where $\dot{\gamma}$ is the shear rate. This indicates that shear plays an important role in the migration of nanoparticles. When the processed material is a polymer blend nanocomposite, random collisions of the dispersed droplets from the minor phase can affect the migration process. During compounding, the floating nanoparticles and the dispersed droplets can collide, which may result in the nanoparticles becoming embedded in the droplets. A similar mechanism might operate when the nanofiller is initially within the droplets of the dispersed phase and needs to migrate through the interface to the matrix polymer due to the velocity gradient inside the droplets, as observed by researchers [161, 169].

The third mechanism involves the entrapment of particles in the interdroplet zone of two droplets undergoing collision. This mechanism can be seen as a special case of the previous one since shear forces are also involved. In this scenario, particles can promote coalescence and ultimately occupy the interior of the newly formed large droplets. This phenomenon is called the "bridging-dewetting" mechanism [170–174], by which nanoparticles are simultaneously absorbed onto the surfaces of two colliding droplets, enhancing coalescence if the particles do not cover the interface. The latter situation is unfavorable for blend stabilization, necessitating fine morphology.

The mechanisms discussed by Elias et al. [161] can explain how nanoparticles migrate in immiscible polymer blends. However, the first mechanism, which involves the migration of particles under static conditions, cannot be considered a valid mechanism since it does not account for hydrodynamic forces, which play a crucial role in the movement of particles over large distances and their penetration into another phase. Therefore, the other two mechanisms seem to be more promising in explaining the migration of nanoparticles in immiscible polymer blends.

However, these mechanisms do not account for the influence of interfacial energy and particle shape. The interfacial energy between the filler particles and the blend components significantly affects the rate and direction of particle migration. Furthermore, for non-spherical particles, their orientation impacts the direction of migration [159]. According to Göldel et al. [157], the penetration rate of anisotropic nanoparticles is greater when they are oriented perpendicular to the interface. This has been attributed to curvature instabilities at the interface. The researchers proposed the Slim-Fast Mechanism (SFM), which states that low aspect ratio nanoparticles are less capable of penetrating blend interfaces than high aspect ratio nanoparticles. SFM was based on work by Krasovitski and Marmur [175], which used a penetration index to quantify the tendency of spherical particles to traverse a liquid/liquid interface. It was observed that increasing the aspect ratio of particles led to a relationship between the penetration index and the contact angle between the particle and the destination fluid approaching a stepwise function. The influence of wetting parameters and particle geometry (size and shape) on particle migration has been discussed in greater depth in Sect. 2.3.4.

The mechanism of particle migration and localization may also involve chemical interactions. Chen et al. [176] produced co-continuous PS/poly(methyl methacrylate) (PMMA) blends, having CNTs selectively localized at the interface, by achieving a balance of π-π interactions between PS and the CNT surfaces and dipole–dipole interactions between PMMA and the CNT surfaces by controlling the concentration of carboxyl groups on the CNT surfaces. Tan et al. [177] grafted a random copolymer PS-PMMA onto graphene sheets to localize the graphene nanosheets at the interface of co-continuous PS/PMMA blends.

Confined free-volume induced migration is another mechanism that is observed at high particle concentration. This type of migration occurs without adequate space when nanoparticles are forced to move to the other phase [178].

It may be summarized that the migration of nanoparticles in immiscible polymer blends occurs through various mechanisms. The mechanism involved depends on several factors, such as shear, viscosity, particle shape and size, and the surface functionality of the filler particles. The sections below discuss the multiple factors involved in particle migration.

2.3.3 *Effects of Mixing Conditions (Dynamic Process)*

2.3.3.1 Mixing Protocol-Sequence

A change in the sequence of adding constituents during the mixing process can alter the final localization of nanoparticles within nanocomposites of immiscible polymer blends [179–182]. There are four possible sequences for mixing components during the melt processing of polymer blend nanocomposites: (i) single-step mixing, where all components are added simultaneously; (ii) adding the nanofiller to the molten blend after the blend components have already been mixed for a period; (iii) premixing the filler with the polymer, which is thermodynamically preferred; and (iv) premixing the filler with the polymer, which is thermodynamically not preferred. Premixing with the thermodynamically less preferred blend component can drive the migration of nanoparticles to the other blend component. However, adequate time must be allowed to achieve thermodynamic equilibrium [156]. Surface modification of the filler can enhance this thermodynamically driven migration by altering the wettability of the filler particles. This section addresses key research findings related to the effect of the sequence of addition of components on particle migration and, consequently, the morphology development of immiscible polymer blend nanocomposites.

Nanoclay

Different research groups have studied the effect of the sequence/mode of addition of components during melt mixing on the morphology and properties of clay nanocomposites of immiscible polymer blends [180, 183–190]. However, the focus has been on the effect on the dispersion quality and the consequent change in properties. Little literature is available on the change in migration and localisation of nanoclay particles.

In this direction, Bandyopadhyay et al. [184] conducted studies on the effect of the mode of nanoclay addition on the morphological characteristics of organoclay (OMMT) reinforced Nylon 6 (N6)/ethyl-vinyl alcohol (EVOH) blends with an 80/20 (wt/wt) N 6/EVOH ratio. For the preparation of the blend nanocomposites, both PP-g-MA/OMMT masterbatch (MB) and OMMT were melt-processed with the blend in a twin-screw extruder (TSE), and the obtained samples were designated as N 6/EVOH/MB and N 6/EVOH/OMMT, respectively. MB-containing nanocomposites of N 6 and EVOH were also prepared and designated as N 6/MB and EVOH/MB, respectively. The silicate content (inorganic portion of OMMT) of all the nanocomposites was around 3 wt%. The 3 D transmission electron microscopy (TEM) images of the nanocomposites are presented in Fig. 2.14. The dispersed clay particles are indicated by the black markers in the 3D distribution of clay platelets (Fig. 2.14a–d). Figure 2.14a′–d′ and a″–d″. The blue markers in Fig. 2.14a′–d′ and a″–d″ indicate the dispersed particles, while the silver, golden, and yellow colors in the background represent the N6, EVOH, and the N 6/EVOH blend, respectively. The intercalated clay particles were oriented in a specific direction. Furthermore, the distribution of intercalated clay particles was better in EVOH/MB (Fig. 2.14b) than in N6/MB (Fig. 2.14a). In N6/EVOH/MB (Fig. 2.14c), the silicate layers were mostly localized in the interphase region, possibly enveloping the EVOH phase. However, such characteristics were absent in N 6/EVOH/OMMT (Fig. 2.14d). Figure 2.14c′, c″, d′, d″ present the dispersion and distribution characteristics of clay particles in the interphase. Clearly, the clay particles were localized only in the interphase in N6/EVOH/MB, forming core–shell particles. This localization of intercalated silicate layers suppressed the coalescence of the dispersed phase and stabilized the blend morphology. Such localization of clay particles was absent in the N6/EVOH/OMMT composite, making it impossible to distinguish between the two phases. The researchers thus demonstrated

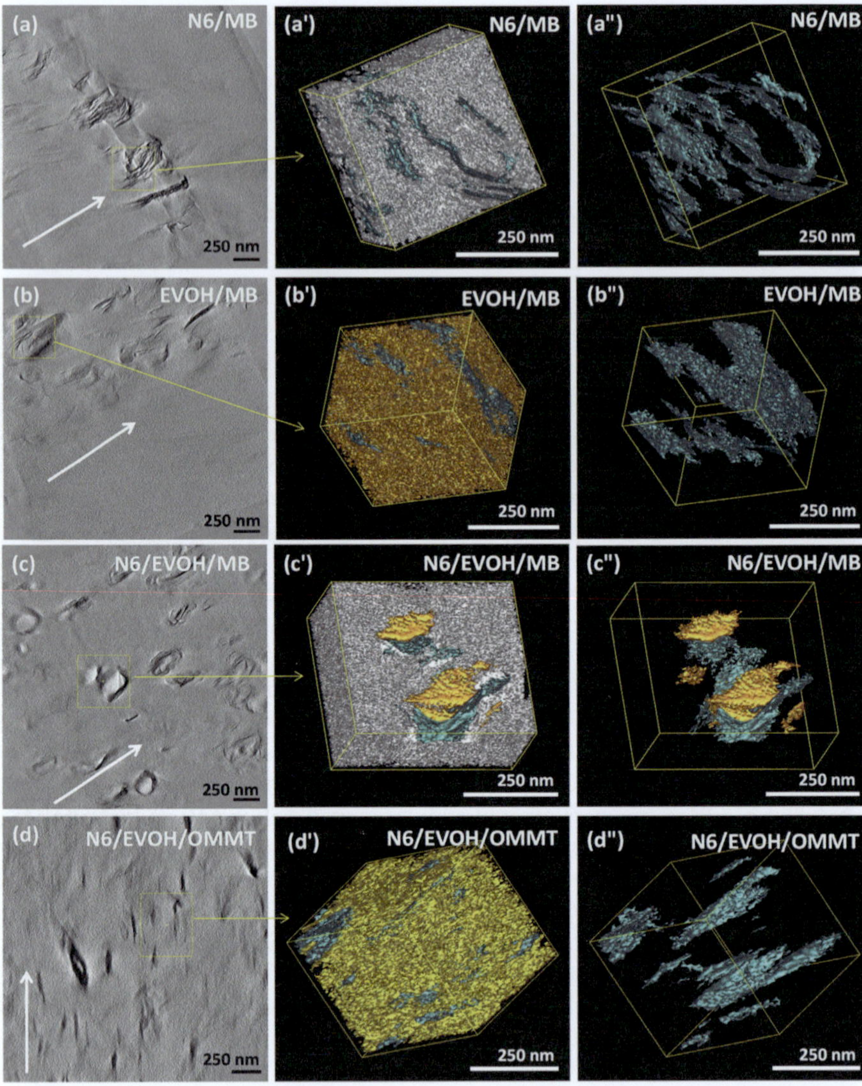

Fig. 2.14 **a–d** The 3D distributions of nanoclay platelets in the blends. The black markers represent the dispersed nanoclay particles in the blend composites. **a′–d′** and **a″–d″** the 3D projection of a selective region of 3D TEM images. The silver, blue, golden and yellow colours in the background represent N6, silicate particles, EVOH, and the N6/EVOH blend, respectively. Arrows indicate flow direction in the injection moulded samples. Reproduced with permission from [184]. Copyright 2017, Elsevier Science Ltd.

the effect of the mode of nanoclay incorporation on the localization of nanoclay in nanoclay-reinforced immiscible polymer blends.

Oliveira et al. [187] investigated how mixing sequence affects the properties and morphology of Cloisite 30B (C30B) reinforced polyamide 6 (PA6)/acrylonitrile–butadiene–styrene (ABS) blends. Styrene/maleic anhydride (SMA) was used as a compatibilizer. The blend composite samples were prepared via melt mixing in a twin-screw extruder. Samples containing 55 wt% PA6, 35 wt% ABS, 5 wt% SMA, and 5 wt% C30B were prepared in a twin-screw extruder using the following four mixing sequences:

(i) *Mixture M1*: All four components were compounded simultaneously.
(ii) *Mixture M2*: PA6 was first compounded with the clay. This was followed by the addition of ABS and SMA in a second extrusion step.
(iii) *Mixture M3*: A pre-mixture of PA6, organoclay, and SMA was prepared in the first step. The prepared composite was blended with ABS in the second step.
(iv) *Mixture M4*: The compatibilized blend was first prepared. The clay was compounded with the blend in a second extrusion step.

Each pre-mixing step was followed by cryogenic grinding of the extruded material. The ground material underwent the second extrusion step. The prepared mixtures were injection molded into test specimens. TEM studies were conducted after selectively staining the ABS phase. The dispersion quality of the ABS domains in the PA6 matrix was significantly influenced by the mixing sequence. M1 and M4 exhibited a much coarser morphology compared to M2 and M3, which showed a significantly reduced domain size. The smallest domain size was found in M3, attributed to the increased viscosity of the PA6 phase, which in the case of M2 resulted from the incorporation of nanoclay particles in PA6 due to the first mixing step. In M3, not only was clay present in PA6, but the compounding with SMA during the first extrusion step provided sufficient time for the grafting reaction between PA6 and SMA to occur, further enhancing the viscosity of the matrix phase and altering the morphology. Other researchers have reported a decrease in domain size of immiscible blend nanocomposites due to the early introduction of nanoclay into the matrix in two-stage mixing [191, 192]. Regardless of the mixing sequence, the nanoclay particles were localized in the PA6 phase and at the interphase, attributed to the greater affinity of C30B for PA6. Thus, the researchers demonstrated that altering the sequence of component addition in the preparation of polymer blend nanocomposites can significantly affect the dispersion quality of the dispersed polymer phase without impacting the localization of the nanofiller particles.

In another recent example, Li et al. [191] conducted studies on the morphology of poly(butylene terephthalate) (PBT)/maleic anhydride grafted poly(ethylene-co-vinyl acetate) (EVA-g-MAH)/C30B nanocomposites having 3 wt% C30B content and PBT: EVA-g-MAH in 75:25 weight ratio, prepared in an internal mixer using the following four different sequences of mixing:

(i) Simultaneous mixing of all three components (Hybrid L-1).
(ii) Mixing PBT with C30B first using 4 wt% epoxy as a compatibilizer, followed by blending of the nanocomposite formed with EVA-g-MAH (Hybrid L-2).
(iii) Mixing EVA-g-MAH with C30B first, followed by blending the nanocomposite formed with PBT (Hybrid L-3).

(iv) Mixing PBT with EVA-g-MAH first, followed by compounding of the blend with C30B (Hybrid L-4).

Injection molded samples of the nanocomposites were subjected to TEM studies. Hybrid L-1 exhibited a layer-like phase morphology with large and irregular EVA-g-MAH domains. C30B was intercalated in both phases, primarily in the EVA-g-MAH phase and the interphase. This was attributed to the stronger polar interaction between the nanoclay particles and EVA-g-MAH compared to PBT. Hybrid L-4 displayed a similar morphology; however, the size of the EVA-g-MAH domains was slightly smaller. Among all the samples, Hybrid L-2 demonstrated the best morphological characteristics, showcasing a fine island-sea morphology of the blend. Nanoclay layers were intercalated within the PBT matrix. The researchers suggested that the epoxy compatibilizer enhanced the interaction between the nanoclay and PBT, thus preventing the migration of the filler particles to the dispersed EVA-g-MAH phase. Another possible explanation for the inhibited migration of nanoclay to the EVA-g-MAH phase, as Borah et al. [193] proposed in similar studies, could be the slow movement of the nanoclay tactoids. When nanoclay is initially mixed with the thermodynamically less favored component, and the more preferred component of the blend is introduced in the second mixing step, the available mixing time in the second step might not be sufficient for the migration of the nanoclay particles to the more favored polymer [156], especially when nanoclay is dispersed as large tactoids. An increase in the size of filler particles may hinder their movement [193]. Hybrid L-3 comprised irregularly shaped EVA-g-MAH domains, with nanoclay mostly intercalated and exfoliated in EVA-g-MAH. The researchers thus demonstrated that a change in the mixing sequence impacts not only the quality of dispersion of the dispersed polymer phase and the filler particles but may also influence the interfacial migration of the filler particles and their localization.

It can thus be summarized that in the preparation of nanoclay reinforced immiscible polymer blends, a change in the mixing sequence may or may not influence the migration and localization of the nanoclay particles. The nanoclay particles tend to localize in the thermodynamically more favorable phase. However, a change in localization may occur if all components are mixed simultaneously or when the nanoclay is first combined with the thermodynamically less preferred blend component and sufficient time is not allowed for its migration to the more preferred component. This is particularly true when the size of the dispersed nanoclay tactoids is large, as an increase in size generally results in slower motion.

Nanosilica

Not much work has been done on the effect of mixing sequences on the migration and localization of nanosilica in immiscible polymer blends. Dil and Favis [194] demonstrated the importance of the mixing sequence in their studies on nanosilica reinforced PLA/poly(butylene adipate-co-terephthalate) (PBAT) blends. Blends containing 3 wt% nanosilica and PLA: PBAT in a 70:30 weight ratio were prepared in an internal batch mixer using the following mixing sequences:

Pr1: A one-step mixing process in which the filler was added after melting the blend components.

Pr2: A two-step process in which the filler was premixed with PLA, followed by blending with PBAT.

AFM and SEM studies revealed that the composition prepared by Pr1 had silica particles selectively localized in the PBAT phase, as predicted by surface energy and interfacial energy calculations. The preferential localization of silica in PBAT was observed to occur in the early stages of mixing. However, when compounding was conducted by Pr2, nanosilica was perfectly localized at the PLA/PBAT interface. This deviation from the theoretical prediction was attributed to the slow migration velocity of the silica particles at the interface, which was due to the low interfacial tension in the system.

A similar observation regarding the effect of mixing sequence on the time available for particle migration was reported by Elias et al. [195] in their studies on nanosilica-reinforced PP/poly(ethylene-co-vinyl acetate) (EVA) blends. Blends containing 3 wt% silica and EVA:PP in a 1:4 ratio were prepared in a twin-screw mini extruder using two sequences of mixing:

(i) The three components were mixed at 200 °C for 5 min.
(ii) The silica was mixed with PP at 200 °C for 5 min, and the resulting nanocomposite was blended with EVA in a second extrusion step.

Compression-molded samples of the compounded material underwent morphological analysis. SEM micrographs revealed that the ternary composition prepared by the first procedure had silica nanoparticles well localized within the EVA droplets, as predicted by thermodynamics, while the composition prepared by procedure 2 had nanosilica located in the EVA phase but close to the interphase. The researchers suggested that the mixing time in procedure 2 was insufficient for achieving thermodynamic equilibrium, thus the final localization of silica was not attained. This was attributed to the low diffusion coefficient of silica in the viscous polymer melt.

The importance of mixing sequence on particle localization was also demonstrated by Yoshida and Trifkovic [181] in their studies on silica-reinforced PP/EVA blends. The researchers observed that by altering the addition sequence of blend components during compounding, hydrophilic silica, which thermodynamically preferred the EVA phase, could be kinetically arrested at the EVA/PP interface of co-continuous blends. In this regard, Saiprasit and Schlarb [196] conducted studies on nano-silica reinforced PLA/PBAT blends prepared through sequential melt mixing. The researchers found that the nano-silica was initially present in the polymer phase where it was incorporated in the masterbatch, as well as at the interface. As the residence time in the extruder increased, more particles migrated to the interface in accordance with thermodynamics, as presented schematically in Fig. 2.15.

It can be summarized that the order of component addition during compounding influences the localization of silica particles in nanosilica-reinforced immiscible

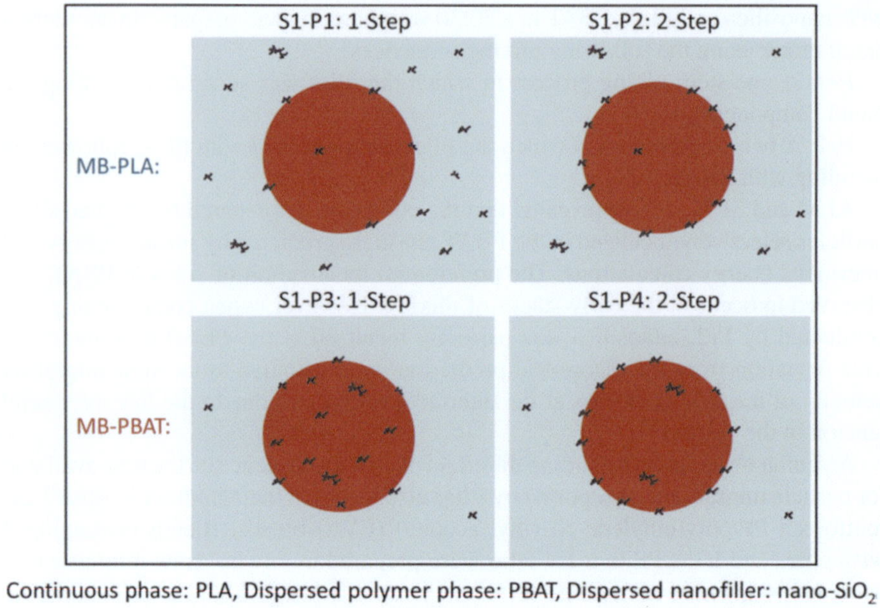

Fig. 2.15 Schematic presentation of the location of nano-SiO_2 with different processing procedures. Reproduced with permission from [196]. Copyright 2021, Polymer Processing Society (Open Access Publication)

polymer blends if sufficient time is not allowed for achieving thermodynamic equilibrium.

Carbon Black

The impact of the mixing sequence of components on the migration and localization of filler particles in carbon black (CB) reinforced immiscible polymer blends has been examined by various research groups [182, 197–200]. For instance, Yang et al. [200] investigated CB-reinforced PP/styrene-butadiene-styrene (SBS) blends with 2 vol% filler content. The ternary composites were processed in a two-roll mixer using three mixing sequences, each consisting of two steps. In the first step, two of the three components were pre-mixed, while the third was blended with the premix in the second step. The researchers noted that the blending sequence where PP was mixed with CB in the first step resulted in the greatest enhancement in conductivity. SBS was the thermodynamically preferred phase for CB, as evident from the wetting parameter. When the two polymers were premixed in the first step, the CB added in the second step favored SBS. However, due to the high viscosity of SBS, the dispersion quality of CB in SBS was poor, leading to agglomeration, as demonstrated in optical micrographs. Additionally, the filler particles that initially entered the PP phase were prevented from migrating and dispersing into the SBS phase because of the high viscosity of the SBS phase. The composite prepared by mixing CB with SBS in the first step demonstrated a poor quality of dispersion of CB particles in SBS due to its high viscosity. In contrast, when CB was first premixed with PP, a superior quality of dispersion was achieved because of the lower viscosity of PP. Furthermore, due to the low percolation threshold of PP, the CB particles formed a complete conductive network in limited PP regions, resulting in a significant improvement in conductivity. The researchers thus illustrated that when the viscosity of the thermodynamically preferred phase is very high and the filler particles are first premixed in the less preferred phase, migration to the more preferred phase may be hindered, causing deviations from expected morphology and changes in properties.

Contrary to the findings of Yang et al. [200], there are also reports [182, 197, 199] indicating that the mixing sequence has no significant effect on the morphology of CB-reinforced immiscible polymer blends. Tchoudakov et al. [197], in their studies on CB-reinforced PP/polyamide (PA) blends with a 70:30 PP:PA ratio and 2 phr CB, observed that regardless of the blending sequence, CB was preferentially localized in the dispersed PA phase. This phenomenon was attributed to the higher affinity of CB for the more polar PA. The overriding influence of nanoparticle affinity for the blend components on particle localization was also shown by Zhang et al. [182], in their morphological analysis of thermoplastic polyurethane (TPU)/20 wt% PA copolymer (COPA)/9 wt% CB composites. The researchers noted that irrespective of the blending sequence, all blends exhibited an island–sea structure, with most of the CB particles located at the TPU/COPA interface. The interfacial localization of the filler particles was attributed to their hydrogen bonding with both blend components, as confirmed by infrared spectroscopy and dynamic mechanical analysis. These studies demonstrate that when the affinity of nanoparticles for one or more blend components is very strong, as in the case of hydrogen bonding, particle localization is directed by this affinity, and the effect of blending sequence may not be evident.

The work of Yu et al. [198] highlights the effect of mixing sequence on mixing time and its subsequent impact on morphology. The researchers conducted studies on how mixing sequence affects the morphology of CB-filled ethylene–vinyl acetate (EVA) copolymer/low-density polyethylene (LDPE) blends processed in a Brabender mixer. They observed that in 2-step mixing, when CB was first mixed with EVA and LDPE was added in the next step, the CB particles migrated to the thermodynamically preferred LDPE phase. However, this migration required sufficient time. Even mixing for 15 min was insufficient for complete migration, as indicated by the oozing of CB and its adherence to the screw of the Brabender. This oozing was not observed at the same mixing time when CB was initially mixed with LDPE and EVA was added subsequently. The researchers thus demonstrated that mixing sequence affects particle localization and morphology of polymer blend nanocomposites only when the filler particles are initially mixed with the thermodynamically less preferred blend component and sufficient time is not available for the particles to migrate to the thermodynamically preferred blend component. When the filler is premixed with the thermodynamically preferred blend component or sufficient time is available for particle migration, mixing sequence does not have an effect.

CNTs

The effect of mixing sequence on the morphology of CNT-reinforced immiscible polymer blends has been studied by various research groups [164, 179, 200–206]. Lee et al. [164], in their studies on the morphology of PA66/[PS/poly(phenylene ether)(PPE)]/CNTs composites prepared using different mixing methods in a TSE, observed that the nanocomposite created by mixing PS/PPE and CNTs in the first step, followed by mixing the (PS/PPE)/CNTs with PA66 in the subsequent step, exhibited better CNT dispersion in the thermodynamically more favorable PA66 phase compared to the nanocomposites prepared using other sequences where the CNTs were initially mixed with PA66 or where all components were blended simultaneously. The researchers proposed a phenomenon called "filler-transfer-induced dispersion" to explain this observation. The transfer of filler particles from a less preferred phase to a more favorable phase is driven by the instability of the curvature at the filler-blend interface [157]. During the migration of filler particles across this interface, the interfacial curvature of low aspect ratio particles, such as CNT aggregates, can relax, resulting in a reduction of the driving force. The driving force for the migration of the higher aspect individual CNT particles remains unchanged. Thus, the individual CNTs migrate faster and more effectively than the aggregates to the more preferred phase, leading to better dispersion of CNTs in that phase compared to a scenario where interfacial migration does not occur. The researchers thus demonstrated the role of instability of the filler-blend interfacial curvature in the migration and dispersion quality of the CNT particles.

The influence of mixing sequence on the migration and dispersion quality of CNT particles was also reported by Gong et al. [206] in their studies on melt processed CNT reinforced polyoxyethylene (PEO)/ethylene-octene random copolymer (ORC) composites. The researchers found that dispersing CNTs in the thermodynamically unfavorable PEO phase before blending with the ORC phase was a promising strategy to enhance the migration process and achieve superior dispersion in the ORC phase.

Sultana et al. [179] also reported similar findings in their investigations of the morphology of CNT-reinforced PS/PMMA blends prepared using various mixing strategies. The different processing strategies and their corresponding nanocomposite designations are presented in Table 2.6.

The TEM images of PS/PMMA (20:80 volume ratio) blends containing 2.7 vol% CNT content, prepared using various mixing strategies, are presented in Fig. 2.16. The CNTs were predominantly located in the PS phase in all the nanocomposites, although some CNTs were also detected in the PMMA phase. However, the degree of dispersion and localization was found to depend on the mixing strategy employed. A greater dispersion of CNTs in larger PS droplets was observed in the PS/PMMA-Solution sample. However, the extent of localization in the PS phase was lower. The PS/PMMA-Melt sample exhibited more CNT localization in smaller PS droplets, but the quality of dispersion of the CNTs was poor. The PSMB(S)/PMMA and PSMB(M)/PMMA samples also displayed better selective localization of CNTs in PS droplets, yet the quality of the dispersion of CNTs remained inadequate. The PS droplets in the PSMB(M)/PMMA sample are smaller than those in the

Table 2.6 Summary of different processing techniques

PS/ PMMA-solution	Blend nanocomposites are prepared by solution mixing, which is also known as one-step solution mixing. First, 2.7 vol% MWCNT is dispersed in chloroform using 30 min of sonication and then transferred to the polymer solution. The mixture is further homogenized for an additional 30 min of sonication. The nanocomposites are extracted from the solution through solvent evaporation and vacuum drying
PS/PMMA-melt	PS and PMMA are initially melt blended for 3 min. MWCNT is then added to the neat PS/PMMA blend and melt blended for an additional 15 min
PSMB(S)/ PMMA	Masterbatch (MB) of PS is prepared by mixing in solution. MB is then melt-blended with PMMA for 5 min
PSMB(M)/ PMMA	MB of PS is prepared through melt blending for 10 min. MB is then melt blended with PMMA for 5 min
PS/ PMMAMB(S)	MB of PMMA is prepared using solution mixing. MB is then melt blended with PS for 5 min
PS/ PMMAMB(M)	MB of PMMA is prepared by melt blending for 10 min. MB is then melt blended further with PS for 5 min

Reproduced with permission from [179]. Copyright 2019, American Chemical Society

PSMB(S)/PMMA sample. Migration of CNTs from the PMMA phase to the PS phase is observed in the PS/PMMAMB(S) and PS/PMMAMB(M) samples. The migration of CNTs into the PS phase resulted in partial continuity of the minor PS phase in the prepared PS/PMMAMB(S) sample. Such partial continuity of the PS phase was not seen in the PS/PMMAMB(M) sample. The researchers clearly demonstrated that the mixing strategy can significantly influence the migration and localization of CNTs as well as the dispersion quality of the dispersed phase and CNTs in CNT-reinforced immiscible polymer blends.

Based on the existing knowledge of the effect of mixing sequence on the migration of CNTs in immiscible polymer blends, Azubuike and Sundararaj [201] advanced research by demonstrating that, with the proper choice of mixing sequence and by conducting reactive compatibilization, CNTs can be trapped at the interface of immiscible polymer blends. This leads to the formation of an interconnected network, resulting in superior electrical conductivity and enhanced EMI shielding effectiveness. The researchers conducted studies on CNT (1.5 vol%) reinforced PS/Polyamide (PA) blends that were reactively compatibilized with polystyrene maleic anhydride (PSMA), having a composition ratio of 25:75 (PS + PSMA: PA). The nanocomposites were prepared by melt mixing in a mini mixer at 200 rpm and 240 °C, using three different mixing sequences, the details of which are provided in Table 2.7.

TEM and optical microscopy (OM) of the nanocomposites revealed that in MO-I, the CNTs were present in the aPA phase, whereas in MO-II and MO-III, they were also found at the interface. The researchers proposed that in MO-II and MO-III, the CNTs were initially blended with the thermodynamically less preferred component PS. In the second mixing step, they migrated to the more favorable component aPA. During this migration process, reactive compatibilization caused their arrest at the

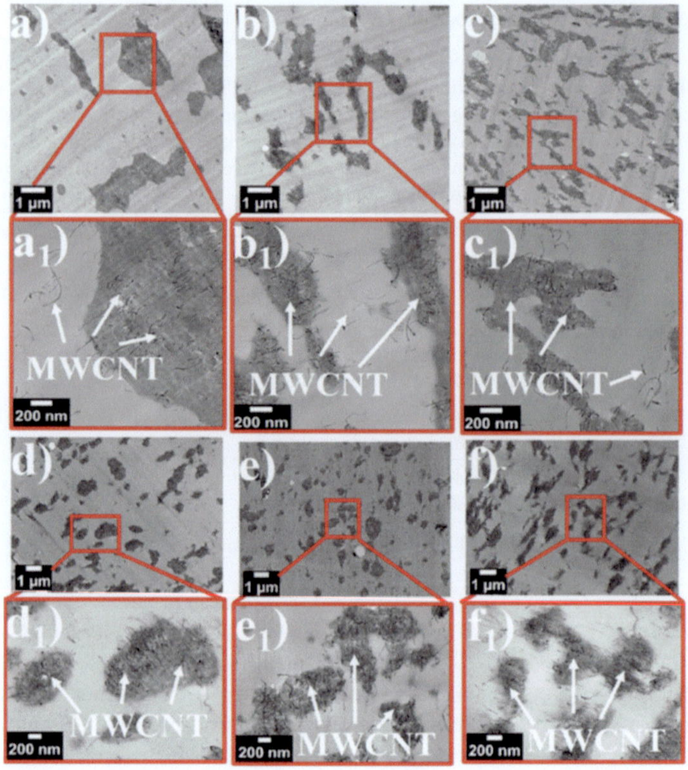

Fig. 2.16 TEM images of PS/PMMA (20:80) vol % blend nanocomposites with 2.7 vol% MWCNT: **a** PS/PMMA-solution, **b** PSMB(S)/PMMA, **c** PS/PMMAMB(S), **d** PS/PMMA-Melt, **e** PSMB(M)/PMMA, and **f** PS/PMMAMB(M). $\mathbf{a_1}$, $\mathbf{b_1}$, $\mathbf{c_1}$, $\mathbf{d_1}$, $\mathbf{e_1}$, and $\mathbf{f_1}$ are high-resolution images of the respective nanocomposites, showing MWCNT with more clarity. Darker domains represent the PS phase, and lighter domains are PMMA. Reproduced with permission from [179]. Copyright 2019, American Chemical Society

Table 2.7 Mixing order of addition

Mixing order	Step 1	Step 2
(MO)	t = 5 min	t = 10 min
I	PS + aPA	PSMA + CNT
II	PS + PSMA + CNT	aPA
III	PS + CNT	aPA+PSMA

Reproduced from [201]. Copyright 2021, the Authors (Open Access Publication)

interface. This interfacial localization resulted in an interconnected network of CNTs, enhancing electrical conductivity and EMI shielding effectiveness. Other research groups also reported that a change in the mixing sequence led to the arrest of CNTs at the interface during their migration from the thermodynamically less preferred phase to the more preferred phase [205, 207].

As reported by Yu et al. [198], in their work on CB-reinforced EVA/LDPE blends, discussed in the previous section, researchers studying CNT-reinforced polymer blends have also demonstrated that the mixing sequence affects particle migration and localization only when the filler is premixed with the thermodynamically less preferred phase and insufficient time is provided for its migration to the more favored phase [204, 208]. In this regard, Goldel et al. [204] noted that inadequate migration time may cause the localization of the CNTs in the thermodynamically less preferred phase or result in incomplete migration, the latter leading to the presence of some CNTs at the interface. Huang et al. [208] also highlighted the significance of adequate mixing time for achieving thermodynamic equilibrium in sequential mixing and the gradual migration of CNTs, which leads to changes in morphology with increased mixing time, in their studies on CNT-reinforced polyvinylidene fluoride (PVDF)/ acrylonitrile butadiene styrene (ABS) blends. This work has been discussed in greater detail in Sect. 2.3.3.2.

Graphene and Graphene Oxide

The impact of mixing sequence on the migration, localization, and quality of dispersion of graphene-based particles in graphene-reinforced immiscible polymer blends has been examined by various research groups [209–216]. For instance, Liebscher et al. [209] performed extensive studies on the influence of mixing sequence and parameters on the morphological characteristics of melt-processed 1 wt% graphene nanoplatelets (GnPs) reinforced co-continuous blends of polycarbonate (PC) and poly(styrene-acrylonitrile) (SAN) with a PC content of 59 wt%. OM, SEM, and TEM studies revealed that single-step mixing resulted in the poorest dispersion quality of the GnPs in the blend. Additionally, the GnPs were not selectively localized in any one component of the blend. In two-step mixing, when the GnPs were premixed with PC, they were selectively localized in PC after the second mixing step. When premixing occurred with SAN, migration of the filler particles to the PC phase was observed. Under mild mixing conditions, the GnPs were present in the SAN phase and at the interphase. However, at higher mixing speeds and longer mixing durations, they were found in the thermodynamically preferred phase as well. The dispersion quality of GnPs in two-step mixing was superior to that achieved in single-step mixing. The researchers demonstrated that the mixing sequence influences the morphological features of GnP-reinforced immiscible polymer blends, achieving better dispersion quality in two-step mixing. GnPs tend to favor the thermodynamically preferred blend component. However, when premixed with the less preferred component during two-step mixing, adequate mixing time and proper mixing conditions are required for their transfer to the more favored phase.

Tiwari et al. [213] added a new dimension to the knowledge about the effect of mixing sequence on the morphology of GO-reinforced immiscible polymer blends by providing an explanation based on interfacial interactions and rheology. The researchers conducted studies on reduced graphene oxide (rGO) reinforced PC/Nylon 66 blends containing 1 wt% rGO and 29.5 wt% PC. They proposed that the sequence of mixing affected blend morphology due to the altered viscosity of the blend component with which the GO was premixed. SEM studies were performed on both the neat blend and the nanocomposites prepared through single-step and two-step mixing. The SEM micrographs and the schematic representation of the morphological changes induced by mixing with rGO are presented in Fig. 2.17. In two-step mixing, when rGO was pre-mixed with PC, the rGO sheets remained localized in the PC phase. This phenomenon was attributed to the strong interactions of rGO with PC combined with PC's higher viscosity. Additionally, selective localization of rGO in the dispersed PC phase resulted in weak interfacial adhesion between the blend components. However, premixing rGO with Nylon 66 led to the dispersion of the filler particles in both phases and enhanced interfacial adhesion between the blend components, as evidenced by the transformation of a droplet-matrix blend morphology into a fibrillar morphology. This morphological change was attributed to the increased viscosity of the Nylon 66 phase caused by the presence of rGO, which improved the viscosity matching of the blend components, facilitating the migration of rGO particles to the PC phase and enhancing interfacial compatibility between the blend components. The researchers thus demonstrated how the mixing sequence influences morphology development in rGO-reinforced immiscible polymer blends by altering the viscosity match of the blend components.

Based on studies of various nanofiller-reinforced immiscible polymer blends, it can be summarized that the mixing sequence affects particle migration and localization only when particles are initially dispersed in the thermodynamically less favorable phase, and there is not enough time for their migration. Furthermore, by tailoring the mixing sequence and employing various strategies, such as reactive compatibilization and incomplete migration, particles can be selectively localized at the interface. In the case of conductive fillers, such interfacial localization results in the formation of an interconnected network of filler particles, leading to an increase in conductivity. The mixing sequence may also affect the dispersion quality of filler particles. Thus, varying the mixing sequence can be seen as an effective strategy for tailoring the morphology and properties of immiscible polymer blend nanocomposites.

2.3.3.2 Mixing Protocol-Time

In studies on kinetically controlled phenomena, the effect of time cannot be ignored. When the thermodynamic and kinetic conditions required for the migration of nanoparticles in an immiscible polymer blend are met, the migration process becomes a matter of time. Too short a mixing time may lead to significantly less migration of the particles to the desired phase. Conversely, too long a mixing time may result in

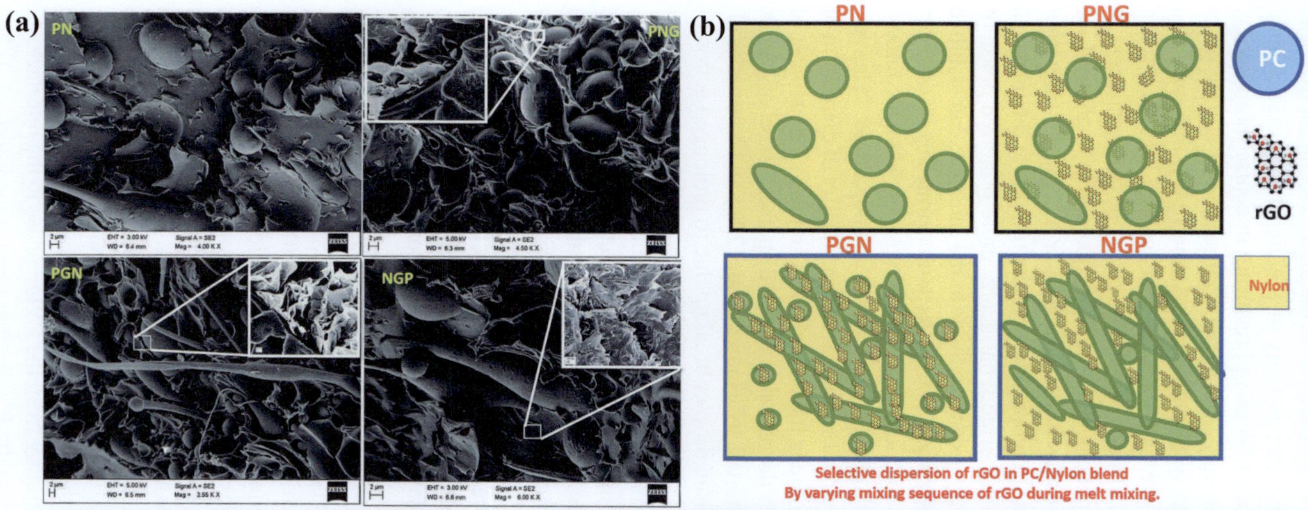

Sample code	Composition and mixing order	Processing conditions
P	Polycarbonates (100%)	Temp 270 °C, RPM = 10
N	Nylon 66 (100%)	Same
P/N66	Nylon 66 (70%) + polycarbonates (30%)	Same
PNG	Polycarbonates (29.5%) + nylon 66 (69.5%) + reduced graphene oxide (1%)	Same
PGN	Polycarbonates (29.5%) + reduced graphene oxide (1%) + nylon 66 (69.5%)	Same
NGP	Nylon 66 (69.5%) + reduced graphene oxide (1%) + polycarbonates (29.5%)	Same

Fig. 2.17 **a** FESEM micrograph of PN, PNG, PGN and NGP at low magnification and inset shows the magnified image of marked sections. **b** Schematic representation of selective dispersion of rGO in blend matrix. Reproduced with permission from [213]. Copyright 2017, Royal Chemical Society

complete migration of the particles to the dispersed droplets if this phase is more thermodynamically preferred. This may be undesirable, as the high concentration of filler in the dispersed droplets will increase the viscosity of the dispersed phase, which, in turn, will not be favorable for droplet break-up, resulting in coarse morphology. Furthermore, extended mixing times may lead to the degradation of certain polymers, such as PLA, which are prone to thermal degradation [217], and may also cause the attrition of nanoparticles like CNTs [218]. In sequential blending, when the nanofiller is premixed with the thermodynamically less preferred component, it migrates to the more preferred component, given that adequate time is allowed for migration. Inadequate mixing time results in no migration or incomplete migration [159]. By controlling the mixing time, nanoparticles can be trapped at the interface before they complete migration to the thermodynamically more preferred phase of the blend [219]. Such kinetic trapping of nanoparticles at the interface can stabilize blend morphology and suppress coarsening over time, as demonstrated by researchers [181, 181] on the localization of nanoparticles in co-continuous blend nanocomposites. Mixing time not only plays a vital role in the migration of nanoparticles in ternary compositions, but also influences blend morphology and the quality of dispersion of the nanofiller [220]. Longer mixing times have resulted in improved dispersion of nanofillers [208, 208]. A decrease in droplet size of the dispersed phase of blends with increased mixing time was reported by Huang et al. [208] and Lencar et al. [221], which was attributed to increased droplet breakup due to prolonged action of shear forces.

In contrast, Navas et al. [222], in their studies on CNT-reinforced PP/PS blends, reported an increase in both domain size and interconnectivity as mixing time increased. The CNTs were found to be preferentially localized in the dispersed PS droplets. The researchers attributed the increase in domain size with increasing mixing time to the migration of CNTs from one PS domain to another, facilitating coalescence. The PS layer covering the migrating CNT formed a thin thread connecting the two domains, which allowed the diffusion of PS chains between them. Thus, the CNTs covered by the PS layer acted as bridges connecting PS domains, leading to their coalescence and enhanced interconnectivity.

Huang et al. [208] demonstrated the role of time in the migration of CNTs during the sequential melt processing of CNT-reinforced PVDF/ABS blends in a Brabender mixer. PVDF was premixed with 3 wt% CNTs for 8 min at 80 rpm and 190 °C. The masterbatch was crushed and subsequently compounded with 70 wt% ABS at the same rotational speed and temperature. The mixing time varied from 2 to 8 min, and the morphological changes over time were monitored. The researchers reported that when the PVDF/CNTs masterbatch was melt processed with ABS, the CNTs migrated from the PVDF phase to the thermodynamically preferred ABS phase. This migration was gradual and consisted of five stages, as illustrated in Fig. 2.18. In the first stage, all CNTs were present in the PVDF phase. After melt processing for 2 to 4 min (Stage 2), the CNTs began migrating from the PVDF phase to the interface. At this stage, the CNTs were found both in the PVDF phase and at the interface. When the sample underwent melt processing for a longer mixing time (6 min), all CNTs were localized at the interphase (Stage 3). A further increase in mixing time resulted

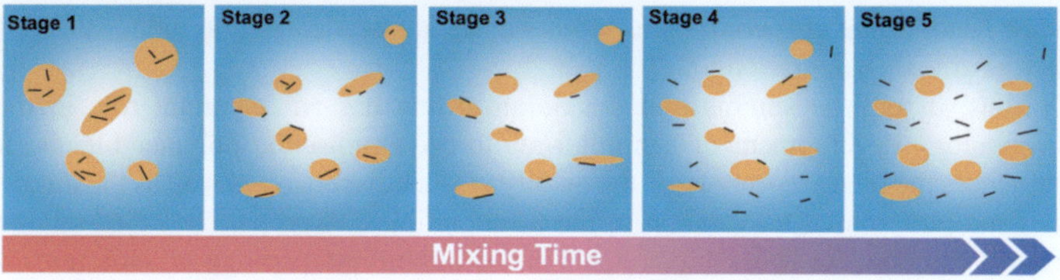

Fig. 2.18 Illustration of the structural evolution of PVDF/ABS/MWCNTs ternary blends along with longer mixing time. Reproduced with permission from [208]. Copyright 2021, Elsevier Science Ltd.

in the gradual migration of the CNTs to the ABS phase. In Stage 4, some CNTs were present at the interface, while others were in ABS. The researchers concluded that with further increases in mixing time, all CNTs would eventually migrate to the ABS phase.

The mixing time needed for particle migration and achieving equilibrium during the morphological development of immiscible blend nanocomposites depends on melt viscosity. Thus, when processed under the same standard conditions, migration rates vary among different blend systems [159]. Mofokeng et al. [220] studied nanoclay-reinforced PP/EVA blends with an 80: 20 ratio of PP to EVA, melt-processed in a batch mixer at 190 °C and a rotor speed of 60 rpm. The researchers found that in sequential mixing, when PP was premixed with nanoclay and EVA was added in the second step, blending the PP/clay mixture with EVA for 1 min was sufficient for the migration of clay particles into the dispersed EVA droplets. However, longer mixing periods for EVA resulted in smaller EVA droplets and improved clay dispersion quality in the EVA droplets. Martins et al. [223] reported a similar short migration time for clay particles in sequential extrusion processing of clay-reinforced PP/EVA blends. Longer migration times have also been reported. Zhou et al. [153] noted a migration time of over 7. 5 min for the migration of CNTs from PMMA-CNTs masterbatch to PS during the sequential melt processing of CNT-reinforced PMMA/PS blends. Huang et al. [208] reported a similar migration time, as mentioned earlier. Extended migration times of CB particles in various blend systems have also been documented by other researchers. For instance, Yu et al. [223], in their studies on CB-reinforced EVA/LDPE blends, observed that in a two-step mixing process, when CB was first mixed with EVA and LDPE was added later, a mixing time of 15 min was inadequate for the complete migration of CB particles from the EVA phase to the LDPE phase. Similarly, Xu et al. [224], in their studies on CNT-reinforced poly(phenylene sulfide) (PPS)/PA 6 blends, reported that the migration of CNTs from the PPS phase to the interphase and the PA 6 phase took 16 min and 30 min, respectively. Such lengthy migration times have been attributed to high viscosity ratios of the blends [225, 226]. All these studies clearly indicate that migration time varies on a case-by-case basis. In some scenarios, a mixing time as short as 1 min may suffice for complete migration, while in other cases, 30 min may be necessary. One factor influencing migration time is the viscosity ratio of the blend components, although other factors may also contribute. More studies are needed to fully understand all factors affecting migration time.

Thus, it may be summarised that mixing time has an important role in the morphology development of nanofiller-reinforced immiscible polymer blends. It influences the blend morphology, the quality of filler dispersion as well as the migration and preferential localisation of the filler particles. A proper choice of mixing time is essential for achieving a stable and desirable morphology of such ternary compositions.

2.3.3.3 Effect of Shear Forces

Shear forces play a crucial role in the development of blend morphology and in the migration and localization of nanoparticles in blend nanocomposites. The size of dispersed droplets in immiscible polymer blends is governed by the fundamental process of breaking up isolated droplets in the flow field of the matrix phase, which is in turn determined by the Capillary number (Ca) equation (2.36):

$$Ca = \frac{\tau R}{\sigma} = \frac{\eta_m \dot{\gamma} R}{\sigma} \tag{2.36}$$

where τ is the shear stress, η_m is the viscosity of the matrix phase, $\dot{\gamma}$ is the shear rate, σ is the interfacial tension, and R is the radius of the droplet. Above a critical value, C_{acrit}, which in turn depends on the viscosity ratio (η_d/η_m, where η_d is the viscosity of the dispersed phase), the shear stress predominates over the interfacial stress, and the droplet is stretched until it breaks. The resulting droplet diameter Bmin is given by equation (2.37) [217]:

$$B_{\min} = \frac{2Ca_{crit}\sigma}{\eta_m \dot{\gamma}} \tag{2.37}$$

Thus, shear forces are crucial to the development of blend morphology. Additionally, during the processing of polymer blend nanocomposites, the nanoparticles and dispersed droplets collide, resulting in the shear-induced migration of nanoparticles toward the thermodynamically favored phase. The frequency of these collisions (C) is given by equation (2.38) [160, 161]:

$$C = \frac{8\dot{\gamma}\varphi}{\pi} \tag{2.38}$$

where φ represents the volume fraction of the dispersed droplets or particles. In addition to raising the frequency of collisions, an increased shear rate can promote particle movement due to the significant forces produced by hydrodynamic shear flow [227].

The effect of shear stresses on the extent of nanoparticle migration has been investigated by various research groups. For example, Taghizadeh and Favis [228] studied the localization and migration of CNTs in CNT-reinforced polycaprolactone (PCL)/thermoplastic starch (TPS) blends. A PCL/CNT masterbatch was first prepared and then blended with 20% TPS. When blending was performed in a TSE, CNTs predominantly localized in the TPS phase and partially at the PCL/TPS interface. The researchers proposed that the CNTs initially migrated to the interface, where they reacted with TPS and were drawn into the low-viscosity TPS phase. X-ray photoelectron spectroscopy revealed a covalent bonding of the CNTs with TPS. However, when blending was conducted in a low-shear stress internal mixer, the CNTs were found in both the PCL phase and at the PCL/TPS interface, indicating that their migration to the thermodynamically preferred TPS phase was not complete under low shear stress conditions. The significance of shear in facilitating the migration of nanoparticles from the thermodynamically less preferred to the more preferred phase has been discussed in depth by Elias et al. [161].

In this direction, Dharaiya and Jana [229] demonstrated that shear forces influence not only particle migration but also blend morphology. The researchers conducted studies on 1 wt% nanoclay reinforced (90/10) PA6/PP blends produced by melt processing in a batch chaotic mixer. Clay was premixed with PP, and this premix was subsequently blended with PA6. Different morphologies were achieved depending on the level of shear forces applied. At lower shear rates, a fibrillar morphology was obtained with the clay particles remaining in the PP phase. At higher shear rates, the PP fibrils disintegrated into droplets, and the clay particles migrated to the PA6 phase. In a different work [230], the same researchers reported the effect of shear forces on the morphology of 1 wt% CB reinforced (90/10) PA6/PP blends. At low strains or shorter mixing times, the CB particles were in the PP dispersed phase, whereas at higher strains, they migrated to the interface.

The effect of shear force on both blend morphology and particle migration was also reported by Li and Shimizu [231] in their studies on CNT-reinforced (50/50) PA6/PVDF blends prepared by melt mixing. The researchers observed that when mixing was conducted at high shear (1000 rpm screw speed in a high-shear extruder), the blends exhibited a co-continuous structure, featuring some CNT-containing PA6 nano-domains within the PVDF phase. The CNTs were localized in the PA6 continuous phase and within the nano-domains. Conversely, when processing was done at low shear (100 rpm screw speed), very few nano-domains were found. The researchers concluded that migration of CNT-filled PA6 inclusions to the PVDF phase occurred at high shear rates.

Therefore, it can be summarized that shear forces are essential for developing blend morphology and affect the migration and localization of nanoparticles in nanofiller-reinforced immiscible polymer blends.

2.3.4 *Effects Associated with Constituents*

2.3.4.1 Influence of Wetting Parameters

The use of wetting parameters in predicting particle localization in nanofiller-reinforced immiscible blends, the discrepancies from predictions, and the reasons behind such discrepancies are discussed in Sect. 2.3.1. Wetting parameters play a crucial role in the migration of nanoparticles from one component to another in immiscible polymer blends. This migration consists of two stages: first, from the bulk of the thermodynamically less preferred component to the interface, and second, from the interface to the thermodynamically more preferred component [159].

Various research groups have studied the role of wetting parameters in particle migration and localization [153, 161, 232–234]. For example, Johansen et al. [232], in their studies on the migration and localization of different types of nanosilica in PS/PC blends, considered the influence of temperature on wetting parameters in their calculations. The researchers calculated using a processing temperature range of 200–260 °C. Hydrophilic silica was predicted to be consistently localized within the PC minor phase, regardless of the processing temperature. In contrast, the predicted localization of hydrophobic silica was found to depend on the processing temperature. At lower (<200 °C) and higher processing temperatures (>200 °C), it was predicted to be localized at the interface and in the PS phase, respectively. Experimental results aligned with theoretical predictions to a significant extent. Hydrophilic silica was confined within the PC phase, while hydrophobic silica was found both at the interface and in the PS phase.

Similarly, Elias et al. [161] demonstrated the role of wetting parameters in the migration and localization of hydrophilic and hydrophobic nanosilica in PP/EVA blends. The blend nanocomposites were produced by melt processing in a mini-extruder for 5 min. In the case of hydrophilic silica, when the blend components and the filler were simultaneously melt mixed in an extruder, the silica particles were found to be localized in the more polar EVA phase, as expected from theoretical predictions using wetting parameters. Furthermore, EVA melted before PP, facilitating the incorporation of silica particles into the thermodynamically preferred phase more quickly. However, the situation differed for hydrophobic silica. The hydrophobic silica particles were mostly located at the interface, to a lesser extent in the PP phase, and partially in the EVA phase. The researchers attributed this observation to kinetic effects. Silica was incorporated into EVA in the early stages of compounding since EVA melted before PP. This silica was subsequently transferred to the interface and eventually to the more preferred PP phase. However, this transfer was slow due to the high viscosity of the PP phase. Thus, thermodynamic equilibrium was not achieved within the 5-min mixing time for hydrophobic silica, unlike in the case of hydrophilic silica.

The significance of the wetting coefficient's role in the morphological evolution of polymer blend composites was also emphasized by Zhang et al. [235]. The researchers demonstrated that the degree of nanoparticle agglomeration affects their wetting coefficient, which consequently influences their migration and localization. Studies were carried out on modified nanosilica-reinforced PMMA/SAN (60/40 wt/wt) blends with varying degrees of nanoparticle agglomeration. The blend nanocomposites were prepared through solution casting. TEM studies (Fig. 2.19) revealed that the isolated silica nanoparticles were situated at the interface, while their agglomerates were found in one phase of the blend. For nanoparticles with the same surface modification, the wetting parameter, as determined from (2.1), should remain constant, regardless of the degree of agglomeration. Thus, the localization of nanoparticles in polymer blends should not be affected by the quality of their dispersion. However, the experimental observation contradicted this expectation. The researchers accounted for this anomaly by introducing Wenzel's roughness parameter R in equation (2.39):

$$\omega_{AB} = \frac{\gamma_{PB} - \gamma_{PA}}{\gamma_{AB}} \times R \tag{2.39}$$

For closely packed spherical particles forming a flat surface, R is approximately 1.7–1.9 [236, 237]. When the particles are not closely packed, or the surface is not flat, R increases or decreases accordingly. Thus, it is easier for $|\omega_{AB}|$ of agglomerates to be greater than 1 in comparison with isolated nanoparticles. The researchers explained the preferential localization of nanoparticles in one phase of the blend, depending on the degree of agglomeration.

Therefore, it can be concluded that the migration and localization of nanoparticles in nanoparticle-reinforced immiscible polymer blends can be predicted using wetting parameters. However, the effects of temperature and the surface roughness of the dispersed particles on wetting parameters must be considered in theoretical predictions. Moreover, experimental observations may vary from theoretical predictions due to processing conditions and kinetic effects.

2.3.4.2 Influence of Particle Shape and Size

The influence of the shape and size of filler particles on their migration and localization in nanofiller-reinforced polymer blends has been examined by various research groups [149, 157, 164, 194, 209, 235, 238–240]. As discussed in Sect. 2.3.4.1, Zhang et al. [235] demonstrated the effects of particle shape and size in their studies on the impact of nanoparticle agglomeration on migration and localization (Fig. 2.19). The researchers explained their findings by proposing that the wetting parameter is influenced by Wenzel's roughness parameter, which in turn depends on particle shape and size. The studies conducted on PA66/(PS/PPE)/CNT nanocomposites by Lee et al. [164] further illuminate this direction. They observed that during the migration of filler particles at the blend interface, the interfacial curvature of low aspect ratio filler particles, such as CNT agglomerates, can relax, leading to a decrease in the driving force for migration. In contrast, the driving force for the migration of high aspect ratio filler particles, such as non-agglomerated CNTs, is stronger. Thus, individual CNT

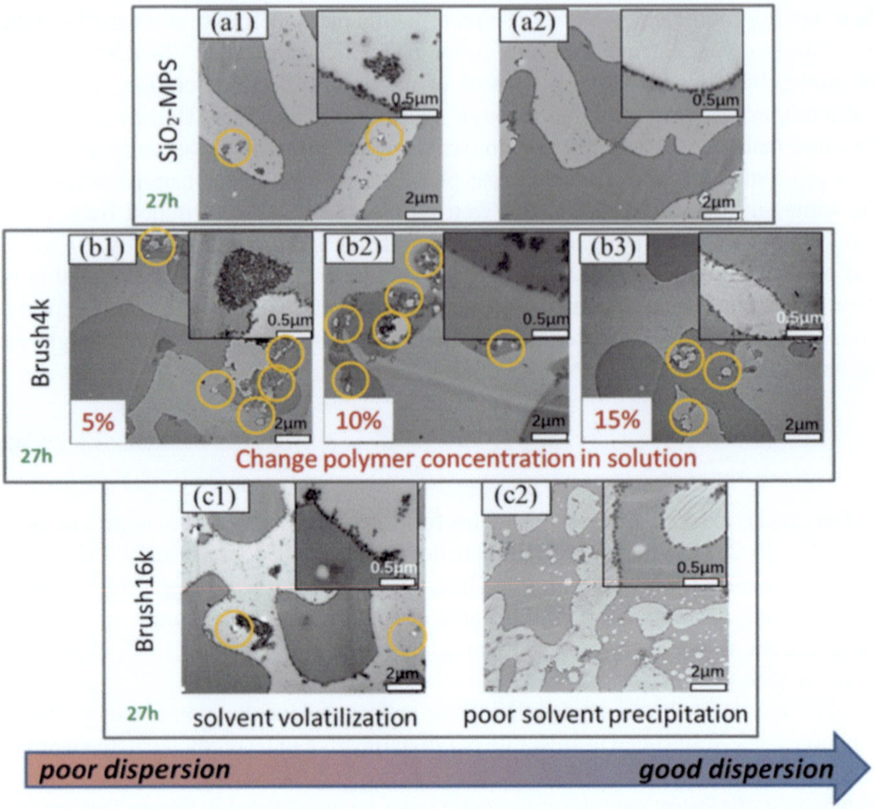

Fig. 2.19 Bright-field TEM micrographs of **a** PMMA/SAN/SiO_2-MPS 60/40/2, **b** PMMA/SAN/Brush4k 60/40/2, and **c** PMMA/SAN/Brush16k 60/40/5 nanocomposites annealed for 27 h at 180 °C. The bright and dark regions refer to the PMMA-rich and SAN-rich phases, respectively. Reproduced with permission from [235]. Copyright 2021, Elsevier Science Ltd.

particles were found to migrate faster and more effectively than CNT agglomerates. The role of particle shape and size in the morphological development of polymer blend nanocomposites was also reported by Liebscher et al. [209] in their studies on GnP-reinforced PC/SAN blends. They observed that the GnPs preferred to localize within PC, which was attributed to thermodynamic reasons. When pre-dispersed in SAN, the GnPs migrated to the thermodynamically preferred PC phase. However, this migration was hindered in the case of large filler agglomerates. The researchers explained that their observations were consistent with SFM [157], attributing the hindered migration of particles to the shape and size of agglomerates.

As discussed in Sect. 2.3.2, SFM was proposed by Göldel et al. [157] in their comparative studies on the migration and localization of filler particles of different shapes and sizes (CNTs and CB) in PC/SAN blends. It was observed that the low aspect ratio filler particles (CNTs) were less capable of penetrating blend interfaces than the high aspect ratio filler particles (CB). Thus, when the filler was pre-dispersed in the thermodynamically less favorable component SAN and the pre-mixture was subsequently blended with PC, most CNTs were found to migrate to PC. In contrast, most of the CB particles were found at the interface. The researchers proposed that interfacial curvature instability is the driving force for the migration of filler particles. An individual 1D CNT perpendicular to the direction of the interface has the maximum instability with constant curvature throughout the penetration process. Relaxation occurs only upon migration. In contrast, in the case of spherical CB, the interface relaxes during penetration, resulting in a decreased driving force for migration. The fact that some CNTs were also found at the interface even after 5 min of melt blending was explained by considering non-ideal cases (agglomerates and randomly coiled structures). In such cases, penetration occurs in the sequence: (i) the sections of CNTs that are perpendicular to the interface, (ii) the sections parallel to the interface, and (iii) coiled portions, resulting in a slower migration rate compared to individual CNTs perpendicular to the interface. The CNTs that migrated to the PC phase followed the SFM route of migration. The researchers also classified fillers of various particle shapes and sizes based on the speed of migration across the interface.

The role of particle shape and size in the morphology development of polymer blend nanocomposites was also discussed by Dil and Favis [194] in their studies on micro- and nano-silica-reinforced PLA/PBAT blends. Two grades of PLA with high and low viscosities were used for the studies. When silica was compounded with a PLA/PBAT melt, the silica particles in the ternary compositions were found to be localized in the thermodynamically preferred PBAT phase, regardless of the filler particle size and the viscosity of PLA. However, when silica was premixed with PLA, and the premixed blend was subsequently combined with PBAT, the results differed. In the case of nano-silica, the filler particles were localized at the interface, irrespective of the viscosity of PLA. For micro-silica, the location of the filler particles depended on the viscosity of PLA. When low-viscosity PLA was employed, the micro-silica particles were localized at the interface. Conversely, when high-viscosity PLA was used, they remained in the PLA phase. This was attributed to the long film-draining time between the large filler particles and the interface. The film drainage time depended on the fourth power of the filler particle radius and was also directly proportional to the viscosity of PLA. Thus, nano-silica exhibited much shorter film drainage times and migrated to the interface, regardless of the viscosity of PLA. The micro-silica particles exhibited a much longer film drainage time and managed to migrate to the interface only when the viscosity of PLA was low.

The influence of filler particle shape on the morphology of immiscible blend nanocomposites is also reflected in the ability of Janus nanoparticles to stabilize co-continuous blend morphologies [158, 241–243]. Janus particles have been found to exhibit a superior capability to stabilize co-continuous morphologies during annealing. This has been attributed to a reduction in the radius of the jamming curvature at the interface. An increase in the shape asymmetry of the Janus particles has been shown to enhance this reduction, leading to greater stabilization of the morphology [241].

Thus, it can be summarized that the shape and size of filler particles influence their migration and localization and may also affect the morphological stability of nanoparticle-reinforced immiscible polymer blends. In general, smaller particle sizes and longer aspect ratios favor faster migration. However, longer aspect ratio CNTs also tend to undergo attrition during compounding. The effect of attrition on the final aspect ratio and its subsequent impact on particle migration and localization must be better understood through more studies [159].

2.3.4.3 Influence of Viscosity Ratio

The role of viscosity ratio in the development of the morphology of multiphase polymeric systems has been extensively reported in the literature [142, 145, 244–246]. Although the final localization of filler particles in nanofiller-reinforced immiscible polymer blends is governed by thermodynamics, the viscosity ratio influences their migration rate. The migration rate is a function of viscosity, and an increase in melt viscosity results in a decrease in the rate of migration. The effect of viscosity ratio on the migration of filler particles has been found to be more pronounced in cases of sequential compounding. If the filler is pre-mixed in the thermodynamically less preferred component, which also has a higher viscosity, insufficient mixing time may cause the filler particles to remain localized within the thermodynamically less preferred component due to their slow migration rate. Furthermore, the viscosity ratio significantly influences droplet deformation. Thus, when the viscosity ratio favors droplet breakup, collisions between these droplets may cause the filler particles to become trapped within the thermodynamically less favored phase, as explained in Sect. 2.3.2 [159]. The role of viscosity ratio in particle migration and localization was investigated by Feng et al. [247] in their studies on CB-reinforced PMMA/PP blends. A ternary composition with a PP:PMMA volume ratio of 2.75 and CB loading of 10 wt% was produced by melt-mixing all components in an internal mixer. Different grades of PMMA were used to adjust the viscosity ratio of the polymers. The researchers observed that when the viscosity of PMMA was significantly higher than that of PP, the CB particles were preferentially localized in PP, even though PMMA was the thermodynamically preferred phase. This was attributed to the difficulty of the filler particles diffusing into the higher viscosity PMMA phase. Lowering the viscosity of PMMA initially resulted in the presence of the CB particles at the interface. When the viscosity of PMMA was further lowered, bringing the viscosity ratio close to 1, the CB particles became localized in the thermodynamically preferred PMMA phase. Hence, the migration and localization of filler particles in the ternary composition were found to be profoundly influenced by the viscosity ratio of the polymers. Similar findings were reported by Yuan et al. [248] in their studies on CNT-reinforced LDPE/PVDF blends, where the CNTs were localized in the thermodynamically less preferred LDPE phase due to the significantly lower viscosity of LDPE. The slow movement of filler particles within and into the high viscosity polymer phase has been demonstrated by various other research groups [225, 249, 250].

The influence of viscosity ratio on particle migration and localization was also demonstrated by Dil and Favis [194] in their studies on micro-silica reinforced PLA/PBAT blends. Their work is discussed in greater detail in Sect. 2.3.4.2.

An in-depth analysis of the role of viscosity ratio on particle migration and localization can be found in the work of Plattier et al. [251]. The researchers conducted extensive studies on CB-filled PP/PCL blends. A co-continuous PP/PCL blend with 60 wt% PP content was prepared by melt mixing at 200 °C for 10 min in a micro-compounder using a screw speed of 20 rpm. 17 phr of CB was added to this molten material, and mixing was continued for another 10 min before extrusion. Viscosity ratio ($K = \eta_{PCL} / \eta_{PP}$) of the blends was varied in the range 0.06–14.7 using different PP and PCL grades. When the viscosity ratio was close to 1, the filler particles were found to be localized at the interface. The researchers proposed that when the filler particles reached the interface under the influence of shear, they experienced similar forces exerted by the two molten polymers, which had comparable viscosity. These forces balanced each other, allowing the particles to remain at the interface. When the viscosity ratio differed significantly from 1, the particles were observed to be localized in the more viscous phase. This was attributed to the greater drag forces generated by the more viscous polymer. The researchers proposed equation (2.40) that quantified the drag forces (F) exerted by the two polymers on a filler particle at the interface:

$$\frac{F_{PCL}}{F_{PP}} = \frac{(1 + \cos\theta)}{(1 - \cos\theta)} K \tag{2.40}$$

where θ is the contact angle at the interface. Thus, a filler particle at the interface experiences a greater drag force from the more viscous polymer, leading to its migration to the more viscous phase. The viscosity ratio of blend components not only influences the migration and localization of filler particles, but the reverse is also true, which may significantly impact the blend morphology in nanofiller-reinforced immiscible polymer blends. Rafeie et al. [252, 253] worked in this direction. The researchers reported that GnP-reinforced linear low-density polyethylene (LLDPE)/PVDF blends with a 1:1 ratio of blend components, produced by melt blending in an internal mixer, exhibited a co-continuous morphology [252]. This contrasted with their earlier observations [253], where an unfilled LLDPE/PVDF blend with the same composition, produced in the same manner, was found to possess a droplet matrix morphology. The co-continuous morphology of the GnP-filled blends was attributed to an enhancement in the viscosity ratio (η_{PVDF}/η_{LLDPE}) because of the selective localisation of the filler particles in the PVDF phase. The researchers, thus, demonstrated the effect of selective localisation of filler particles on the viscosity ratio of blend components and its consequent impact on the morphology of nanofiller-reinforced polymer blends. Other researchers have also reported the effect of filler localisation on viscosity ratio and blend morphology [254–260].

Thus, it can be summarized that the viscosity ratio of blend components influences the migration and localization of filler particles in polymer blend composites. Filler particles experience a slower rate of motion within the more viscous polymer phase. The effect of viscosity ratio on the migration and localization of filler particles becomes more evident when compounding a ternary composition sequentially. The selective localization of filler particles in one of the phases of a polymer blend also alters the viscosity ratio of the blend components, which may, in turn, influence the morphology of the blend.

2.4 Conclusions

The properties of an immiscible polymer blend are dictated by the blend components and morphology, the latter being significantly influenced by the volume fractions and viscosities of the phases, interfacial tension, shear rate used during mixing, and mixing time. Morphology develops through a sheeting mechanism. Compatibilization lowers interfacial tension and assists in achieving and maintaining the desired level of dispersion with improved adhesion between the phases. Reactive compatibilization is more effective than compatibilization through addition.

The morphology and other properties of immiscible blend nanocomposites are largely influenced by the location of the filler particles. While nanoparticles tend to gather in the thermodynamically preferred phase of polymer blends, it is possible to tailor this localization by modifying certain parameters. For instance, because thermodynamic equilibrium is influenced by the wetting parameter, which is governed by interfacial tension, altering the characteristics of the filler particles (shape and size) can aid in adjusting its localization. By introducing functional groups on the filler surface and inducing interfacial reactions, filler particles may become trapped at the interface. The use of nanofillers in interfacial reactions has been discussed in greater detail in Chap. 5.

Kinetics and processing conditions play a crucial role in achieving thermodynamic equilibrium. The processing time may be adjusted so that sufficient time is not available for the nanoparticles to reach thermodynamic equilibrium. This can be particularly accomplished in sequential blending when the filler is first dispersed in the thermodynamically less favored phase or in single-step blending when the thermodynamically less favored phase melts first, causing the nanoparticles to disperse within it. The time required for migration may be modified by altering the viscosity ratio, as an increase in melt viscosity leads to a decrease in the migration rate. In summary, the migration and localization of nanoparticles in immiscible polymer blends are influenced by various factors (thermodynamic and kinetic), none of which can be overlooked. This complexity results in the migration process and associated mechanisms, where a change in any single parameter can produce a markedly different morphology of the ternary composite. Therefore, extensive morphological characterization of polymer blend nanocomposites is essential for gaining a deeper understanding of the structure–property-processing relationship of these materials, which is vital for material development. The techniques used for characterizing polymer blends are discussed in Chap. 4.

Now a question arises: why is tailoring the migration of nanoparticles necessary? The answer can be found in several studies where researchers have demonstrated that by tailoring the localization of nanoparticles, one can achieve the desired properties of blend composites. For example, the key to achieving conductive polymer composites with a low percolation threshold is developing an interconnected network of conductive filler particles. This can be readily accomplished by trapping the conductive filler particles at the interface. Similarly, interfacially located filler particles with strong interfacial adhesion have been found to result in superior mechanical properties. The properties of polymer blend nanocomposites and their dependence on filler migration and localization have been discussed in Chap. 6.

References

1. M. Nofar, R. Salehiyan, S.S. Ray, Influence of nanoparticles and their selective localization on the structure and properties of polylactide-based blend nanocomposites. Compos. Part B Eng. **215**, 108845 (2021). https://doi.org/10.1016/J.COMPOSITESB.2021.108845
2. L.A. Utracki, Introduction to polymer blends, in *Polymer Blends Handbook*, ed. by L.A. Utracki, vol. 1 (Kluwer Academic Publishers, Montreal, 1999), pp. 1–122
3. L.A. Utracki, *Commercial Polymer Blends* (Springer Science & Business Media, New York, 2013)
4. C.W. Macosko, Morphology development and control in immiscible polymer blends. Macromol. Symp. **149**, 171–184 (2000). https://doi.org/10.1002/1521-3900(200001)149:1%3c171::AID-MASY171%3e3.0.CO;2-8
5. L.A. Utracki, Compatibilization of polymer blends. Can. J. Chem. Eng. **80**, 1008–1016 (2002). https://doi.org/10.1002/cjce.5450800601
6. A. Ajji, L.A. Utracki, Interphase and compatibilization of polymer blends. Polym. Eng. Sci. **36**, 1574–1585 (1996). https://doi.org/10.1002/pen.10554
7. R.C. Willemse, A. Posthuma de Boer, J. van Dam, A.D. Gotsis, Co-continuous morphologies in polymer blends: a new model. Polymer (Guildf.) **39**, 5879–5887 (1998). https://doi.org/10.1016/S0032-3861(97)10200-2
8. R.J. Shields, D. Bhattacharyya, S. Fakirov, Fibrillar polymer–polymer composites: morphology, properties and applications. J. Mater. Sci. **43**, 6758–6770 (2008). https://doi.org/10.1007/s10853-008-2693-z
9. H. Veenstra, J. Van Dam, A. Posthuma de Boer, On the coarsening of co-continuous morphologies in polymer blends: effect of interfacial tension, viscosity and physical cross-links. Polymer (Guildf.) **41**, 3037–3045 (2000). https://doi.org/10.1016/S0032-3861(99)00455-3
10. J. Lyngaae-Jørgensen, L.A. Utracki, Dual phase continuity in polymer blends. Makromol. Chemie Macromol. Symp. **48–49**, 189–209 (1991). https://doi.org/10.1002/masy.19910480117
11. L.A. Utracki, *Commercial Polymer Blends* (Chapman & Hall, London, 1998)
12. E. Helfand, Y. Tagami, Theory of the interface between immiscible polymers. J. Polym. Sci. Part B Polym. Lett. **9**, 741–746 (1971). https://doi.org/10.1002/pol.1971.110091006
13. E. Helfand, Y. Tagami, Theory of the interface between immiscible polymers II. J. Chem. Phys. **56**, 3592–3601 (1972)
14. S. Yukioka, T. Inoue, Ellipsometric studies on immiscible polymer-polymer interfaces. Polymer (Guildf.) **34**, 1256–1259 (1993). https://doi.org/10.1016/0032-3861(93)90782-6
15. J. Kressler, N. Higashida, T. Inoue, W. Heckmann, F. Seitz, Study of polymer-polymer interfaces: a comparison of ellipsometric and TEM data of PMMA/PS and PMMA/SAN systems. Macromolecules **26**, 2090–2094 (1993). https://doi.org/10.1021/ma00060a043

16. A. Luciani, L.A. Utracki, The extensional flow mixer. EFM. Int. Polym. Process **11**, 299–309 (1996). https://doi.org/10.3139/217.960299
17. A. Luciani, M.F. Champagne, L.A. Utracki, Interfacial tension determination by retraction of an ellipsoid. J. Polym. Sci. Part B Polym. Phys. **35**, 1393–1403 (1997)
18. A. Luciani, M.F. Champagne, L.A. Utracki, Interfacial tension in polymer blends. Macromol. Symp. **126**, 307–321 (1998). https://doi.org/10.1002/masy.19981260125
19. L. Leibler, Emulsifying effects of block copolymers in incompatible polymer blends. Makromol. Chemie Macromol. Symp. **16**, 1–17 (1988). https://doi.org/10.1002/masy.19880160103
20. S. Wu, *Polymer Interface and Adhesion* (M. Dekker, New York, 1982)
21. A.W. Adamson, *Physical Chemistry of Surfaces* (Wiley, New York, 1982)
22. S.H. Anastasiadis, J.K. Chen, J.T. Koberstein, A.F. Siegel, J.E. Sohn, J.A. Emerson, The determination of interfacial tension by video image processing of pendant fluid drops. J. Colloid Interface Sci. **119**, 55–66 (1987). https://doi.org/10.1016/0021-9797(87)90244-X
23. N.R. Demarquette, M.R. Kamal, Interfacial tension in polymer melts. I: An improved pendant drop apparatus. Polym. Eng. Sci. **34**, 1823–1833 (1994). https://doi.org/10.1002/pen.760342407
24. M.R. Kamal, R. Lai-Fook, N.R. Demarquette, Interfacial tension in polymer melts. Part II: Effects of temperature and molecular weight on interfacial tension. Polym. Eng. Sci. 34, 1834–1839 (1994). https://doi.org/10.1002/pen.760342408
25. L.M. Coucoulas, R.A. Dawe, The calculation of interfacial tension from sessile drops. J. Colloid Interface Sci. **103**, 230–236 (1985). https://doi.org/10.1016/0021-9797(85)90095-5
26. H.T. Patterson, K.H. Hu, T.H. Grindstaff, Measurement of interfacial and surface tensions in polymer systems. J. Polym. Sci. Part C Polym. Symp. **34**, 31–43 (2007). https://doi.org/10.1002/polc.5070340106
27. D.D. Joseph, M.S. Arney, G. Gillberg, H. Hu, D. Hultman, C. Verdier et al., A spinning drop tensioextensometer. J. Rheol. (N. Y. N. Y.) **36**, 621–662 (1992). https://doi.org/10.1122/1.550311
28. J.J. Elmendorp, A.K. Van Der Vegt, A study on polymer blending microrheology: Part IV. The influence of coalescence on blend morphology origination. Polym. Eng. Sci. 26, 1332–1338 (1986). https://doi.org/10.1002/pen.760261908
29. P.H.M. Elemans, J.M.H. Janssen, H.E.H. Meijer, The measurement of interfacial tension in polymer/polymer systems: the breaking thread method. J. Rheol. (N. Y. N. Y.) **34**, 1311–1325 (1990). https://doi.org/10.1122/1.550087
30. K. Cho, H.K. Jeon, C.E. Park, J. Kim, K.U. Kim, The effect of end-sulfonated polystyrene on the interfacial tension of nylon-6/polystyrene blends. Polymer (Guildf.) **37**, 1117–1122 (1996). https://doi.org/10.1016/0032-3861(96)80837-8
31. C.J. Carriere, A. Cohen, C.B. Arends, Estimation of interfacial tension using shape evolution of short fibers. J. Rheol. (N. Y. N. Y.) **33**, 681–689 (1989). https://doi.org/10.1122/1.550033
32. A. Cohen, C.J. Carriere, Analysis of a retraction mechanism for imbedded polymeric fibers. Rheol. Acta **28**, 223–232 (1989). https://doi.org/10.1007/BF01332854
33. T. Rundqvist, A. Cohen, C. Klason, The imbedded disk retraction method for measurement of interfacial tension between polymer melts. Rheol. Acta **35**, 458–469 (1996). https://doi.org/10.1007/BF00368996
34. P. Xing, M. Bousmina, D. Rodrigue, M.R. Kamal, Critical experimental comparison between five techniques for the determination of interfacial tension in polymer blends: model system of polystyrene/polyamide-6. Macromolecules **33**, 8020–8034 (2000). https://doi.org/10.1021/ma000537x
35. D.F. Siqueira, D.W. Schubert, V. Erb, M. Stamm, J.P. Amato, Interface thickness of the incompatible polymer system PS/PnBMA as measured by neutron reflectometry and ellipsometry. Colloid Polym. Sci. **273**, 1041–1048 (1995). https://doi.org/10.1007/BF00657671
36. F. Bruder, R. Brenn, B. Stühn, G.R. Strobl, Interdiffusion in the partially miscible polymer blend of deuterated polystyrene and poly(styrene-co-bromostyrene). Macromolecules **22**, 4434–4437 (1989). https://doi.org/10.1021/ma00202a009

37. J. Kressler, T. Inoue, Thermodynamics and interfaces of polymer blends. Front. Polym. Adv. Mater. (Springer US) 561–574 (1994). https://doi.org/10.1007/978-1-4615-2447-2_52
38. S.A. Schwarz, B.J. Wilkens, M.A.A. Pudensi, M.H. Rafailovich, J. Sokolov, X. Zhao et al., Studies of surface and interface segregation in polymer blends by secondary ion mass spectrometry. Mol. Phys. **76**, 937–950 (1992). https://doi.org/10.1080/00268979200101771
39. M. Stamm, Polymer interfaces on a molecular scale: comparison of techniques and some examples. Adv. Polym. Sci. **100**, 356–400 (1991). https://doi.org/10.1007/bfb0051640
40. S. Horiuchi, D. Yin, T. Ougizawa, Nanoscale analysis of polymer interfaces by energy-filtering transmission electron microscopy. Macromol. Chem. Phys. **206**, 725–731 (2005). https://doi.org/10.1002/macp.200400519
41. R. Schnell, M. Stamm, F. Rauch, Segregation of diblock copolymers to the interface between weakly incompatible polymers. Macromol. Chem. Phys. **200**, 1806–1812 (1999). https://doi.org/10.1002/(SICI)1521-3935(19990701)200:7%3c1806::AID-MACP1806%3e3.0.CO;2-9
42. C. Yeung, A.C. Shi, Formation of interfaces in incompatible polymer blends: a dynamical mean field study. Macromolecules **32**, 3637–3642 (1999). https://doi.org/10.1021/ma981648n
43. J.T. Koberstein, B. Morra, R.S. Stein, The determination of diffuse-boundary thicknesses of polymers by small-angle X-ray scattering. J. Appl. Crystallogr. **13**, 34–45 (1980). https://doi.org/10.1107/s0021889880011478
44. G.C. Gemeinhardt, R.B. Moore, Characterization of ionomer-compatibilized blend morphology using synchrotron small-angle X-ray scattering. Macromolecules **38**, 2813–2819 (2005). https://doi.org/10.1021/ma0479823
45. D.W. Schubert, M. Stamm, Influence of chain length on the interface width of an incompatible polymer blend. Europhys. Lett. **35**, 419–424 (1996). https://doi.org/10.1209/epl/i1996-00130-3
46. M. Stamm, G. Reiter, K. Kunz, The use of X-ray and neutron reflectometry for the investigation of polymeric thin films. Phys. B Phys. Condens. Matter. **173**, 35–42 (1991). https://doi.org/10.1016/0921-4526(91)90032-A
47. D.W. Schubert, M. Stamm, Neutron reflectometry and small-angle neutron scattering—two complementary techniques for polymer blend investigations. Phys. B Condens. Matter **234–236**, 286–288 (1997). https://doi.org/10.1016/S0921-4526(96)00959-3
48. M. Hayashi, T. Hashimoto, H. Hasegawa, M. Takenaka, H. Grüll, A.R. Esker et al., Interface between a polysulfone and polyamide as studied by combined neutron reflectivity and small-angle neutron scattering techniques. Macromolecules **33**, 8375–8387 (2000). https://doi.org/10.1021/ma000661g
49. P. Müller-Buschbaum, S.A. O'Neill, S. Affrossman, M. Stamm, Phase separation and dewetting of weakly incompatible polymer blend films. Macromolecules **31**, 5003–5009 (1998). https://doi.org/10.1021/ma971802q
50. M. Tirrell, E.E. Parsonage, Structure and properties of polymers. Mater. Sci. Technol. **12**, 653 (1993)
51. M. Stamm, D.W. Schubert, Interfaces between incompatible polymers. Annu. Rev. Mater. Sci. **25**, 325–356 (1995). https://doi.org/10.1146/annurev.ms.25.080195.001545
52. H.R. Brown, Effect of a diblock copolymer on the adhesion between incompatible polymers. Macromolecules **22**, 2859–2860 (1989). https://doi.org/10.1021/ma00196a059
53. S.B. Brown, Reactive compatibilization of polymer blends, in *Polymer Blends Handbook*, ed. by L.A. Utracki (Kluwer Academic Press, Dordrecht, 2002), pp. 339–415
54. C.E. Scott, C.W. Macosko, Model experiments concerning morphology development during the initial stages of polymer blending. Polym. Bull. **26**, 341–348 (1991). https://doi.org/10.1007/BF00587979
55. C.E. Scott, C.W. Macosko, Morphology development during the initial stages of polymer-polymer blending. Polymer (Guildf.) **36**, 461–470 (1995). https://doi.org/10.1016/0032-3861(95)91554-K
56. J.T. Lindt, A.K. Ghosh, Fluid mechanics of the formation of polymer blends. Part I: Formation of lamellar structures. Polym. Eng. Sci. **32**, 1802–1813 (1992). https://doi.org/10.1002/pen.760322403

57. U. Sundararaj, C.W. Macosko, Drop breakup and coalescence in polymer blends: the effects of concentration and compatibilization. Macromolecules **28**, 2647–2657 (1995). https://doi.org/10.1021/ma00112a009
58. U. Sundararaj, Y. Dori, C.W. Macosko, Visualization of morphology development in polymer blends, in *Proceedings of the 52nd Annual Technical Conference ANTEC 94* (Society of Plastics Engineers, San Francisco, 1994), pp. 2448–2449
59. U. Sundararaj, Y. Dori, C.W. Macosko, Sheet formation in immiscible polymer blends: model experiments on initial blend morphology. Polymer (Guildf.) **36**, 1957–1968 (1995). https://doi.org/10.1016/0032-3861(95)91438-D
60. H.P. Grace, Dispersion phenomena in high viscosity immiscible fluid systems and application of static mixers as dispersion devices in such systems. Chem. Eng. Commun. **14**, 225–277 (1982). https://doi.org/10.1080/00986448208911047
61. J.J. Elmendrop, *A Study on Polymer Blending Microrheologie*. Doctoral dissertation, Delft University of Technology, 1986
62. J.M.H. Janssen, *Dynamics of Liquid–Liquid Mixing.*. Doctoral dissertation, Eindhoven University of Technology, 1993
63. R.C. Willemse, E.J.J. Ramaker, J. Van Dam, A. Posthuma De Boer, Morphology development in immiscible polymer blends: initial blend morphology and phase dimensions. Polymer (Guildf.) 40, 6651–6659 (1999). https://doi.org/10.1016/S0032-3861(99)00038-5
64. P.H.M. Elemans, *Modelling of the Processing of Incompatible Polymer Blends*. Doctoral dissertation, Eindhoven University of Technology, 1989
65. W. Kuhn, Spontane Aufteilung von Flüssigkeitszylindern in kleine Kugeln. Kolloid-Zeitschrift **132**, 84–99 (1953). https://doi.org/10.1007/BF01513708
66. T. Mikami, R.G. Cox, S.G. Mason, Breakup of extending liquid threads. Int. J. Multiph. Flow **2**, 113–138 (1975). https://doi.org/10.1016/0301-9322(75)90003-8
67. S. Tomotika, On the instability of a cylindrical thread of a viscous liquid surrounded by another viscous fluid. Proc. R. Soc. Lond. Ser. A Math. Phys. Sci. **150**, 322–337 (1935). https://doi.org/10.1098/rspa.1935.0104
68. A.P. Plochocki, S.S. Dagli, R.D. Andrews, The interface in binary mixtures of polymers containing a corresponding block copolymer: effects of industrial mixing processes and of coalescence. Polym. Eng. Sci. **30**, 741–752 (1990). https://doi.org/10.1002/pen.760301207
69. C.-K. Shih, Mixing and morphological transformations in the compounding process for polymer blends: the phase inversion mechanism. Polym. Eng. Sci. **35**, 1688–1694 (1995). https://doi.org/10.1002/pen.760352105
70. U. Sundararaj, C.W. Macosko, R.J. Rolando, H.T. Chan, Morphology development in polymer blends. Polym. Eng. Sci. **32**, 1814–1823 (1992). https://doi.org/10.1002/pen.760322404
71. D.G. Baird, D.I. Collias, *Polymer Processing* (Butterworth-Heinemann, Boston, 1995)
72. W.J. Schrenk, N.L. Bradley, T. Alfrey, H. Maack, Interfacial flow instability in multilayer coextrusion. Polym. Eng. Sci. **18**, 620–623 (1978). https://doi.org/10.1002/pen.760180803
73. C.W. Macosko, P. Guégan, A.K. Khandpur, A. Nakayama, P. Marechal, T. Inoue, Compatibilizers for melt blending: premade block copolymers. Macromolecules **29**, 5590–5598 (1996). https://doi.org/10.1021/ma9602482
74. M. Maric, Compatibilization of model poly (styrene)/poly (dimethylsiloxane) blends, in *ANTEC*, Tech. Pap. 45 (Soc. Plast. Eng., New York, 1999)
75. M.S. Lee, T.P. Lodge, C.W. Macosko, Can random copolymers serve as effective polymeric compatibilizers? J. Polym. Sci. Part B Polym. Phys. **35**, 2835–2842 (1997). https://doi.org/10.1002/(SICI)1099-0488(199712)35:17%3c2835::AID-POLB8%3e3.0.CO;2-P
76. M.S. Lee, T.P. Lodge, C.W. Macosko, Blend encapsulation by random copolymers: C/B/A-ran-B and C/D/A-ran-B systems. Macromol. Chem. Phys. **199**, 1555–1559 (1998). https://doi.org/10.1002/(SICI)1521-3935(19980801)199:8%3c1555::AID-MACP1555%3e3.0.CO;2-O
77. Y. Koseki, M.S. Lee, C.W. Macosko, Encapsulation in ternary elastomer blends. Rubber Chem. Technol. **72**, 109–118 (1999). https://doi.org/10.5254/1.3538780
78. P. Guegan, A.K. Khandpur, C.W. Macosko, ABC triblocks as compatibilizers for mechanically mixed immiscible polymer blends. ACS Polym. Prepr. **36**, 188–189 (1995)

79. P. Guégan, J.J. Cernohous, A.K. Khandpur, T.R. Hoye, C.W. Macosko, Anionic synthesis, modification, and characterization of ABC triblock copolymers. Macromolecules **29**, 4605–4612 (1996). https://doi.org/10.1021/ma951193r
80. C. Scott, C. Macosko, Model experiments for the interfacial reaction between polymers during reactive polymer blending. J. Polym. Sci. Part B Polym. Phys. **32**, 205–213 (1994). https://doi.org/10.1002/polb.1994.090320201
81. A.R. Padwa, Y. Sasaki, K.A. Wolske, C.W. Macosko, Kinetics of amine-cyclic anhydride reactions in moderately polar solutions. J. Polym. Sci. Part A Polym. Chem. **33**, 2165–2174 (1995). https://doi.org/10.1002/pola.1995.080331310
82. C.A. Orr, A. Adedeji, A. Hirao, F.S. Bates, C.W. Macosko, Flow-induced reactive self-assembly. Macromolecules **30**, 1243–1246 (1997). https://doi.org/10.1021/ma961574k
83. C.E. Scott, C.W. Macosko, Morphology development during reactive and non-reactive blending of an ethylene-propylene rubber with two thermoplastic matrices. Polymer (Guildf.) **35**, 5422–5433 (1994). https://doi.org/10.1016/S0032-3861(05)80005-9
84. C.E. Scott, C.W. Macosko, Compounding and morphology of nylon/ethylene-propylene rubber reactive and nonreactive blends. Int. Polym. Process. **10**, 36–45 (2013). https://doi.org/10.3139/217.950036
85. P. Guégan, C.W. Macosko, T. Ishizone, A. Hirao, S. Nakahama, Kinetics of chain coupling at melt interfaces. Macromolecules **27**, 4993–4997 (1994). https://doi.org/10.1021/ma00096a022
86. C.E. Scott, C.W. Macosko, Processing and morphology of polystyrene/ethylene-propylene rubber reactive and nonreactive blends. Polym. Eng. Sci. **35**, 1938–1948 (1995). https://doi.org/10.1002/pen.760352405
87. A. Nakayama, T. Inoue, P. Guégan, C.W. Macosko, Compatibilizers for melt blending: premade vs. reactively formed block copolymers. Polym. Prepr. **34**, 840–841 (1993)
88. B. Majumdar, D.R. Paul, A.J. Oshinski, Evolution of morphology in compatibilized vs. uncompatibilized polyamide blends. Polymer (Guildf.) **38**, 1787–1808 (1997). https://doi.org/10.1016/S0032-3861(96)00702-1
89. P.A. Bhadane, A.H. Tsou, J. Cheng, B.D. Favis, Morphology development and interfacial erosion in reactive polymer blending. Macromolecules **41**, 7549–7559 (2008). https://doi.org/10.1021/ma801390s
90. H. Li, U. Sundararaj, Morphology development of polymer blends in extruder: the effects of compatibilization and rotation rate. Macromol. Chem. Phys. **210**, 852–863 (2009). https://doi.org/10.1002/macp.200800543
91. S. Kim, J.K. Kim, C.E. Park, Effect of molecular architecture of in situ reactive compatibilizer on the morphology and interfacial activity of an immiscible polyolefin/polystyrene blend. Polymer (Guildf.) **38**, 1809–1815 (1997). https://doi.org/10.1016/S0032-3861(96)00714-8
92. P. Pötschke, D.R. Paul, Formation of co-continuous structures in melt-mixed immiscible polymer blends. J. Macromol. Sci. Polym. Rev. **43**, 87–141 (2003). https://doi.org/10.1081/MC-120018022
93. W.P. Gergen, R.G. Lutz, S. Davison, Hydrogenated block copolymers in thermoplastic elastomer interpenetrating polymer networks, in *Thermoplastic Elastomers*, ed. by G. Holden, N.R. Legge, R. Quirk, H.E. Schroeder (Hanser Publishers, Munich, 1996)
94. D.R. Paul, S. Newman, *Polymer Blends*, vol. 1 (Academic Press, New York, 1978)
95. J. Lyngaae-Jørgensen, K.L. Rasmussen, E.A. Chtcherbakova, L.A. Utracki, Flow induced deformation of dual-phase continuity in polymer blends and alloys. Part I. Polym. Eng. Sci. **39**, 1060–1071 (1999). https://doi.org/10.1002/pen.11494
96. I.S. Miles, A. Zurek, Preparation, structure, and properties of two-phase co-continuous polymer blends. Polym. Eng. Sci. **28**, 796–805 (1988)
97. N. Mekhilef, B.D. Favis, P.J. Carreau, Morphological stability, interfacial tension, and dual-phase continuity in polystyrene-polyethylene blends. J. Polym. Sci. Part B Polym. Phys. **35**, 293–308 (1997). https://doi.org/10.1002/(SICI)1099-0488(19970130)35:2%3c293::AID-POLB7%3e3.0.CO;2-T

98. J. He, W. Bu, J. Zeng, Co-phase continuity in immiscible binary polymer blends. Polymer (Guildf.) **38**, 6347–6353 (1997). https://doi.org/10.1016/S0032-3861(97)00212-7
99. J. Lyngaae-Jorgensen, A. Kuta, K. Sondergaard, K.V. Poulsen, Structure and properties of polymer blends with dual phase continuity. Polym. Netw. Blends **3**, 1–13 (1993)
100. L.A. Utracki, On the viscosity-concentration dependence of immiscible polymer blends. J. Rheol. (N. Y. N. Y.) **35**, 1615–1637 (1991). https://doi.org/10.1122/1.550248
101. F. Gubbels, S. Blacher, E. Vanlathem, R. Jérôme, R. Deltour, F. Brouers et al., Design of electrical conductive composites: key role of the morphology on the electrical properties of carbon black filled polymer blends. Macromolecules **28**, 1559–1566 (1995). https://doi.org/10.1021/ma00109a030
102. G.M. Jordhamo, J.A. Manson, L.H. Sperling, Phase continuity and inversion in polymer blends and simultaneous interpenetrating networks. Polym. Eng. Sci. **26**, 517–524 (1986). https://doi.org/10.1002/pen.760260802
103. L. Utracki, *Polymer Alloys and Blends: Thermodynamics and Rheology* (Hanser Gardner Publications, Munich, 1989)
104. C.D. Han, *Multiphase Flow in Polymer Processing* (Academic Press, New York, 1981)
105. G.N. Avgeropoulos, F.C. Weissert, P.H. Biddison, G.G.A. Boehm, Heterogeneous blends of polymers. Rheology and morphology. Rubber Chem. Technol. **49**, 93–104 (1976). https://doi.org/10.5254/1.3534954
106. C.J. Nelson, G.N. Avgeropoulos, F.C. Weissert, G.G.A. Böhm, The relationship between rheology, morphology and physical properties in heterogeneous blends. Angew. Makromol. Chemie **60**, 49–86 (1977). https://doi.org/10.1002/apmc.1977.050600103
107. N. Mekhilef, H. Verhoogt, Phase inversion and dual-phase continuity in polymer blends: theoretical predictions and experimental results. Polymer (Guildf.) **37**, 4069–4077 (1996). https://doi.org/10.1016/0032-3861(96)00254-6
108. P.T. Hietaoja, R.M. Holsti-Miettinen, J.V. Seppälä, O.T. Ikkala, The effect of viscosity ratio on the phase inversion of polyamide 66/polypropylene blends. J. Appl. Polym. Sci. **54**, 1613–1623 (1994). https://doi.org/10.1002/app.1994.070541104
109. L.N. Andradi, G.P. Hellmann, Morphologies of mechanically mixed amorphous blends before and after annealing. Polym. Eng. Sci. **35**, 693–702 (1995). https://doi.org/10.1002/pen.760350809
110. J.M. Willis, V. Caldas, B.D. Favis, Processing-morphology relationships of compatibilized polyolefin/polyamide blends—Part II The emulsifying effect of an ionomer compatibilizer as a function of blend composition and viscosity ratio. J. Mater. Sci. **26**, 4742–4750 (1991). https://doi.org/10.1007/BF00612413
111. B.D. Favis, J.P. Chalifoux, Influence of composition on the morphology of polypropylene/polycarbonate blends. Polymer (Guildf.) **29**, 1761–1767 (1988). https://doi.org/10.1016/0032-3861(88)90388-6
112. S. Steinmann, W. Gronski, C. Friedrich, Cocontinuous polymer blends: influence of viscosity and elasticity ratios of the constituent polymers on phase inversion. Polymer (Guildf.) **42**, 6619–6629 (2001). https://doi.org/10.1016/S0032-3861(01)00100-8
113. R.M. Ho, C.H. Wu, A.C. Su, Morphology of plastic/rubber blends. Polym. Eng. Sci. **30**, 511–518 (1990). https://doi.org/10.1002/pen.760300903
114. N. Kitayama, H. Keskkula, D.R. Paul, Reactive compatibilization of nylon 6/styrene-acrylonitrile copolymer blends: Part 1. Phase inversion behavior. Polymer (Guildf.) **41**, 8041–8052 (2000). https://doi.org/10.1016/S0032-3861(00)00156-7
115. V. Everaert, L. Aerts, G. Groeninckx, Phase morphology development in immiscible PP/(PS/PPE) blends influence of the melt-viscosity ratio and blend composition. Polymer (Guildf.) **40**, 6627–6644 (1999). https://doi.org/10.1016/S0032-3861(99)00048-8
116. V. Metelkin, V. Blekht, Formation d'une phase continue dans des mélanges hétérogènes de polymères. Kolloidn Žurnal **46**, 476–480 (1984)
117. S. Steinmann, W. Gronski, C. Friedrich, Quantitative rheological evaluation of phase inversion in two-phase polymer blends with cocontinuous morphology. Rheol. Acta **41**, 77–86 (2002). https://doi.org/10.1007/s003970200007

118. T.H. Chen, A.C. Su, Morphology of poly(p-phenylene sulfide)polyethylene blends. Polymer (Guildf.) **34**, 4826–4831 (1993). https://doi.org/10.1016/0032-3861(93)90004-T
119. D. Bourry, B.D. Favis, Cocontinuity and phase inversion in HDPE/PS blends: influence of interfacial modification and elasticity. J. Polym. Sci. Part B Polym. Phys. **36**, 1889–1899 (1998). https://doi.org/10.1002/(SICI)1099-0488(199808)36:11%3c1889::AID-POLB10%3e3.0.CO;2-3
120. R.C. Willemse, A. Posthuma De Boer, J. Van Dam, A.D. Gotsis, Co-continuous morphologies in polymer blends: the influence of the interfacial tension. Polymer (Guildf.) **40**, 827–834 (1999). https://doi.org/10.1016/S0032-3861(98)00307-3
121. W.Y. Hsu, S. Wu, Percolation behavior in morphology and modulus of polymer blends. Polym. Eng. Sci. **33**, 293–302 (1993). https://doi.org/10.1002/pen.760330509
122. A. Margolina, S. Wu, Percolation model for brittle-tough transition in nylon/rubber blends. Polymer (Guildf.) **29**, 2170–2173 (1988). https://doi.org/10.1016/0032-3861(88)90108-5
123. S. Feng, P.N. Sen, Percolation on elastic networks: new exponent and threshold. Phys. Rev. Lett. **52**, 216–219 (1984). https://doi.org/10.1103/PhysRevLett.52.216
124. D. Deptuck, J.P. Harrison, P. Zawadzki, Measurement of elasticity and conductivity of a three-dimensional percolation system. Phys. Rev. Lett. **54**, 913–916 (1985). https://doi.org/10.1103/PhysRevLett.54.913
125. J. Lyngaae-Jørgensen, Structuring of interface-modified polymer blends. Int. Polym. Process. **14**, 213–220 (1999). https://doi.org/10.3139/217.1553
126. D. Quintens, G. Groeninckx, M. Guest, L. Aerts, Phase morphology coarsening and quantitative morphological characterization of a 60/40 blend of polycarbonate of bisphenol a (pc) and poly(styrene-bco-acrylonitrile) (SAN). Polym. Eng. Sci. **30**, 1484–1490 (1990). https://doi.org/10.1002/pen.760302209
127. R.C. Willemse, Co-continuous morphologies in polymer blends: stability. Polymer (Guildf.) **40**, 2175–2178 (1999). https://doi.org/10.1016/S0032-3861(98)00430-3
128. N.D.B. Lazo, C.E. Scott, Isolating the effect of reaction on the phase inversion of model PA/PS blends, in *ANTEC—Proceedings 58th Annual Technical Conference and Exhibition* (Society of Plastics Engineers, Orlando, 2000), pp. 2500–2504
129. H. Van Oene, Rheology of polymer blends and dispersions, in *Polymer Blends*, ed. by D.R. Paul, S. Newman, vol. 2 (Academic Press, New York, n.d.), pp. 295–352
130. H. Veenstra, B. Norder, J. Van Dam, A. Posthuma De Boer, Stability of co-continuous polystyrene/poly(ether-ester) blends in shear flow. Polymer (Guildf.) **40**, 5223–5226 (1999). https://doi.org/10.1016/S0032-3861(99)00043-9
131. S. Raha, N. Kao, S.N. Bhattacharya, Modification of styrene-ethylene/butylene-styrene copolymer microstructure by polystyrene homopolymer and evolution of a cocontinuous blend morphology. Polym. Eng. Sci. **52**, 2559–2572 (2012)
132. W.P. Gergen, *Uniqueness of Hydrogenated Block Copolymers for Elastomer Applications* (Kautschuk Und Gummi, Kunststoffe, 1984), p. 37
133. H. Veenstra, B.J.J. Van Lent, J. Van Dam, A. Posthuma De Boer, Co-continuous morphologies in polymer blends with SEBS block copolymers. Polymer (Guildf.) **40**, 6661–6672 (1999). https://doi.org/10.1016/S0032-3861(98)00875-1
134. H. Veenstra, J. Van Dam, A. Posthuma De Boer, Formation and stability of co-continuous blends with a poly(ether-ester) block copolymer around its order-disorder temperature. Polymer (Guildf.) **40**, 1119–1130 (1999). https://doi.org/10.1016/S0032-3861(98)00342-5
135. E.N. Kresge, Rubbery thermoplastic blends, in *Polymer Blends*, ed. by D.R. Paul, S. Newman, vol. 2 (Academic Press, New York, 1978), pp. 293–318
136. X. Xu, X. Yan, T. Zhu, C. Zhang, J. Sheng, Phase morphology development of polypropylene/ethylene-octene copolymer blends: effects of blend composition and processing conditions. Polym. Bull. **58**, 465–478 (2007). https://doi.org/10.1007/s00289-006-0678-0
137. H. Vanoene, Modes of dispersion of viscoelastic fluids in flow. J. Colloid Interface Sci. **40**, 448–467 (1972). https://doi.org/10.1016/0021-9797(72)90355-4
138. Z.H. Shi, L.A. Utracki, Development of polymer blend morphology during compounding in a twin-screw extruder. Part II: Theoretical derivations. Polym. Eng. Sci. **32**, 1834–1845 (1992)

139. J.M. Willis, B.D. Favis, J. Lunt, Reactive processing of polystyrene-co-maleic anhydride/elastomer blends: processing-morphology-property relationships. Polym. Eng. Sci. **30**, 1073–1084 (1990)
140. C.W. Macosko, H.K. Jeon, T.R. Hoye, Reactions at polymer–polymer interfaces for blend compatibilization. Prog. Polym. Sci. **30**, 939–947 (2005)
141. H. Potente, S. Krawinkel, M. Bastian, M. Stephan, P. Pötschke, Investigation of the melting behaviour and morphology development of polymer blends in the melting zone of twin-screw extruders. J. Appl. Polym. Sci. **82**, 1986–2002 (2001)
142. R. Banerjee, S.S. Ray, A.K. Ghosh, Investigations on blending and foaming behavior of styrene-ethylene-butylene-styrene/polystyrene blends. Int. Polym. Process. **32**, 434–445 (2017). https://doi.org/10.3139/217.3362
143. A.P. Plochocki, P.S. Hope, J.E. Curry, Fabrication of a standard industrial polyblend/alloy: repeatability of the compounding process, in *ANTEC* (CRC Press, Montreal, 1991)
144. G.E. Schoolenberg, F. During, G. Ingenbleek, Coalescence and interfacial tension measurements for polymer melts: experiments on a PS-PE model system. Polymer (Guildf.) **39**, 765–772 (1998)
145. J.K. Lee, C.D. Han, Evolution of polymer blend morphology during compounding in an internal mixer. Polymer (Guildf.) **40**, 6277–6296 (1999). https://doi.org/10.1016/S0032-3861(99)00022-1
146. C.D. Han, K.Y. Lee, N.C. Wheeler, Plasticating single-screw extrusion of amorphous polymers: development of a mathematical model and comparison with experiment. Polym. Eng. Sci. **36**, 1360–1376 (1996). https://doi.org/10.1002/pen.10531
147. M.D. Elkovitch, D.L. Tomasko, Effect of supercritical carbon dioxide on morphology development during polymer blending. Polym. Eng. Sci. **40**, 1850–1861 (2000). https://doi.org/10.1002/pen.11317
148. M. Sumita, K. Sakata, S. Asai, K. Miyasaka, H. Nakagawa, Dispersion of fillers and the electrical conductivity of polymer blends filled with carbon black. Polym. Bull. **25**, 265–271 (1991). https://doi.org/10.1007/BF00310802
149. A. Taguet, P. Cassagnau, J.M. Lopez-Cuesta, Structuration, selective dispersion and compatibilizing effect of (nano)fillers in polymer blends. Prog. Polym. Sci. **39**, 1526–1563 (2014). https://doi.org/10.1016/J.PROGPOLYMSCI.2014.04.002
150. N.R. Demarquette, A. Martinelli Catelli de Souza, G. Palmer, P.H.P. Macaubas, Comparison between five experimental methods to evaluate interfacial tension between molten polymers. Polym. Eng. Sci. **43**, 670–683 (2003). https://doi.org/10.1002/PEN.10055
151. D.K. Owens, R.C. Wendt, Estimation of the surface free energy of polymers. J. Appl. Polym. Sci. **13**, 1741–1747 (1969). https://doi.org/10.1002/APP.1969.070130815
152. R. Salehiyan, M. Nofar, S.S. Ray, V. Ojijo, Kinetically controlled localization of carbon nanotubes in polylactide/poly(vinylidene fluoride) blend nanocomposites and their influence on electromagnetic interference shielding, electrical conductivity, and rheological properties. J. Phys. Chem. C **123**, 19195–19207 (2019). https://doi.org/10.1021/ACS.JPCC.9B04494/SUPPL_FILE/JP9B04494_SI_001.PDF
153. L. Zhou, Y. Tian, P. Xu, H. Wei, Y. Li, H.X. Peng et al., Effect of the selective localization of carbon nanotubes and phase domain in immiscible blends on tunable microwave dielectric properties. Compos. Sci. Technol. **213**, 108919 (2021). https://doi.org/10.1016/J.COMPSCITECH.2021.108919
154. A. Taguet, B. Otazaghine, M. Batistella, Kinetic and thermodynamic parameters guiding the localization of regioselectively modified kaolin platelets into a PS/PA6 co-continuous blend. Polymer (Guildf.) **191**, 122277 (2020). https://doi.org/10.1016/J.POLYMER.2020.122277
155. A. Rostami, M. Vahdati, H. Nazockdast, Unraveling the localization behavior of MWCNTs in binary polymer blends using thermodynamics and viscoelastic approaches. Polym. Compos. **39**, 2356–2367 (2018). https://doi.org/10.1002/PC.24216
156. X. Qi, J. Yang, N. Zhang, T. Huang, Z. Zhou, I. Kühnert et al., Selective localization of carbon nanotubes and its effect on the structure and properties of polymer blends. Prog. Polym. Sci. **123**, 101471 (2021). https://doi.org/10.1016/J.PROGPOLYMSCI.2021.101471

157. A. Göldel, A. Marmur, G.R. Kasaliwal, P. Pötschke, G. Heinrich, Shape-dependent localization of carbon nanotubes and carbon black in an immiscible polymer blend during melt mixing. Macromolecules **44**, 6094–6102 (2011). https://doi.org/10.1021/MA200793A/ASSET/IMAGES/MEDIUM/MA-2011-00793A_0001.GIF
158. F.I. Seyni, B.P. Grady, Janus particles as immiscible polymer blend compatibilizers: a review. Colloid Polym. Sci. **299**, 585–593 (2021). https://doi.org/10.1007/S00396-021-04820-X/METRICS
159. R. Salehiyan, S.S. Ray, Tuning the conductivity of nanocomposites through nanoparticle migration and interface crossing in immiscible polymer blends: a review on fundamental understanding. Macromol. Mater. Eng. **304**, 1800431 (2019). https://doi.org/10.1002/MAME.201800431
160. F. Fenouillot, P. Cassagnau, J.C. Majesté, Uneven distribution of nanoparticles in immiscible fluids: morphology development in polymer blends. Polymer (Guildf.) **50**, 1333–1350 (2009). https://doi.org/10.1016/J.POLYMER.2008.12.029
161. L. Elias, F. Fenouillot, J.C. Majesté, G. Martin, P. Cassagnau, Migration of nanosilica particles in polymer blends. J. Polym. Sci. Part B Polym. Phys. **46**, 1976–1983 (2008). https://doi.org/10.1002/POLB.21534
162. M.M. Rueda, M.C. Auscher, R. Fulchiron, T. Périé, G. Martin, P. Sonntag et al., Rheology and applications of highly filled polymers: a review of current understanding. Prog. Polym. Sci. **66**, 22–53 (2017). https://doi.org/10.1016/J.PROGPOLYMSCI.2016.12.007
163. R. Salehiyan, H.Y. Song, M. Kim, W.J. Choi, K. Hyun, Morphological evaluation of PP/PS blends filled with different types of clays by nonlinear rheological analysis. Macromolecules **49**, 3148–3160 (2016). https://doi.org/10.1021/ACS.MACROMOL.6B00268/ASSET/IMAGES/MEDIUM/MA-2016-002687_0022.GIF
164. C.J. Lee, R. Salehiyan, D.S. Ham, S.K. Cho, S.J. Lee, K.J. Kim et al., Influence of carbon nanotubes localization and transfer on electrical conductivity in PA66/(PS/PPE)/CNTs nanocomposites. Polymer (Guildf.) **84**, 198–208 (2016). https://doi.org/10.1016/J.POLYMER.2015.12.055
165. L. Bai, S. He, J.W. Fruehwirth, A. Stein, C.W. Macosko, X. Cheng, Localizing graphene at the interface of cocontinuous polymer blends: morphology, rheology, and conductivity of cocontinuous conductive polymer composites. J. Rheol. (N. Y. N. Y.) **61**, 575–587 (2017). https://doi.org/10.1122/1.4982702
166. H. Yoon, K. Okamoto, M. Yamaguchi, Carbon nanotube imprinting on a polymer surface. Carbon N. Y. **47**, 2840–2846 (2009). https://doi.org/10.1016/J.CARBON.2009.06.028
167. V.A. Doan, S. Nobukawa, S. Ohtsubo, T. Tada, M. Yamaguchi, Selective migration of silica particles between rubbers. J. Polym. Res. **20**, 1–6 (2013). https://doi.org/10.1007/S10965-013-0145-1/METRICS
168. R. Wiwattananukul, Y. Hachiya, T. Endo, S. Nobukawa, M. Yamaguchi, Anomalous transfer phenomenon of carbon nanotube in the blend of polyethylene and polycarbonate. Compos. Part B Eng. **78**, 409–414 (2015). https://doi.org/10.1016/J.COMPOSITESB.2015.04.009
169. J.R. Austin, M. Kontopoulou, Effect of organoclay content on the rheology, morphology, and physical properties of polyolefin elastomers and their blends with polypropylene. Polym. Eng. Sci. **46**, 1491–1501 (2006). https://doi.org/10.1002/PEN.20622
170. P. Thareja, K. Moritz, S.S. Velankar, Interfacially active particles in droplet/matrix blends of model immiscible homopolymers: particles can increase or decrease drop size. Rheol. Acta **49**, 285–298 (2010). https://doi.org/10.1007/S00397-009-0421-5/METRICS
171. S. Nagarkar, S.S. Velankar, Rheology and morphology of model immiscible polymer blends with monodisperse spherical particles at the interface. J. Rheol. (N. Y. N. Y.) **57**, 901–926 (2013). https://doi.org/10.1122/1.4801757
172. S.P. Nagarkar, S.S. Velankar, Morphology and rheology of ternary fluid–fluid–solid systems. Soft Matter **8**, 8464–8477 (2012). https://doi.org/10.1039/C2SM25758K
173. P. Thareja, S. Velankar, Rheology of immiscible blends with particle-induced drop clusters. Rheol. Acta **47**, 189–200 (2008). https://doi.org/10.1007/S00397-007-0231-6/METRICS

174. Z.M. Zou, Z.Y. Sun, L.J. An, Effect of fumed silica nanoparticles on the morphology and rheology of immiscible polymer blends. Rheol. Acta **53**, 43–53 (2014). https://doi.org/10.1007/S00397-013-0740-4/METRICS
175. B. Krasovitski, A. Marmur, Particle adhesion to drops. J. Adhes. **81**, 869–880 (2005). https://doi.org/10.1080/00218460500189489
176. J. Chen, X. Cui, Y. Zhu, W. Jiang, K. Sui, Design of superior conductive polymer composite with precisely controlling carbon nanotubes at the interface of a co-continuous polymer blend via a balance of π-π interactions and dipole-dipole interactions. Carbon N. Y. **114**, 441–448 (2017). https://doi.org/10.1016/J.CARBON.2016.12.048
177. Y. Tan, L. Fang, J. Xiao, Y. Song, Q. Zheng, Grafting of copolymers onto graphene by miniemulsion polymerization for conductive polymer composites: improved electrical conductivity and compatibility induced by interfacial distribution of graphene. Polym. Chem. **4**, 2939–2944 (2013). https://doi.org/10.1039/C3PY00164D
178. A. Maani, P.J. Carreau, Rheological and morphological properties of thermoplastic olefin blends containing nanosilica. J. Nonnewton. Fluid Mech. **233**, 95–106 (2016). https://doi.org/10.1016/J.JNNFM.2016.01.017
179. S.M.N. Sultana, S.P. Pawar, U. Sundararaj, Effect of processing techniques on EMI SE of immiscible PS/PMMA blends containing MWCNT: enhanced intertube and interphase scattering. Ind. Eng. Chem. Res. **58**, 11576–11584 (2019). https://doi.org/10.1021/ACS.IECR.8B05957/SUPPL_FILE/IE8B05957_SI_001.PDF
180. G. Xie, J. Yu, S. Qin, J. Sun, Z. Yang, L. Wei et al., Selective localization of organic montmorillonite nanoparticles in multilayered high-density polyethylene/polyamide 6 composites. Adv. Polym. Technol. **37**, 1526–1536 (2018). https://doi.org/10.1002/ADV.21811
181. S. Yoshida, M. Trifkovic, Unraveling the effect of 3D particle localization on coarsening dynamics and rheological properties in cocontinuous polymer blend nanocomposites. Macromolecules **52**, 7678–7687 (2019). https://doi.org/10.1021/ACS.MACROMOL.9B01437/SUPPL_FILE/MA9B01437_SI_002.MP4
182. Q. Zhang, J. Wang, J. Yu, Z.X. Guo, Improved electrical conductivity of TPU/carbon black composites by addition of COPA and selective localization of carbon black at the interface of sea-island structured polymer blends. Soft Matter **13**, 3431–3439 (2017). https://doi.org/10.1039/C7SM00346C
183. T. Baouz, E. Acik, F. Rezgui, U. Yilmazer, Effects of mixing protocols on impact modified poly(lactic acid) layered silicate nanocomposites. J. Appl. Polym. Sci. **132** (2015). https://doi.org/10.1002/APP.41518
184. J. Bandyopadhyay, S.S. Ray, R. Salehiyan, V. Ojijo, M. Khoza, J. Wesley-Smith, Effect of the mode of nanoclay inclusion on morphology development and rheological properties of nylon6/ethyl–vinyl-alcohol blend composites. Polymer (Guildf.) **126**, 96–108 (2017). https://doi.org/10.1016/j.polymer.2017.08.032
185. R. Salehiyan, W.J. Choi, J.H. Lee, K. Hyun, Effects of mixing protocol and mixing time on viscoelasticity of compatibilized PP/PS blends. Korea Aust. Rheol. J. **26**, 311–318 (2014). https://doi.org/10.1007/S13367-014-0035-Z/METRICS
186. L. As'habi, S.H. Jafari, H.A. Khonakdar, R. Boldt, U. Wagenknecht, G. Heinrich, Tuning the processability, morphology and biodegradability of clay incorporated PLA/LLDPE blends via selective localization of nanoclay induced by melt mixing sequence. Express Polym. Lett. **7**, 21–39 (2013). https://doi.org/10.3144/expresspolymlett.2013.3
187. A.D. Oliveira, N.M. Larocca, D.R. Paul, L.A. Pessan, Effects of mixing protocol on the performance of nanocomposites based on polyamide 6/acrylonitrile–butadiene–styrene blends. Polym. Eng. Sci. **52**, 1909–1919 (2012). https://doi.org/10.1002/PEN.23152
188. M. Nofar, M. Mohammadi, P.J. Carreau, Nanoclay migration and the rheological response of PBAT/LDPE blends. Int. Polym. Process. **36**, 287–296 (2021). https://doi.org/10.1515/IPP-2020-4057/MACHINEREADABLECITATION/RIS
189. S.B. Kazemabad, A. Khavandi, D. Fox, Y. Chen, H. Zhang, B. Chen, Effects of blending sequence on morphology and mechanical properties of polypropylene/ethylene-octene copolymer/clay nanocomposites. Manuf. Sci. Technol. **3**, 236–239 (2015)

190. F.R. Passador, A. Ruvolo-Filho, L.A. Pessan, Effect of blending protocol on the rheological properties and morphology of HDPE/LLDPE blend-based nanocomposites. Int. Polym. Process. **27**, 378–385 (2012). https://doi.org/10.3139/217.2567/MACHINEREADABLECITATION/RIS
191. X. Li, H.M. Park, J.O. Lee, C.S. Ha, Effect of blending sequence on the microstructure and properties of PBT/EVA-g-MAH/organoclay ternary nanocomposites. Polym. Eng. Sci. **42**, 2156–2164 (2002). https://doi.org/10.1002/PEN.11105
192. A. Dasari, Z.Z. Yu, Y.W. Mai, Effect of blending sequence on microstructure of ternary nanocomposites. Polymer (Guildf.) **46**, 5986–5991 (2005). https://doi.org/10.1016/J.POLYMER.2005.05.145
193. J.S. Borah, N. Karak, T.K. Chaki, Effect of organoclay platelets on morphology and properties of LLDPE/EMA blends. Mater. Sci. Eng. A **528**, 2820–2830 (2011). https://doi.org/10.1016/J.MSEA.2010.12.067
194. E. Jalali Dil, B.D. Favis, Localization of micro- and nano-silica particles in heterophase poly(lactic acid)/poly(butylene adipate-co-terephthalate) blends. Polymer (Guildf.) **76**, 295–306 (2015). https://doi.org/10.1016/J.POLYMER.2015.08.046
195. L. Elias, F. Fenouillot, J.C. Majesté, P. Alcouffe, P. Cassagnau, Immiscible polymer blends stabilized with nano-silica particles: rheology and effective interfacial tension. Polymer (Guildf.) **49**, 4378–4385 (2008). https://doi.org/10.1016/J.POLYMER.2008.07.018
196. P. Saiprasit, A.K. Schlarb, The effect of the compounding procedure on the morphology and mechanical properties of PLA/PBAT-based nanocomposites. Int. Polym. Process. **36**, 219–227 (2021). https://doi.org/10.1515/IPP-2020-4032/MACHINEREADABLECITATION/RIS
197. R. Tchoudakov, O. Breuer, M. Narkis, A. Siegmann, Conductive polymer blends with low carbon black loading: polypropylene/polyamide. Polym. Eng. Sci. **36**, 1336–1346 (1996). https://doi.org/10.1002/PEN.10528
198. G. Yu, M.Q. Zhang, H.M. Zeng, Y.H. Hou, H.B. Zhang, Conductive polymer blends filled with carbon black: positive temperature coefficient behavior. Polym. Eng. Sci. **39**, 1678–1688 (1999). https://doi.org/10.1002/PEN.11562
199. S.L. Scherzer, E. Pavlova, J.D. Esper, Z. Starý, Phase structure, rheology and electrical conductivity of co-continuous polystyrene/polymethylmethacrylate blends filled with carbon black. Compos. Sci. Technol. **119**, 138–147 (2015). https://doi.org/10.1016/J.COMPSCITECH.2015.10.003
200. B. Yang, J. Zhang, J. Bao, Effect of polymeric ratio in matrix and blending sequence on electrical resistivity of carbon black-filled multi-phase composites. J. Mater. Sci. Mater. Electron. **34**, 625 (2023). https://doi.org/10.1007/S10854-023-10027-3/METRICS
201. L. Azubuike, U. Sundararaj, Interface strengthening of PS/aPA polymer blend nanocomposites via in situ compatibilization: enhancement of electrical and rheological properties. Materials **14**, 4813 (2021). https://doi.org/10.3390/MA14174813
202. A.C. Baudouin, J. Devaux, C. Bailly, Localization of carbon nanotubes at the interface in blends of polyamide and ethylene–acrylate copolymer. Polymer (Guildf.) **51**, 1341–1354 (2010). https://doi.org/10.1016/J.POLYMER.2010.01.050
203. A.C. Baudouin, C. Bailly, J. Devaux, Interface localization of carbon nanotubes in blends of two copolymers. Polym. Degrad. Stab. **95**, 389–398 (2010). https://doi.org/10.1016/J.POLYMDEGRADSTAB.2009.11.007
204. A. Göldel, G. Kasaliwal, P. Pötschke, Selective localization and migration of multiwalled carbon nanotubes in blends of polycarbonate and poly(styrene-acrylonitrile). Macromol. Rapid Commun. **30**, 423–429 (2009). https://doi.org/10.1002/MARC.200800549
205. Y. Li, Y. Zheng, P. Zhan, G. Zheng, K. Dai, C. Liu et al., Vapor sensing performance as a diagnosis probe to estimate the distribution of multi-walled carbon nanotubes in poly(lactic acid)/polypropylene conductive composites. Sens. Actuators B Chem. **255**, 2809–2819 (2018). https://doi.org/10.1016/J.SNB.2017.09.098
206. T. Gong, M.Q. Liu, H. Liu, S.P. Peng, T. Li, R.Y. Bao et al., Selective distribution and migration of carbon nanotubes enhanced electrical and mechanical performances in polyolefin elastomers. Polymer (Guildf.) **110**, 1–11 (2017). https://doi.org/10.1016/J.POLYMER.2016.12.056

207. L. Zonder, A. Ophir, S. Kenig, S. McCarthy, The effect of carbon nanotubes on the rheology and electrical resistivity of polyamide 12/high density polyethylene blends. Polymer (Guildf.) **52**, 5085–5091 (2011). https://doi.org/10.1016/J.POLYMER.2011.08.048
208. Z.X. Huang, M.L. Zhao, G.Z. Zhang, J. Song, J.P. Qu, Controlled localizing multi-wall carbon nanotubes in polyvinylidene fluoride/acrylonitrile butadiene styrene blends to achieve balanced dielectric constant and dielectric loss. Compos. Sci. Technol. **212**, 108874 (2021). https://doi.org/10.1016/J.COMPSCITECH.2021.108874
209. M. Liebscher, M.O. Blais, P. Pötschke, G. Heinrich, A morphological study on the dispersion and selective localization behavior of graphene nanoplatelets in immiscible polymer blends of PC and SAN. Polymer (Guildf.) **54**, 5875–5882 (2013). https://doi.org/10.1016/J.POLYMER.2013.08.009
210. C. Tu, K. Nagata, S. Yan, Influence of melt-mixing processing sequence on electrical conductivity of polyethylene/polypropylene blends filled with graphene. Polym. Bull. **74**, 1237–1252 (2017). https://doi.org/10.1007/S00289-016-1774-4/METRICS
211. S.K. Tiwari, G. Hatui, R. Oraon, A. De Adhikari, G.C. Nayak, Mixing sequence driven controlled dispersion of graphene oxide in PC/PMMA blend nanocomposite and its effect on thermo-mechanical properties. Curr. Appl. Phys. **17**, 1158–1168 (2017). https://doi.org/10.1016/J.CAP.2017.05.007
212. L. Bai, R. Sharma, X. Cheng, C.W. Macosko, Kinetic control of graphene localization in co-continuous polymer blends via melt compounding. Langmuir **34**, 1073–1083 (2018). https://doi.org/10.1021/ACS.LANGMUIR.7B03085/SUPPL_FILE/LA7B03085_SI_001.PDF
213. S.K. Tiwari, K. Verma, P. Saren, R. Oraon, A. De Adhikari, G.C. Nayak et al., Manipulating selective dispersion of reduced graphene oxide in polycarbonate/nylon 66 based blend nanocomposites for improved thermo-mechanical properties. RSC Adv. **7**, 22145–22155 (2017). https://doi.org/10.1039/C7RA02044A
214. Y. Cao, J. Zhang, J. Feng, P. Wu, Compatibilization of immiscible polymer blends using graphene oxide sheets. ACS Nano **5**, 5920–5927 (2011). https://doi.org/10.1021/NN201717A/SUPPL_FILE/NN201717A_SI_001.PDF
215. A. Paydayesh, A. Arefazar, A. Jalaliarani, A morphological study on the migration and selective localization of graphene in the PLA/PMMA blends. J. Appl. Polym. Sci. **133** (2016). https://doi.org/10.1002/APP.43799
216. I. Kelnar, J. Kratochvíl, L. Kaprálková, Z. Špitálsky, M. Ujčič, A. Zhigunov et al., Effect of graphene oxide on structure and properties of impact-modified polyamide 6. Polym. Plast. Technol. Eng. **57**, 827–835 (2018). https://doi.org/10.1080/03602559.2017.1354223
217. S.S. Ray, R. Banerjee, *Sustainable Polylactide-Based Blends* (Elsevier, Amsterdam, 2022)
218. P.B. Tambe, A.R. Bhattacharyya, A.R. Kulkarni, The influence of melt-mixing process conditions on electrical conductivity of polypropylene/multiwall carbon nanotubes composites. J. Appl. Polym. Sci. **127**, 1017–1026 (2013). https://doi.org/10.1002/app.37889
219. D.Y. Kim, J. Bin Lee, D.Y. Lee, Selective localization of nanofiller on mechanical properties of poly(lactic acid)/poly(butylene adipate-co-terephthalate) nanocomposites via the surface energy and melt blending technique. Macromolecules **55**, 3287–3300 (2022). https://doi.org/10.1021/ACS.MACROMOL.1C02074/SUPPL_FILE/MA1C02074_SI_001.PDF
220. T.G. Mofokeng, S.S. Ray, V. Ojijo, R. Salehiyan, Tuning the nano/micro-structure and properties of melt-processed ternary composites of polypropylene/ethylene vinyl acetate blend and nanoclay: the influence of kinetic and thermodynamic parameters. J. Appl. Polym. Sci. **135**, 45585 (2018). https://doi.org/10.1002/APP.45585
221. C. Lencar, S. Ramakrishnan, E. Erfanian, U. Sundararaj, The role of phase migration of carbon nanotubes in melt-mixed PVDF/PE polymer blends for high conductivity and EMI shielding applications. Molecules **27**, 933 (2022). https://doi.org/10.3390/MOLECULES27030933
222. I.O. Navas, M. Arjmand, U. Sundararaj, Effect of carbon nanotubes on morphology evolution of polypropylene/polystyrene blends: understanding molecular interactions and carbon nanotube migration mechanisms. RSC Adv. **7**, 54222–54234 (2017). https://doi.org/10.1039/C7RA11390K

223. C.G. Martins, N.M. Larocca, D.R. Paul, L.A. Pessan, Nanocomposites formed from polypropylene/EVA blends. Polymer (Guildf.) **50**, 1743–1754 (2009). https://doi.org/10.1016/J.POLYMER.2009.01.059
224. L. Xu, B.Y. Zhang, Z.Y. Xiong, Z.X. Guo, J. Yu, Preparation of conductive polyphenylene sulfide/polyamide 6/multiwalled carbon nanotube composites using the slow migration rate of multiwalled carbon nanotubes from polyphenylene sulfide to polyamide 6. J. Appl. Polym. Sci. **132** (2015). https://doi.org/10.1002/APP.42353
225. F. Gubbels, R. Jerome, E. Vanlathem, R. Deltour, S. Blacher, F. Brouers, Kinetic and thermodynamic control of the selective localization of carbon black at the interface of immiscible polymer blends. Chem. Mater. **10**, 1227–1235 (1998). https://doi.org/10.1021/CM970594D
226. M. Liebscher, L. Tzounis, P. Pötschke, G. Heinrich, Influence of the viscosity ratio in PC/SAN blends filled with MWCNTs on the morphological, electrical, and melt rheological properties. Polymer (Guildf.) **54**, 6801–6808 (2013). https://doi.org/10.1016/J.POLYMER.2013.10.040
227. J.S. Hong, Y.K. Kim, K.H. Ahn, S.J. Lee, Shear-induced migration of nanoclay during morphology evolution of PBT/PS blend. J. Appl. Polym. Sci. **108**, 565–575 (2008). https://doi.org/10.1002/APP.27330
228. A. Taghizadeh, B.D. Favis, Carbon nanotubes in blends of polycaprolactone/thermoplastic starch. Carbohydr. Polym. **98**, 189–198 (2013). https://doi.org/10.1016/J.CARBPOL.2013.05.024
229. D.P. Dharaiya, S.C. Jana, Nanoclay-induced morphology development in chaotic mixing of immiscible polymers. J. Polym. Sci. Part B Polym. Phys. **43**, 3638–3651 (2005). https://doi.org/10.1002/POLB.20657
230. D.P. Dharaiya, S.C. Jana, S.F. Lyuksyutov, Production of electrically conductive networks in immiscible polymer blends by chaotic mixing. Polym. Eng. Sci. **46**, 19–28 (2006). https://doi.org/10.1002/PEN.20445
231. Y. Li, H. Shimizu, Conductive PVDF/PA6/CNTs nanocomposites fabricated by dual formation of cocontinuous and nanodispersion structures. Macromolecules **41**, 5339–5344 (2008). https://doi.org/10.1021/MA8006834/SUPPL_FILE/MA8006834-FILE001.PDF
232. L.H.B. Johansen, L.B. Canto, S.V. Canevarolo, Development of an immiscible polymer/polymer/nanoparticle system to assess the location of nanoparticles by quantitative optical microscopy. Int. Polym. Process. **31**, 607–617 (2016). https://doi.org/10.3139/217.3241/MACHINEREADABLECITATION/RIS
233. L. Elias, F. Fenouillot, J.C. Majested, P. Cassagnauab, Morphology and rheology of immiscible polymer blends filled with silica nanoparticles. Polymer (Guildf.) **48**, 6029–6040 (2007). https://doi.org/10.1016/J.POLYMER.2007.07.061
234. X. Yan, A. Cayla, E. Devaux, F. Salaün, Microstructure evolution of immiscible PP-PVA blends tuned by polymer ratio and silica nanoparticles. Polymers (Basel) **10**, 1031 (2018). https://doi.org/10.3390/POLYM10091031
235. H. Zhang, M. Zuo, X. Zhang, X. Shi, L. Yang, S. Sun et al., Effect of agglomeration on the selective distribution of nanoparticles in binary polymer blends. Compos. Part A Appl. Sci. Manuf. **149**, 106590 (2021). https://doi.org/10.1016/J.COMPOSITESA.2021.106590
236. P.M. Hansson, L. Skedung, P.M. Claesson, A. Swerin, J. Schoelkopf, P.A.C. Gane et al., Robust hydrophobic surfaces displaying different surface roughness scales while maintaining the same wettability. Langmuir **27**, 8153–8159 (2011). https://doi.org/10.1021/LA201121P/ASSET/IMAGES/MEDIUM/LA-2011-01121P_0007.GIF
237. P.S. Tsai, Y.M. Yang, Y.L. Lee, Fabrication of hydrophobic surfaces by coupling of Langmuir-Blodgett deposition and a self-assembled monolayer. Langmuir **22**, 5660–5665 (2006). https://doi.org/10.1021/LA053152M/ASSET/IMAGES/MEDIUM/LA053152MN00001.GIF
238. R. Salehiyan, M. Nofar, K. Malkappa, S.S. Ray, Effect of nanofillers characteristics and their selective localization on morphology development and rheological properties of melt-processed polylactide/poly(butylene adipate-co-terephthalate) blend composites. Polym. Eng. Sci. **60**, 2749–2760 (2020). https://doi.org/10.1002/PEN.25505
239. R. Salehiyan, M. Nofar, S.S. Ray, V. Ojijo, Kinetically controlled localization of carbon nanotubes in polylactide/poly(vinylidene fluoride) blend nanocomposites and their influence

on electromagnetic interference shielding, electrical conductivity, and rheological properties. J. Phys. Chem. C **123**, 19195–19207 (2019). https://doi.org/10.1021/acs.jpcc.9b04494
240. R. Salehiyan, M. Nofar, D. Makwakwa, S.S. Ray, Shear-induced carbon nanotube migration and morphological development in polylactide/poly(vinylidene fluoride) blend nanocomposites and their impact on dielectric constants and rheological properties. J. Phys. Chem. C **124**, 9536–9547 (2020). https://doi.org/10.1021/ACS.JPCC.0C00891/ASSET/IMAGES/MEDIUM/JP0C00891_0013.GIF
241. W. You, W. Yu, Onset reduction and stabilization of cocontinuous morphology in immiscible polymer blends by snowmanlike janus nanoparticles. Langmuir **34**, 11092–11100 (2018). https://doi.org/10.1021/ACS.LANGMUIR.8B02503/SUPPL_FILE/LA8B02503_SI_001.PDF
242. J. Lin, C. Chi, Y. Zhang, J. Lin, Q. Chen, Increased continuity of the PA6 phase from the PS matrix induced by migrating janus particles and its application in thermal conductivity. Ind. Eng. Chem. Res. **60**, 13905–13913 (2021). https://doi.org/10.1021/ACS.IECR.1C03202/SUPPL_FILE/IE1C03202_SI_002.MP4
243. J. Guan, Y. Yang, B. Tang, X. Shen, Y. Li, The synthesis of functional Janus nanosheets as compatibilizers for the immiscible polyamide 6/polystyrene (PA6/PS): formation of the nanosilica monolayer at the interface. Colloids Surfaces A Physicochem. Eng. Asp. **643**, 128788 (2022). https://doi.org/10.1016/J.COLSURFA.2022.128788
244. B.D. Favis, The effect of processing parameters on the morphology of an immiscible binary blend. J. Appl. Polym. Sci. **39**, 285–300 (1990)
245. R. Banerjee, S.S. Ray, Role of rheology in morphology development and advanced processing of thermoplastic polymer materials: a review. ACS Omega **8**, 27969–28001 (2023). https://doi.org/10.1021/ACSOMEGA.3C03310/ASSET/IMAGES/LARGE/AO3C03310_0020.JPEG
246. S.S. Ray, R. Banerjee, *Sustainable Polylactide-Based Composites* (Elsevier, Amsterdam, 2023)
247. J. Feng, C.M. Chan, J.X. Li, A method to control the dispersion of carbon black in an immiscible polymer blend. Polym. Eng. Sci. **43**, 1058–1063 (2003). https://doi.org/10.1002/PEN.10089
248. J.K. Yuan, S.H. Yao, A. Sylvestre, J. Bai, Biphasic polymer blends containing carbon nanotubes: heterogeneous nanotube distribution and its influence on the dielectric properties. J. Phys. Chem. C **116**, 2051–2058 (2012). https://doi.org/10.1021/JP210872W/SUPPL_FILE/JP210872W_SI_001.PDF
249. S. Maiti, S.K. De, A.K. Bhowmick, Quantitative estimation of filler distribution in immiscible rubber blends by mechanical damping studies. Rubber Chem. Technol. **65**, 293–302 (1992). https://doi.org/10.5254/1.3538612
250. A.E. Zaikin, R.R. Karimov, V.P. Arkhireev, A study of the redistribution conditions of carbon black particles from the bulk to the interface in heterogeneous polymer blends. Colloid J. **63**, 53–59 (2001). https://doi.org/10.1023/A:1009490108111
251. J. Plattier, L. Benyahia, M. Dorget, F. Niepceron, J.F. Tassin, Viscosity-induced filler localisation in immiscible polymer blends. Polymer (Guildf.) **59**, 260–269 (2015). https://doi.org/10.1016/j.polymer.2014.12.044
252. O. Rafeie, M.K. Razavi Aghjeh, A. Tavakoli, M. Salami Kalajahi, O.A. Jameie, Conductive poly(vinylidene fluoride)/polyethylene/graphene blend-nanocomposites: relationship between rheology, morphology, and electrical conductivity. J. Appl. Polym. Sci. **135**, 46333 (2018). https://doi.org/10.1002/APP.46333
253. A. Jameie Oskooie, M.K. Razavi Aghjeh, O. Rafeie, A. Tavakoli, Composition and compatibilization induced morphology alteration in PVDF/LLDPE blends: correlation between rheology and morphology. J. Polym. Res. **24**, 1–11 (2017). https://doi.org/10.1007/S10965-017-1183-X/METRICS
254. M.R. Aghjeh, V. Asadi, P. Mehdijabbar, H.A. Khonakdar, S.H. Jafari, Application of linear rheology in determination of nanoclay localization in PLA/EVA/Clay nanocomposites: correlation with microstructure and thermal properties. Compos. Part B Eng. **86**, 273–284 (2016). https://doi.org/10.1016/J.COMPOSITESB.2015.09.064

255. M. Gahleitner, B. Kretzschmar, G. Van Vliet, J. Devaux, D. Pospiech, K. Bernreitner et al., Rheology/morphology interactions in polypropylene/polyamide-6 nanocomposites. Rheol. Acta **45**, 322–330 (2006). https://doi.org/10.1007/S00397-005-0029-3/METRICS
256. T. Malwela, S.S. Ray, Study of morphology and crystal growth behaviour of nanoclay-containing biodegradable polymer blend thin films using atomic force microscopy. Polymer (Guildf.) **53**, 2705–2716 (2012). https://doi.org/10.1016/J.POLYMER.2012.04.025
257. V. Ojijo, T. Malwela, S.S. Ray, R. Sadiku, Unique isothermal crystallization phenomenon in the ternary blends of biopolymers polactide and poly [(butylene succinate)-co-adipate] and nano-clay. Polymer (Guildf.) **53**, 505–518 (2012)
258. M.S. Thompson, S. Agarwal, R.K. Gupta, Effects of extensional flow and nanofiller incorporation on dispersed phase size in the blending of high-viscosity-ratio immiscible vinyl polymers. J. Vinyl Addit. Technol. **28**, 104–114 (2022). https://doi.org/10.1002/VNL.21870
259. O.J. Botlhoko, S.S. Ray, J. Ramontja, Influence of functionalized exfoliated reduced graphene oxide nanoparticle localization on mechanical, thermal and electronic properties of nanobiocomposites. Eur. Polym. J. **102**, 130–140 (2018). https://doi.org/10.1016/J.EURPOLYMJ.2018.03.012
260. T. Gcwabaza, S. Sinha Ray, W.W. Focke, A. Maity, Morphology and properties of nanostructured materials based on polypropylene/poly(butylene succinate) blend and organoclay. Eur. Polym. J. **45**, 353–367 (2009). https://doi.org/10.1016/J.EURPOLYMJ.2008.11.007

Chapter 3
Processing Technologies for Polymer Blend Nanocomposites

Abstract The chapter opens with a summary of essential techniques for processing polymer blends and nanocomposites, explaining how these methods can be employed separately or together to create blend nanocomposites with diverse morphologies. This is due to the frequent use of multiple processing techniques in combination, which allows for the customization of these combinations to control the morphological properties of multiphase materials.

3.1 Introduction

McKelvey [1] has described polymer processing as "an engineering speciality concerned with the operation carried out on polymeric materials or systems to increase their utility". He has further categorised the area into three classes: (i) modification, which includes mixing or compounding; (ii) forming, which includes processes used for imparting shape to the product; and (iii) bonding or finishing [2, 3]. In the context of polymer blends, nanocomposites, and blend nanocomposites, compounding refers to the techniques used for the preparation of these biphasic/multiphase materials. Though the basic processing techniques of polymer blends, nanocomposites, and blend nanocomposites are the same, the processing of polymer blend nanocomposites is more complex. This is because various processing techniques are often used in combination, and by tailoring this combination, one can control the morphological characteristics of these multiphase materials. Beginning with an overview of the fundamental techniques used in processing polymer blends and nanocomposites, the chapter discusses how these techniques are used individually or in combination to produce blend nanocomposites of various morphologies.

S. S. Ray and R. Banerjee, *Polymer Blend Nanocomposites*, Springer Series in Materials Science 354, https://doi.org/10.1007/978-3-032-00309-6_3

3.2 Processing of Polymer Blends and Nanocomposites

Polymer blends and/or nanocomposites are generally produced using three main techniques [4]:

(a) *In situ polymerisation*—This process is used for producing polymer nanocomposites. In this process, the nanoparticles are pre-mixed with the liquid monomer or monomer solution. This is followed by polymerization accomplished with the help of heat, radiation, or suitable initiators.

(b) *Solution/solvent casting*—This method is used to produce both blends and nanocomposites. This method involves the dissolution of the film ingredients in a suitable carrier, followed by the evaporation of the solvent. In the case of polymer blends, co-continuous or dispersed phase morphology may be produced by tailoring the rates of phase separation and solvent evaporation. It is not necessary to use a single solvent to produce a blend by this method. Often, a mixture of solvents with different evaporation rates and miscibility with the blend components is employed. Furthermore, the films can be annealed and selected blend components extracted to form the desired size and shape of domains. Another method of controlling the blend morphology is by using a mixture of solvents with different affinities for the polymeric components. When the non-solvent has a lower volatility than the good solvent, evaporation leads to the formation of a two-phase structure [5].

    In the case of nanocomposites, a solution of the polymer in a suitable solvent is mixed with a dispersion of the nanoparticles in the same or a different solvent. In the production of nanoclay-containing polymer nanocomposites, the solvent is usually used to pre-swell the nanoclay, and the dispersion of the swollen nanoclay is mixed with the polymer solution [6]. The polymer chains intercalate, displacing the solvent from the nanoclay galleries and getting absorbed on the silicate surface. The impetus for this process is the entropy gain due to the desorption of the solvent molecules, which compensates for the decrease in entropy due to chain confinement in the intercalated structure. Evaporation of the solvent leads to the production of a nanocomposite with an intercalated structure.

(c) Melt processing is the most widely used process for producing blends and nanocomposites. In the case of polymer blends, the blend components are mixed in the molten state, whereas in the case of nanocomposites, the nanoparticles are mixed with the polymer melt. The process is accomplished at a temperature higher than the softening point of the polymer(s), typically under shear [7, 8]. Compared to direct blending with nanofillers, compounding by masterbatch dilution has been more effective in achieving superior dispersion in the melt processing of polymer nanocomposites. This is because two processing cycles have been reported to facilitate higher exfoliation. This is coupled with the ease of handling masterbatches [9–12].

Melt processing has significant advantages over solution casting and in situ intercalative polymerisation techniques. First, no solvent is required, which makes the process environmentally friendly. Second, this process is compatible with processing techniques used in the polymer industry, such as extrusion and injection moulding.

However, there are some advantages of solvent casting over melt mixing [13]:

(a) More uniform and thinner films;
(b) Absence of pinholes and gel marks;
(c) Purity and clarity;
(d) Absence of residual stresses; and
(e) Possibility of production of patterns and dull finishes.

Though melt mixing is predominantly used in industry, in small-scale production, and in research, the choice of technique is also determined by other factors, such as the exact product requirements and the infrastructure available.

As in the case of polymer blends and nanocomposites, melt mixing is predominantly used in the production of polymer blend nanocomposites [14–27]. However, in situ polymerisation [28] and solvent casting [29] have been used as well.

Yan and Yang [28] prepared 1 and 1.5 wt% multi-walled carbon nanotube (MWNT) reinforced polyamide 6 (PA6)/polystyrene (PS) blends having different PA6/PS weight ratios, by successive in situ polymerisation. As presented schematically in Fig. 3.1, the process comprised two steps: polymerisation of styrene in the presence of MWCNTs and ε-caprolactam (CL), followed by polymerisation of CL. Unfilled PA6/PS blends were also prepared similarly for comparative studies on the morphology of the filled and unfilled blends.

Scanning electron microscopy (SEM) studies, after selectively etching out the PS phase using tetrahydrofuran, revealed that increased PS content gradually changed the morphology from PS dispersed/PA6 matrix to PA6 dispersed/PS matrix. For the same PS content, the blend nanocomposites had a smaller size of dispersed domains than the unfilled blend. For the unfilled blends, phase inversion from PS dispersed phase to PS continuous phase was found to occur at a PS loading of 15 wt%. In contrast, in the nanocomposite of 1 wt% MWNT content, phase inversion did not occur until the PS content was as high as 30 wt%. When the MWNT loading was increased to 1.5 wt%, phase inversion didn't happen even at 30 wt% PS content. SEM and transmission electron microscopy (TEM) studies revealed that, irrespective of

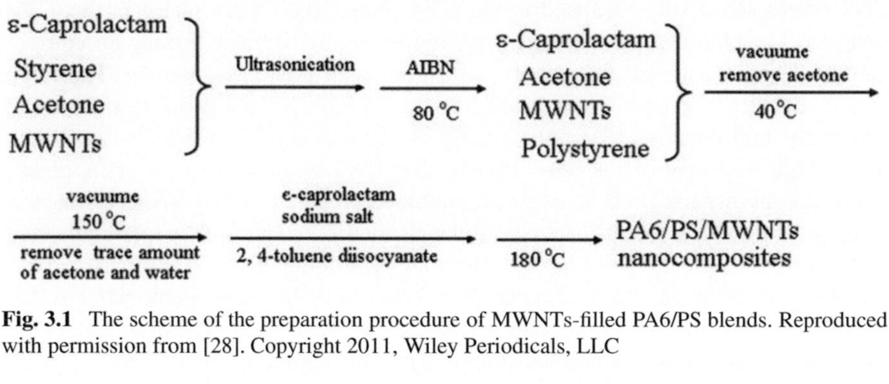

Fig. 3.1 The scheme of the preparation procedure of MWNTs-filled PA6/PS blends. Reproduced with permission from [28]. Copyright 2011, Wiley Periodicals, LLC

whether the PS phase was the dispersed or continuous phase, the MWNTs were selectively localised at the blend interface. This was ascribed to the poor wettability of the carbon nanotubes (CNTs) by both the blend components, which led to their exclusion at the interface. The lowering of domain size with the addition of CNTs was ascribed to the fact that the CNTs localised at the interface served as compatibilizers. Furthermore, the PA6/PS interface was continuous in the (70/30) PA6/PS blends. The selectively localised CNTs at the continuous interface formed an interconnected network, leading to high electrical conductivity values. The volume resistivity of (70/30) PA6/PS blends having 0.5 and 1 wt% CNT content was nine orders of magnitude lower than that of the unfilled blend of the same composition. Thus, the researchers successfully prepared conductive polymer blend nanocomposites having interfacially localised CNTs by in situ polymerisation.

Masterbatch dilution [15, 30–33] is popular in melt mixing. In this direction, Lopes Pereira et al. [34] demonstrated that by the proper choice of mixing sequence in masterbatch dilution, it may be possible to achieve the desired morphology and properties. The researchers conducted studies on poly(vinylidene fluoride) (PVDF)/ethylene–vinyl acetate copolymer (EVA)/CNT nanocomposites having PVDF/EVA volume ratio of 50/50 and CNT content of 0.56 vol%. The blend nanocomposites were prepared by melt mixing in an internal mixer at 190 °C and 60 rpm rotor speed for 7 min, using different sequences of mixing:

(a) Mixing 1—All components were mixed in a single step.
(b) Mixing 2—A PVDF/CNT (97:3 volume ratio) masterbatch was prepared by dry mixing both components, followed by melt processing at 210 °C for 10 min. The PVDF masterbatch was diluted with an additional amount of PVDF at 190 °C for 1 min, after which EVA was added and melt processing was continued for another 7 min.
(c) Mixing 3—An EVA/CNT (97:3 volume ratio) masterbatch was prepared by dry mixing both components, followed by melt processing at 180 °C for 10 min. This masterbatch, along with an additional amount of EVA was blended with PVDF which had been heated to 190 °C for 1 min. This mixture, which had a PVDF: EVA volume ratio of 50:50, was blended for 7 min.

TEM images (Fig. 3.2) revealed that irrespective of the mixing sequence, the CNTs were selectively localised in the EVA phase (the lighter phase in the TEM images). The blend nanocomposite prepared by simultaneously mixing all components (Mixing 1) had well-dispersed CNTs, along with a few agglomerates. The blend nanocomposite prepared by the dilution of the PVDF/CNT masterbatch (Mixing 2) had many agglomerates. The poor quality of dispersion of the CNTs was ascribed to the high viscosity of the masterbatch. The CNTs migrated to the EVA phase, but could not disperse well. In contrast, the blend prepared using EVA/CNT masterbatch (Mixing 3) had a well-dispersed and well-distributed physically interconnected network of CNTs, leading to a much higher electrical conductivity in comparison with the other blend nanocomposites (Fig. 3.3). The researchers thus demonstrated that choosing the proper mixing sequence in masterbatch dilution is crucial for achieving the desired morphology of polymer blend nanocomposites.

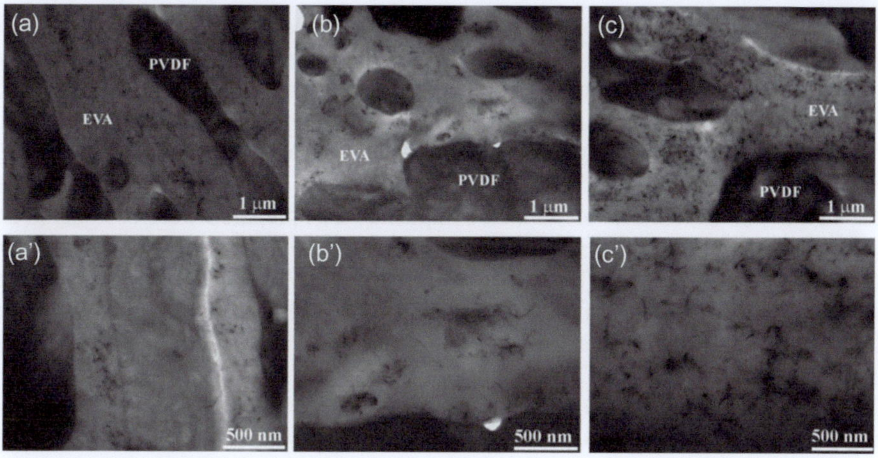

Fig. 3.2 Transmission electron microscopy (TEM) images of PVDF/EVA composites prepared by Mixing 1 (**a**), Mixing 2 (**b**), and Mixing 3 (**c**). Reproduced with permission from [34]. Copyright 2021, Wiley Periodicals, LLC

Fig. 3.3 Dependence of the ac electrical conductivity with frequency of (a) PVDF/CNT, (b) EVA/CNT, and the PVDF/EVA composites prepared by (c) Mixing 1, (d) Mixing 2, and (e) Mixing 3. Reproduced with permission from [34]. Copyright 2021, Wiley Periodicals, LLC

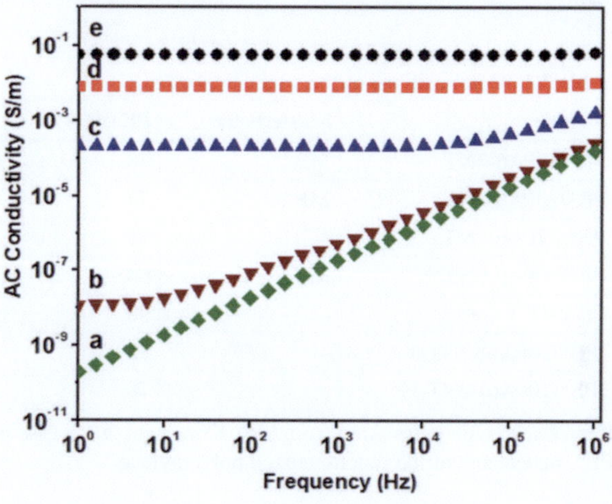

Périé et al. [35] used masterbatches of reactive polymers to produce low molecular-weight amino-terminated PA6/ maleic anhydride functionalized polyethylene (PE*)/ MWNT nanocomposites by reactive processing. The blend nanocomposites were produced in a 2-step process. In the first step, the CNTs and the PA6 were melt processed at 220 °C and 200 rpm screw speed for 8 min in a micro-compounder to produce PA6—CNT masterbatches having 9.3 wt% (MB1) or 16.5 wt% (MB2) CNT content. In the second step, the PE* was melt processed with the masterbatch at 220 °C and 200 rpm screw speed for 4 min. The blend nanocomposites thus prepared were designated as PE*/xPA6-yNT, where x and y were the weight fractions of PA6 and CNT, respectively.

For comparative studies, blend nanocomposites having 78 wt% PE*, 20 wt% PA6, and 2 wt% CNT content were produced using a different mixing sequence, without masterbatch dilution. In this case, the PE* and the PA6 were melt processed at 220 °C and 200 rpm for 4 min, before the introduction of the CNTs. The mixing time of the CNTs was varied. The blend nanocomposites thus prepared were designated as [PE*/20PA6]-2NTz, where z was the mixing time (in min) employed for the melt processing of the PE*/PA6 blend with the CNTs. A PE*-2NT composite having 2 wt% CNT content was also prepared similarly, using different processing conditions (190 °C, 200 rpm, 8 min). Details of the various compositions produced have been presented in Table 3.1.

The formation of graft copolymer during reactive processing was confirmed by selective extraction of the unreacted polymers from the extruded strands. The unreacted PE* was first extracted using 1,2,4-trichlorobenzene (TCB) at 140 °C for 4 h. This was followed by extracting the unreacted PA6 using hexafluoro-2-propanol (HFIP) at room temperature for 24 h. Fractions of the unreacted PE*, unreacted PA6, and the remaining graft copolymer were determined from mass balance and have been presented in Table 3.2.

Table 3.1 Compositions of the nanocomposites

Sample	Masterbatch	PE (wt%)	PA6 (wt%)	CNT (wt%)
PE*/20PA6-2NT	MB1	78	20	2
PE*/19PA6-4NT	MB2	77	19	4
PE*/37PA6-4NT	MB1	59	37	4
PE*/30PA6-6NT	MB2	64	30	6
PE-2NT	–	98	–	2
[PE*/20PA6]-2NT-4′	–	78	20	2
[PE*/20PA6]-2NT-12′	–	78	20	2

Reproduced with permission from [35]. Copyright 2012, Elsevier Science Ltd.
PE*, maleic anhydride functionalized polyethylene

Table 3.2 Weight fraction of the unreacted PE* (x_{PE*}) and the unreacted PA6 (x_{PA6}) after solvent extractions, respectively, during 4 h at 140 °C in TCB and 24 h at 25 °C in HFIP, and weight fraction of the remaining graft copolymer ($x_{PE*\text{-}g\text{-}PA6}$)

Sample	x_{PE*}	x_{PA6}	$x_{PE*\text{-}g\text{-}PA6}$
PE*/20PA6-2NT	0.45	0.12	0.41
PE*/19PA6-4NT	0.46	0.11	0.39
PE*/30PA6-6NT	0.27	0.20	0.47
PE*/37PA6-4NT	0.15	0.27	0.54

Reproduced with permission from [35]. Copyright 2012, Elsevier Science Ltd.

Figure 3.4 presents the TEM images of the PE*/PA6-NT nanocomposites having 20, 30, and 37 wt% of PA6 after staining the PA6 with phosphotungstic acid. Irrespective of the scale and the PA6 loadings, a microstructure was observed without any macrophase separation between the blend components. Furthermore, the PA6 micro-domains were interconnected, forming a sponge-like mesophase, thus indicating a co-continuous morphology. The width of the PA6 ligaments was in the range of 10–50 nm, which was the same range as that of the CNT diameter.

Due to the lack of contrast between the CNT and the stained blend, it was challenging to discern the localisation of the filler particles in the microstructure. However, focus on a crack in a film of PE*/20PA6-2NT revealed a CNT completely wrapped in a PA6 micro-domain. Localisation of the CNTs in the PA6 phase, as predicted by thermodynamics [36, 37], was also observed in compositions having a higher content of PA6. The TEM micrograph of PE*/30PA6-6NT at a high magnification (Fig. 3.5b) revealed that many CNTs were localised in PA6 and a few were bridging the two phases. The presence of a few CNTs as bridges between the two phases was ascribed to the fact that the PA6 micro-domains were too small and tortuous for entirely confining all the CNTs.

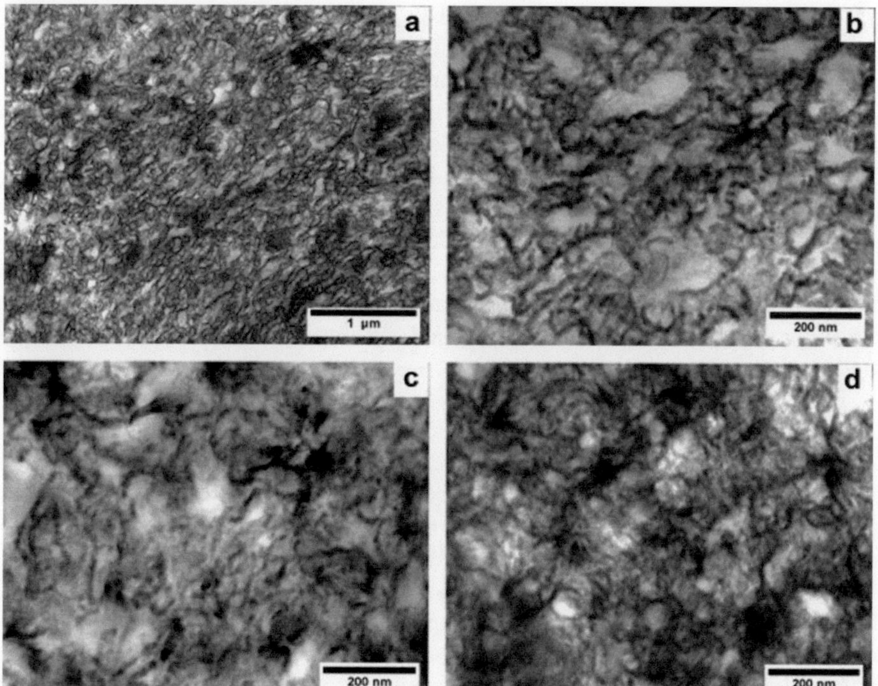

Fig. 3.4 TEM micrographs. **a**, **b** PE*/20PA6-2NT, **c** PE*/30PA6-6NT, **d** PE*/37PA6-4NT. The PA6 is stained in black by phosphotungstic acid. Reproduced with permission from [35]. Copyright 2012, Elsevier Science Ltd.

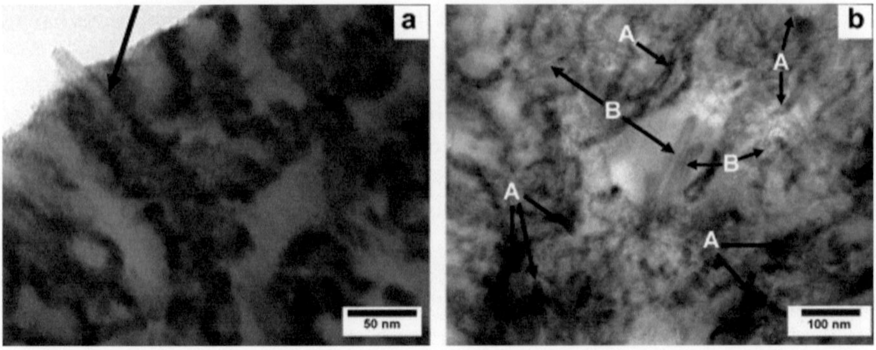

Fig. 3.5 **a** PE*/20PA6-2NT: CNT wrapped in a PA6 micro-domains; **b** PE*/30PA6-6NT, A: CNT embedded in PA6, B: CNT bridging the different phases. The PA6 is stained in black by phosphotungstic acid. Reproduced with permission from [35]. Copyright 2012, Elsevier Science Ltd.

The presence of interconnected CNTs dispersed in the co-continuous nanostructured blend led to a significant improvement in electrical conductivity and mechanical properties. PE*/20PA6-2NT had a resistivity >1 × 10^{13} Ω/square. A higher CNT loading resulted in a remarkable decrease (8 orders of magnitude) in resistivity. The compositions having 4 wt% CNT content exhibited a resistivity of 2–4 × 10^5 Ω/square, irrespective of the PA6 content. This remarkable decrease in resistivity was ascribed to forming a conductive network of CNTs at higher CNT loadings. PE*-2NT and PE*/20PA6-2NT had similar values of elongation at break (ε_b ~ 300%). However, the yield stress (Y_s) and stress at break (σ_b) of PE*/20PA6-2NT were ~8.6 MPa and ~12 MPa, respectively, whereas the corresponding values for PE*-2NT were ~4.3 MPa and 5.3 MPa, respectively. The significant improvement in mechanical properties was ascribed to the synergistic effect of the presence of the nanofiller and the nanostructure of the blend. An increase in the PA6 content led to higher values of Y_s and σ_b.

Thus, the researchers achieved a unique morphology and an improved combination of properties by combining masterbatch blending and reactive processing.

Researchers have also used different techniques to tailor the morphology of these complex-structured materials. In this direction, Yang et al. [38] conducted comparative studies on PA6/PS/organoclay nanocomposites prepared by two methods, namely single-step melt processing and in situ polymerisation followed by melt processing. Two types of organoclay were used for studies, namely C_{18}–MMT and VC_{18}–MMT. To carry out in situ polymerisation, a suspension of the organoclay in styrene was stirred for 4 h and sonicated for 15 min in an ice/water bath. The initiator AIBN (0.2 wt% of the styrene) was added to this suspension, and polymerisation was carried out at 70 °C for 15 h. The product was heated to 100 °C for 5 h for completion of the reaction. The nanocomposites were milled into a fine powder, washed with hot alcohol, and dried. They were designated as PS/C_{18}–MMT and PS/VC_{18}–MMT. The blend nanocomposites were prepared by the melt processing of PS/organoclay nanocomposites with PA6 or of the three components simultaneously in an internal mixer at 220 °C for 8 min, using a rotor speed of 80 rpm. Details of the compositions prepared using the two methods have been presented in Table 3.3.

TEM micrographs of the ternary compositions have been presented in Fig. 3.6. The contrast between the components made them easily distinguishable: the white ellipsoids, the grey continuous phase, and the black lines being the dispersed PS domains, the PA6 matrix, and the clay layers, respectively. The distributions of C_{18}–MMT in N2 and N4 were almost identical. Most of the C_{18}–MMT layers were found to be localised in the PA6 component, while some were localised at the interface. The preferential localisation of the nanoclay particles in PA6 was ascribed to its greater affinity for PA6 than PS. Though the localisation of nanoclay particles was similar in N2 and N4, the size of the dispersed PS domains was smaller in N2, and the quality of nanoclay particles dispersion was also better. In contrast to N2 and N4, the nanoclay particles in N3 were primarily localised in the PS phase. A small amount was localised at the interface. The researchers ascribed the localisation of VC_{18}–MMT in PS to the fact that during polymerisation of styrene, the VC_{18}–MMT layers reacted with styrene, leading to the growth of PS chains directly from the nanoclay layers. The strong chemical interaction between the nanoclay layers and PS increased their compatibility, resulting in localisation of the nanoclay particles in the PS phase. Thus, the researchers achieved an improvement in the morphological characteristics of blend nanocomposites by using a combination of in situ polymerisation and melt processing. Furthermore, they could tailor the localisation of the clay particles by changing the surface properties of the nanoclay particles.

Mohamadi and Sharifi-Sanjani [39] used a combination of masterbatch dilution and solution casting to produce PVDF/PMMA/graphene blend nanocomposites. However, unlike most studies where masterbatches were produced by melt processing and studies were conducted on immiscible blends, the researchers produced PMMA/graphene masterbatch by in situ polymerisation and selected a miscible system, PVDF/PMMA [40] for their studies, which further added novelty to their work. To prepare the masterbatch, 300 mg vinyl group-modified graphite (MG) was ultrasonicated in 15 mL MMA for 30 min. This was followed by polymerisation of the MMA

Table 3.3 The formulations of compounds blended in haake-mixture

Samples	Blending ingredient (wt%)				
	PA6	PS	PS/C_{18}–MMT	PS/VC_{18}–MMT	C_{18}–MMT
N1	70	30	–	–	–
N2	70	–	30 (6.7)[a]	–	–
N3	70	–	–	30 (6.9)[a]	–
N4	70	28	–		2

[a] The numbers in the brackets represent the actual organoclay content (%) in the nanocomposites

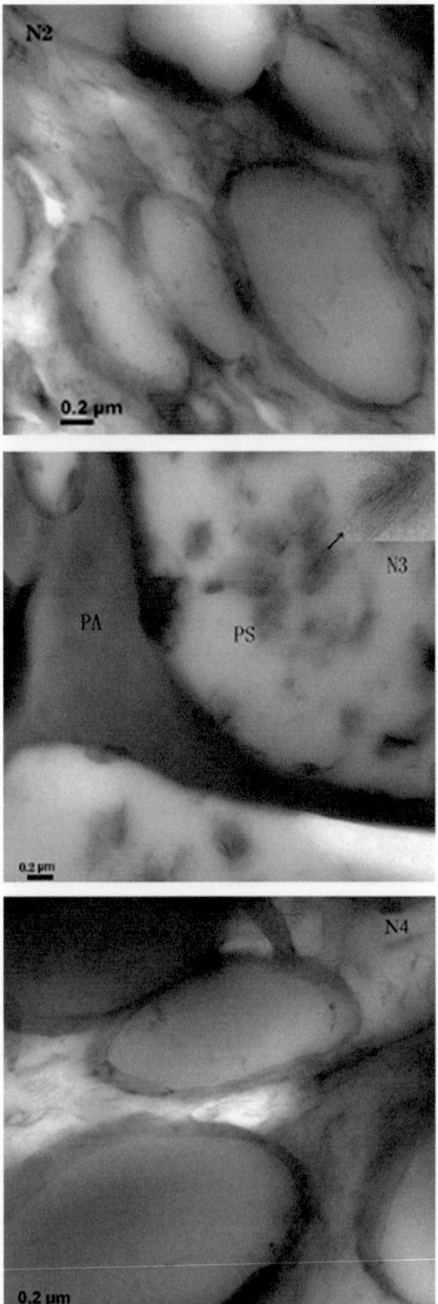

Fig. 3.6 The typical TEM micrographs of ternary nanocomposites. Reproduced with permission from [38]. Copyright 2008, Wiley Periodicals, LLC

in the presence of 1.1 wt% benzoyl peroxide initiator at 90 °C until the entire reaction mixture was solidified. The product was dried in a vacuum at 100 °C to remove residual monomer.

PVDF and PMMA-graphene masterbatch (PMMA-G master) solutions in dimethylformamide were separately prepared at 50 °C by stirring for a day. The two solutions were mixed, and the final concentration of the mixture was adjusted to 12% with respect to the solvent. The mixture was sonicated for 0.5 h, stirred for 2 days, and solution-cast films were prepared by evaporating the solvent at 60 °C in an oven for 20 h. Samples having PVDF/PMMA-G master ratios 80:20 and 70:30 were thus prepared.

TEM images of PMMA-G master and the nanocomposites have been presented in Fig. 3.7. In the PMMA-G master, the separated graphene layers were easily discernible, but the layers were not individual graphene sheets, as the thickness was greater than that of an individual sheet (1 nm). Each layer comprised multiple sheets stacked together. In the 80:20 blend nanocomposite, individual sheets were found to coexist with the stacked sheets. A good quality of dispersion of graphene was observed in the 70:30 blend nanocomposite. The crumple feature of graphene, comprising ripples corresponding to either bending or buckling, was also evident. The researchers thus achieved a superior quality of dispersion of graphene in a miscible polymer blend, preparing a masterbatch with one of the blend components by means of in situ polymerisation and using this masterbatch to produce the blend nanocomposite by means of solution casting.

Studies conducted by Liu et al. [41] also show deviation from the general method of preparing masterbatch by melt mixing, wherein the researchers prepared a PA6/4 wt% masterbatch by solution mixing and then melt mixed this masterbatch with acrylonitrile–butadiene–styrene (ABS). The masterbatch was produced by dissolving 96 g PA6 in 400 mL formic acid at 60 °C. To this solution, 4 g MWNT was added, and the mixture was subjected to ultrasonication, along with mechanical stirring at 60 °C for 2.5 h. The suspension was poured into a large amount of water, and the flocculant solid precipitate was washed, dried, and pulverised into a powder. The masterbatch thus produced was used to produce (70/30 and 50/50) PA6/ABS/MWNT nanocomposites having 0, 0.1, 0.25, 0.5 and 1 wt% MWNT content, by melt mixing in a co-rotating twin-screw, using a temperature profile of 210–230 °C and a screw speed of 150 rpm.

SEM micrographs of the cryofractured injection moulded samples of the unfilled blends and the blend nanocomposites have been presented in Fig. 3.8. As observed in Fig. 3.8A, both the unfilled and filled (70/30) PA6/ABS blends had an island-sea morphology. However, the presence of the MWNTs resulted in a finer morphology with a more regular shape and reduced size of the dispersed ABS domains. The number average diameter of the ABS domains in the unfilled blend was 22.07 μm, while that of the nanocomposites was 2–4 μm. Furthermore, interfacial adhesion was improved with the addition of the nanoparticles, as evident from the characteristics of the cryofractured surface. The cryofractured surface of the specimen of unfilled blend was rough with many concave holes, indicating that the ABS domains had peeled off during cryofracturing. Also, the surfaces of these holes were smooth,

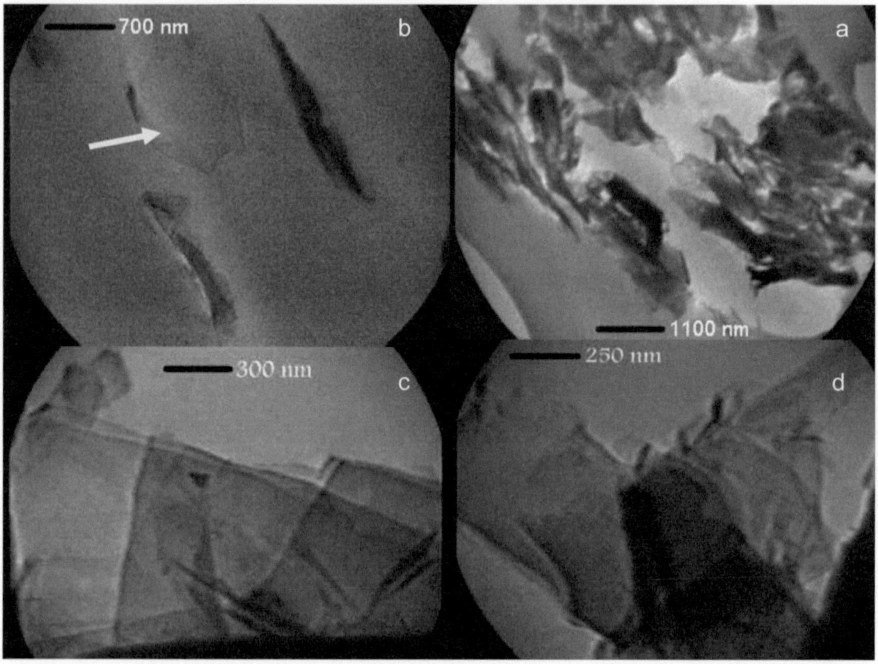

Fig. 3.7 TEM images of PMMA-G master (**a**), 80:20 (**b**), and 70:30 polymer blend nanocomposites (**c**) and (**d**). Reproduced with permission from [39]. Copyright 2011, Wiley Periodicals, LLC

indicating that interfacial adhesion between ABS and PA6 was poor, leading to the domains getting easily peeled off. In contrast, the cryofractured surfaces of the nanocomposite samples had a smoother appearance, with fewer holes. The interface was also less distinct. All these features indicated an improvement in the interfacial adhesion of the bend components due to the presence of the filler particles. A similar improvement in morphology brought about by the presence of nanoparticles was also observed in the 50/50 blend (Fig. 3.8B). The unfilled and filled 50/50 blends had a co-continuous morphology. But the morphology became more refined, and the interfacial adhesion improved with the addition of the nanoparticles. The MWNTs were found to be selectively distributed in the PA6 phase. Addition of only 0.25 wt% MWNTs resulted in a significant improvement (20.8% for the 70/30 blend and 14.3% for the 50/50 blend) in the yield strength of the blends. The improvement in yield strength was ascribed to the improvement in blend morphology and interfacial adhesion due to the addition of the nanoparticles.

The researchers thus successfully produced polymer blend nanocomposites of superior morphology by combining solution blending, masterbatch dilution, and melt processing.

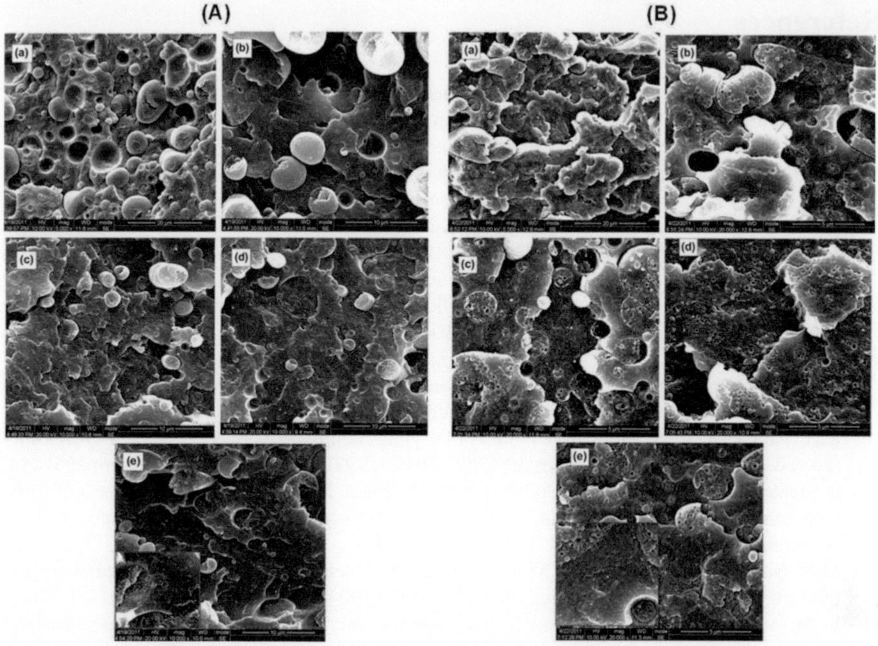

Fig. 3.8 **A** SEM micrographs of cryo-fractured surface morphologies of **a** PA6/ABS (70/30 wt) blends and PA6/ABS (70/30 wt) blends with; **b** 0.1 wt% MWNTs; **c** 0.25 wt% MWNTs; **d** 0.5 wt% MWNTs; and **e** 1 wt% MWNTs. **B** SEM micrographs of cryo-fractured **a** PA6/ABS (50/50 wt) blend and PA6/ABS (50/50 wt) blends with; **b** 0.1 wt% MWNTs; **c** 0.25 wt% MWNTs; **d** 0.5 wt% MWNTs; and **e** 1 wt% MWNTs. Reproduced with permission from [41]. Copyright 2012, Elsevier Science Ltd.

3.3 Conclusions

Polymer blend nanocomposites of superior morphological characteristics may be produced by in situ polymerisation, solution casting, and melt mixing. Of these, melt mixing is the most popular. However, it is often combined with in situ polymerisation or solvent casting. Furthermore, masterbatch dilution is widely used to achieve a superior quality of nanoparticle dispersion. By the proper choice of processing technique(s) and mixing sequence, it is possible to tailor not only the quality of dispersion of the nanoparticles, but also their localisation and the morphological characteristics of the blend. Structural and morphological characterization of these multiphase materials is crucial for understanding the structure–property-processing relationship, which is essential in material design. This topic has been discussed in depth in Chap. 4.

References

1. J. McKelvey, *Polymer Processing* (Wiley, New Jersey, 1962)
2. M.M. Dumoulin, Polymer blends forming, in *Polymer Blends Handbook*, ed. by L.A. Utracki, vol. 2 (Kluwer Academic Publishers, Dordrecht, 2002)
3. S.S. Ray, R. Banerjee, *Sustainable Polylactide-Based Blends* (Elsevier, Amsterdam, 2022)
4. V. Ojijo, S.S. Ray, Processing strategies in bionanocomposites. Prog. Polym. Sci. **38**, 1543–1589 (2013). https://doi.org/10.1016/j.progpolymsci.2013.05.011
5. L.A. Utracki, G.Z.H. Shi, Compounding polymer blends, in *Polymer Blends Handbook*, ed. by L.A. Utracki, vol. 1 (Kluwer Academic Publishers, Dordrecht, 2002)
6. S.S. Ray, M. Okamoto, Polymer/layered silicate nanocomposites: a review from preparation to processing. Prog. Polym. Sci. **28**, 1539–1641 (2003)
7. R.A. Vaia, K.D. Jandt, E.J. Kramer, E.P. Giannelis, Microstructural evolution of melt intercalated polymer-organically modified layered silicates nanocomposites. Chem. Mater. **8**, 2628–2635 (1996). https://doi.org/10.1021/cm960102h
8. R.A. Vaia, E.P. Giannelis, Lattice model of polymer melt intercalation in organically-modified layered silicates. Macromolecules **30**, 7990–7999 (1997). https://doi.org/10.1021/ma9514333
9. P. Eteläaho, K. Nevalainen, R. Suihkonen, J. Vuorinen, K. Hanhi, P. Järvelä, Effects of direct melt compounding and masterbatch dilution on the structure and properties of nanoclay-filled polyolefins. Polym. Eng. Sci. **49**, 1438–1446 (2009)
10. M.N. Bureau, F. Perrin-Sarazin, M.T. Ton-That, Polyolefin nanocomposites: essential work of fracture analysis. Polym. Eng. Sci. **44**, 1142–1151 (2004)
11. L. Xu, M. Xanthos, Effects of process conditions and mixing protocols on structure of extruded polypropylene nanocomposites. J. Appl. Polym. Sci. **93**, 1891–1899 (2004)
12. T. Villmow, P. Pötschke, S. Pegel, L. Häussler, B. Kretzschmar, Influence of twin-screw extrusion conditions on the dispersion of multi-walled carbon nanotubes in a poly(lactic acid) matrix. Polymer (Guildf.) **49**, 3500–3509 (2008). https://doi.org/10.1016/j.polymer.2008.06.010
13. S. Ricklin, Solvent-cast films. Plast. Eng. **39**, 29–33 (1983)
14. T. Baouz, E. Acik, F. Rezgui, U. Yilmazer, Effects of mixing protocols on impact modified poly(lactic acid) layered silicate nanocomposites. J. Appl. Polym. Sci. **132** (2015). https://doi.org/10.1002/APP.41518.
15. J. Bandyopadhyay, S.S. Ray, R. Salehiyan, V. Ojijo, M. Khoza, J. Wesley-Smith, Effect of the mode of nanoclay inclusion on morphology development and rheological properties of nylon6/ethyl–vinyl-alcohol blend composites. Polymer (Guildf.) **126**, 96–108 (2017). https://doi.org/10.1016/j.polymer.2017.08.032
16. T. Gong, M.Q. Liu, H. Liu, S.P. Peng, T. Li, R.Y. Bao et al., Selective distribution and migration of carbon nanotubes enhanced electrical and mechanical performances in polyolefin elastomers. Polymer (Guildf.) **110**, 1–11 (2017). https://doi.org/10.1016/J.POLYMER.2016.12.056
17. C.J. Lee, R. Salehiyan, D.S. Ham, S.K. Cho, S.J. Lee, K.J. Kim et al., Influence of carbon nanotubes localization and transfer on electrical conductivity in PA66/(PS/PPE)/CNTs nanocomposites. Polymer (Guildf.) **84**, 198–208 (2016). https://doi.org/10.1016/J.POLYMER.2015.12.055
18. M. Liebscher, M.O. Blais, P. Pötschke, G. Heinrich, A morphological study on the dispersion and selective localization behavior of graphene nanoplatelets in immiscible polymer blends of PC and SAN. Polymer (Guildf.) **54**, 5875–5882 (2013). https://doi.org/10.1016/J.POLYMER.2013.08.009
19. O. Rafeie, M.K. Razavi Aghjeh, A. Tavakoli, M. Salami Kalajahi, O.A. Jameie, Conductive poly(vinylidene fluoride)/polyethylene/graphene blend-nanocomposites: relationship between rheology, morphology, and electrical conductivity. J. Appl. Polym. Sci. **135**, 46333 (2018). https://doi.org/10.1002/APP.46333
20. R. Salehiyan, W.J. Choi, J.H. Lee, K. Hyun, Effects of mixing protocol and mixing time on viscoelasticity of compatibilized PP/PS blends. Korea Aust. Rheol. J. **26**, 311–318 (2014). https://doi.org/10.1007/S13367-014-0035-Z/METRICS

21. G. Xie, J. Yu, S. Qin, J. Sun, Z. Yang, L. Wei et al., Selective localization of organic montmorillonite nanoparticles in multilayered high-density polyethylene/polyamide 6 composites. Adv. Polym. Technol. **37**, 1526–1536 (2018). https://doi.org/10.1002/ADV.21811
22. L. As'habi, S.H. Jafari, H.A. Khonakdar, R. Boldt, U. Wagenknecht, G. Heinrich, Tuning the processability, morphology and biodegradability of clay incorporated PLA/LLDPE blends via selective localization of nanoclay induced by melt mixing sequence. Express Polym. Lett. **7**, 21–39 (2013). https://doi.org/10.3144/expresspolymlett.2013.3
23. A.D. Oliveira, N.M. Larocca, D.R. Paul, L.A. Pessan, Effects of mixing protocol on the performance of nanocomposites based on polyamide 6/acrylonitrile–butadiene–styrene blends. Polym. Eng. Sci. **52**, 1909–1919 (2012). https://doi.org/10.1002/PEN.23152
24. M. Nofar, M. Mohammadi, P.J. Carreau, Nanoclay migration and the rheological response of PBAT/LDPE blends. Int. Polym. Process. **36**, 287–296 (2021). https://doi.org/10.1515/IPP-2020-4057/MACHINEREADABLECITATION/RIS
25. S.B. Kazemabad, A. Khavandi, D. Fox, Y. Chen, H. Zhang, B. Chen, Effects of blending sequence on morphology and mechanical properties of polypropylene/ethylene-octene copolymer/clay nanocomposites. Manuf. Sci. Technol. **3**, 236–239 (2015)
26. F.R. Passador, A. Ruvolo-Filho, L.A. Pessan, Effect of blending protocol on the rheological properties and morphology of HDPE/LLDPE blend-based nanocomposites. Int. Polym. Process. **27**, 378–385 (2012). https://doi.org/10.3139/217.2567/MACHINEREADABLECITATION/RIS
27. L. Azubuike, U. Sundararaj, Interface strengthening of PS/aPA polymer blend nanocomposites via in situ compatibilization: enhancement of electrical and rheological properties. Materials **14**, 4813 (2021). https://doi.org/10.3390/MA14174813
28. D. Yan, G. Yang, Effect of multiwalled carbon nanotubes on the morphology and electrical properties of polyamide 6/polystyrene blends prepared via successive polymerization. J. Appl. Polym. Sci. **125**, E167–E174 (2012). https://doi.org/10.1002/APP.35680
29. R.J. Sengwa, S. Choudhary, Structural characterization of hydrophilic polymer blends/montmorillonite clay nanocomposites. J. Appl. Polym. Sci. **131**, 40617 (2014). https://doi.org/10.1002/APP.40617
30. P. Saiprasit, A.K. Schlarb, The effect of the compounding procedure on the morphology and mechanical properties of PLA/PBAT-based nanocomposites. Int. Polym. Process. **36**, 219–227 (2021). https://doi.org/10.1515/IPP-2020-4032/MACHINEREADABLECITATION/RIS
31. Z.X. Huang, M.L. Zhao, G.Z. Zhang, J. Song, J.P. Qu, Controlled localizing multi-wall carbon nanotubes in polyvinylidene fluoride/acrylonitrile butadiene styrene blends to achieve balanced dielectric constant and dielectric loss. Compos. Sci. Technol. **212**, 108874 (2021). https://doi.org/10.1016/J.COMPSCITECH.2021.108874
32. C.G. Martins, N.M. Larocca, D.R. Paul, L.A. Pessan, Nanocomposites formed from polypropylene/EVA blends. Polymer (Guildf.) **50**, 1743–1754 (2009). https://doi.org/10.1016/J.POLYMER.2009.01.059
33. A. Taghizadeh, B.D. Favis, Carbon nanotubes in blends of polycaprolactone/thermoplastic starch. Carbohydr. Polym. **98**, 189–198 (2013). https://doi.org/10.1016/J.CARBPOL.2013.05.024
34. E.C. Lopes Pereira, B.G. Soares, A.A. Silva, G.M.O. Barra, Master batch approach for developing PVDF/EVA/CNT nanocomposites with co-continuous morphology and improved electrical conductivity. J. Appl. Polym. Sci. **138**, 51164 (2021). https://doi.org/10.1002/APP.51164
35. T. Périé, A.C. Brosse, S. Tencé-Girault, L. Leibler, Co-continuous nanostructured nanocomposites by reactive blending of carbon nanotube masterbatches. Polymer (Guildf.) **53**, 984–992 (2012). https://doi.org/10.1016/J.POLYMER.2012.01.007
36. P. Pötschke, S. Pegel, M. Claes, D. Bonduel, A novel strategy to incorporate carbon nanotubes into thermoplastic matrices. Macromol. Rapid Commun. **29**, 244–251 (2008). https://doi.org/10.1002/MARC.200700637
37. A.C. Baudouin, J. Devaux, C. Bailly, Localization of carbon nanotubes at the interface in blends of polyamide and ethylene–acrylate copolymer. Polymer (Guildf.) **51**, 1341–1354 (2010). https://doi.org/10.1016/J.POLYMER.2010.01.050

38. J. Yang, L. Sun, S. Xiang, J. He, L. Gu, M. Zhong, Influence of organoclay and preparation technique on the morphology of polyamide6/polystyrene/organoclay nanocomposites. J. Appl. Polym. Sci. **110**, 276–282 (2008). https://doi.org/10.1002/APP.28178
39. S. Mohamadi, N. Sharifi-Sanjani, Investigation of the crystalline structure of PVDF in PVDF/PMMA/graphene polymer blend nanocomposites. Polym. Compos. **32**, 1451–1460 (2011). https://doi.org/10.1002/PC.21175
40. T. Nishi, T.T. Wang, Melting point depression and kinetic effects of cooling on crystallization in poly (vinylidene fluoride)-poly (methyl methacrylate) mixtures. Macromolecules **8**, 909–915 (1975)
41. X.Q. Liu, W. Yang, B.H. Xie, M.B. Yang, Influence of multiwall carbon nanotubes on the morphology, melting, crystallization and mechanical properties of polyamide 6/acrylonitrile–butadiene–styrene blends. Mater. Des. **34**, 355–362 (2012). https://doi.org/10.1016/J.MATDES.2011.08.028

Chapter 4
Techniques for Structural and Morphological Characterisation of Polymer Blend Nanocomposites

Abstract To manufacture a polymer blend nanocomposite with exceptional properties, it is essential to manage the localization of nanoparticles and achieve the desired morphology of the blend, ensuring an excellent dispersion quality. Therefore, it is vital to characterize the structure and morphology of these multiphase materials to assess the effectiveness of the processing. This understanding contributes to the relationship between structure and properties, being crucial for both quality control and research aimed at developing high-performance polymer blend nanocomposites. This chapter offers an overview of the techniques utilized for structuring and morphologically characterizing polymer blend nanocomposites.

4.1 Optical Microscopy

Optical microscopy (OM) is a very adaptable imaging method extensively employed to investigate the crystal growth characteristics and kinetics of polymeric materials [1]. This technique involves illuminating an object and collecting the light it scatters or transmits through a lens system, forming an image [2]. The surface structure of a sample is analysed using reflected light mode, while the bulk structure is examined with transmitted light. In bright-field microscopy, illumination is mostly parallel to the main optical axis. Conversely, dark-field microscopy utilises illumination at a significantly larger angle to the main optical axis. Both microscopy methods—bright-field and dark-field—allow studies in either reflected or transmitted light mode [3, 4].

Optical microscopy (OM) is useful for investigating spherulite structures in polymer blend nanocomposites. However, when the blend components are immiscible and both are semi-crystalline, their spherulites might overlap, making them difficult to differentiate. To gain better insights, it's essential to compare the microstructure of the homopolymers with that of the multi-phase system [2]. Figure 4.1 presents polarized optical microscopy (POM) images of both an unfilled sample and a 5 wt% organically modified vermiculite nanoclay (OVMT) reinforced poly(L-lactic acid) (PLLA)/poly(methyl methacrylate) (PMMA)/poly(ethylene oxide) (PEO) blend

S. S. Ray and R. Banerjee, *Polymer Blend Nanocomposites*, Springer Series in Materials Science 354, https://doi.org/10.1007/978-3-032-00309-6_4

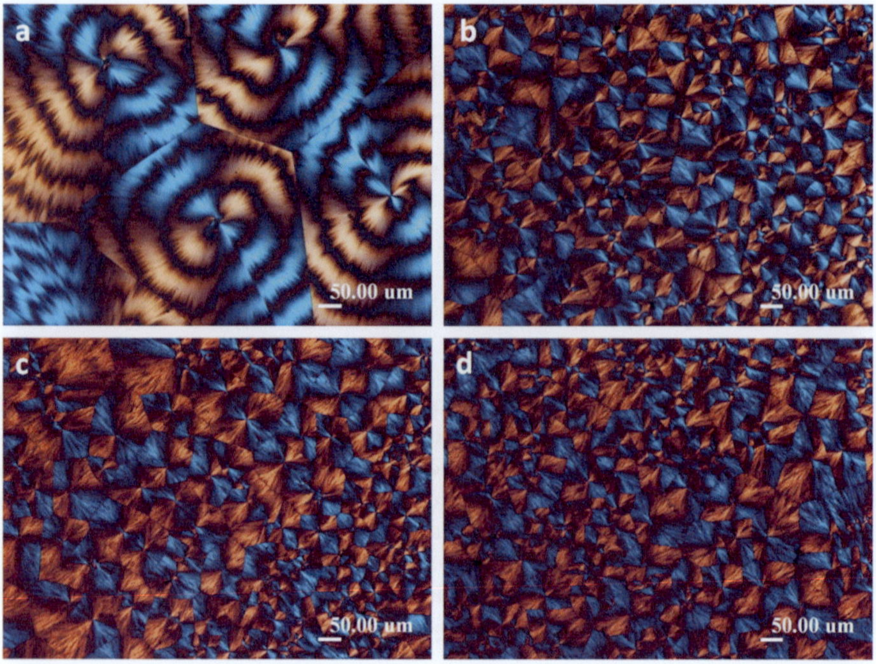

Fig. 4.1 POM images of **a** PLLA/PMMA/PEO 80/10/10 with **b** 1 wt%; **c** 3 wt%; and **d** 5 wt% OVMT held at 128 °C. Reproduced with permission from [5]. Copyright 2011, Wiley Periodicals LLC

nanocomposite. POM studies were performed on thin film samples of the unfilled blend and nanocomposites sandwiched between two glass slides, with phase observations conducted at a scanning rate of 2 °C/min. The growth of spherulites was tracked at various isothermal crystallization temperatures. The spherulites from the unfilled blend displayed a larger size and regular ring patterns, whereas the spherulites of the nanocomposites were significantly smaller and exhibited a narrower size distribution. This variation in spherulite size and distribution was linked to the enhanced nucleation density of crystals caused by the presence of filler particles.

OM has been employed to estimate the alterations in the domain size of the dispersed phase in the blend due to the incorporation of nanoparticles in polymer blend nanocomposites. Figure 4.2 presents OM images of (70/30 wt ratio) (iPP)/PEO blends that are reinforced with 5 wt% of two varieties of organically modified nanoclay, specifically Cloisite 30B (C30B) and 1.30 T from Nanocor Inc. The blend nanocomposites were produced through melt processing under consistent conditions. For OM analysis, the molten samples were placed between two glass slides, heated to 200 °C for 10 min, and subsequently cooled to room temperature. OM analysis indicated that C30B led to a notably greater alteration in the blend domain size than 1.30 T, which is believed to result from the varying interactions between the clay surfactant groups and the polymer entities [6].

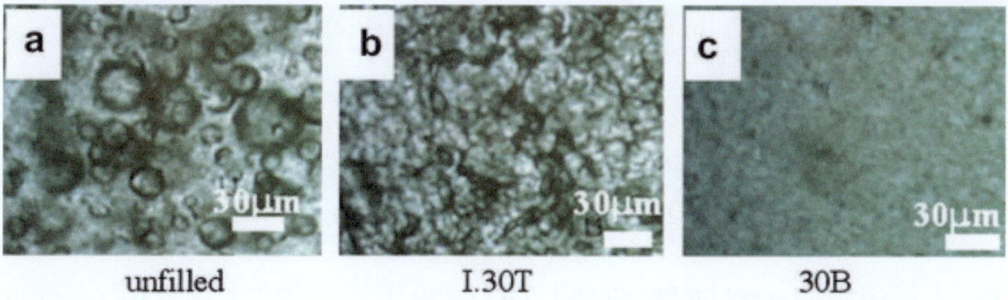

Fig. 4.2 Optical micrographs of the unfilled blend and the 5 wt% nanocomposite samples as indicated. Reproduced with permission from [6]. Copyright 2008, Elsevier Science Ltd.

OM indicates how dispersed the filler particles are within the polymer melt. When particles achieve nanolevel dispersion, turbidity is eliminated because the particle size drops below λ/4, where λ is the wavelength of light. Figure 4.3 shows OM images of microtomed extruded strands from amino-terminated polyamide-6 (PA6)/maleic anhydride functionalized polyethylene (PE*)/carbon nanotube (CNT) nanocomposites, composed of 78 wt% PE*, 20 wt% PA6, and 2 wt% MWNT, which were created through melt processing with different mixing sequences. The observed variation in CNT dispersion levels in the OM images resulted from the superior effectiveness of masterbatch dilution in achieving better quality dispersion of the nanofiller [7].

Nevertheless, the lack of turbidity in optical microscopy (OM) images does not automatically indicate a dispersion of particles at the nano level. This is due to the fact that OM cannot resolve aggregates smaller than λ/4. Consequently, the resolution of OM is constrained by the wavelength of visible light. However, recent advancements in near-field scanning techniques (NSOM) can considerably mitigate this limitation [1, 8].

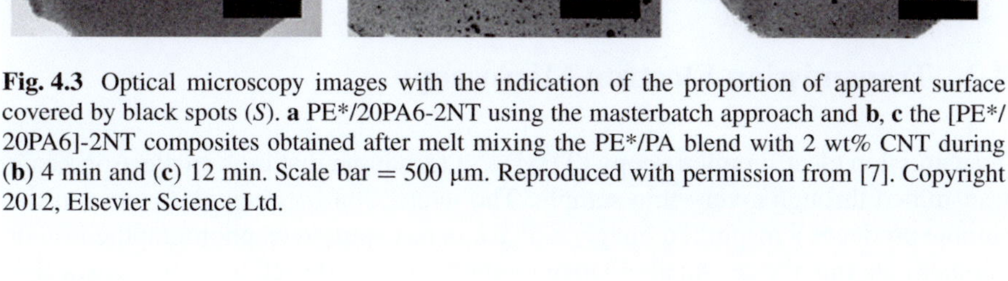

Fig. 4.3 Optical microscopy images with the indication of the proportion of apparent surface covered by black spots (*S*). **a** PE*/20PA6-2NT using the masterbatch approach and **b**, **c** the [PE*/20PA6]-2NT composites obtained after melt mixing the PE*/PA blend with 2 wt% CNT during (**b**) 4 min and (**c**) 12 min. Scale bar = 500 μm. Reproduced with permission from [7]. Copyright 2012, Elsevier Science Ltd.

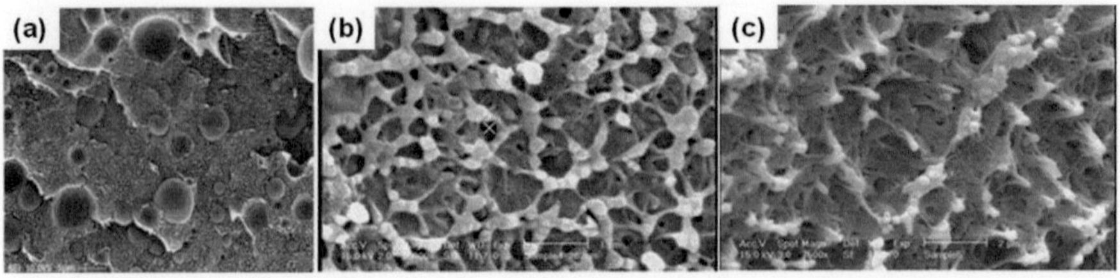

Fig. 4.4 Micrograph of the fracture surface of **a** C0-P10-N150, **b** C1-P10-N150, and **c** C3-P10-N150. Reproduced with permission from [9]. Copyright 2012, Taylor & Francis

4.2 Scanning Electron Microscopy

Scanning electron microscopy (SEM) is utilised to investigate the surface morphology of polymeric materials. An image is created by scanning a probe over the sample. This probe comprises a focused electron beam that interacts with a thin layer on the surface of the sample, typically just a few micrometers thick [2]. To prepare polymer samples for SEM analysis, they are often coated with a thin layer of electrically conductive material, such as gold, using low vacuum sputter coating or high vacuum evaporation. SEM instruments have an inherent resolution of less than 1 nm, making it quite challenging to image nanolevel dispersions of filler particles. Nevertheless, SEM remains widely used for the morphological characterization of polymer blend nanocomposites, allowing for the examination of both the morphological properties of the blend and the fracture surface morphology [1].

Figure 4.4 shows SEM micrographs of an unfilled linear low-density polyethylene (LLDPE)/maleic anhydride grafted LLDPE (LLDPE-g-MA)/polyamide 6 (PA6) blend alongside LLDPE/LLDPE-g-MA/PA6/Cloisite 15A nanocomposites. In the C1-P10-N150 nanocomposite, which contains 1 wt% clay, 10 wt% PA6, and 22 wt% LLDPE-g-MA, the PA6 domain size ranged from 29 to 262 nm, representing a reduction to 30 times smaller than that observed in C0-P10-N150, the unfilled blend with the same PA6 and LLDPE-g-MA proportions. Increasing the filler content resulted in a further reduction of the PA6 domain size; for instance, in C3-P10-N150 (the nanocomposite with 3 wt% clay), the domain size measured between 58 and 77 nm [9].

4.3 Transmission Electron Microscopy

Transmission electron microscopy (TEM) is a technique that uses an electron beam transmitted through a very thin sample. The interaction between electrons and the sample produces a magnified image, which can be captured on photographic film or another imaging device. An alternative method involves detection with a sensor like a charge-coupled device (CCD) camera. In standard TEM, the image is formed from transmitted or certain forward-scattered beams. On the other hand, high-resolution TEM (HRTEM) generates an interference image using both transmitted and scattered beams.

Sample preparation techniques vary based on the material analyzed and the information needed. Polymer samples are sliced with a microtome. For oriented samples, including extruded strands or injection molded specimens, it is crucial to maintain the orientation throughout the sectioning process. Samples should be cut at a 45° angle from any orientation to reduce the likelihood of misinterpreting features induced by preparation, such as knife marks [1].

TEM is essential for the morphological analysis of polymer blend nanocomposites, providing a visual tool that helps clarify the spatial distribution of the present phases and defects. Additionally, for nanostructured blends, it enables a deeper insight into the morphological features of the blend [4, 8]. When the components of the blend have similar chemical compositions, it may be necessary to selectively stain one component for better visualisation of the phase arrangement [10]. Figure 4.4 displays TEM micrographs of both unstained and stained images of a polystyrene (PS)/acrylonitrile–butadiene–styrene (ABS)/clay nanocomposite. The unstained image (Fig. 4.4a) shows that clay is predominantly intercalated and partially exfoliated, although its precise location remains undetermined. Staining with OsO_4 enhanced clarity regarding this issue. It revealed that the clay platelets are positioned along, within, and outside the rubber phase boundaries of ABS, marked by the dark grey rings (indicated by the arrows in Fig. 4.5b) [11].

Staining can be labour-intensive and involves toxic dyes. These challenges can be mitigated using energy-filtered transmission electron microscopy (EFTEM), which allows for molecular mapping in the low energy-loss spectral zone [12]. The spectral variations observed among polymers in this low-energy-loss region are linked to differences in their molecular structures, making it possible to distinguish between phases formed by molecules consisting of the same elements. Even subtle changes in molecular structure result in alterations in the molecular spectra intensity, which

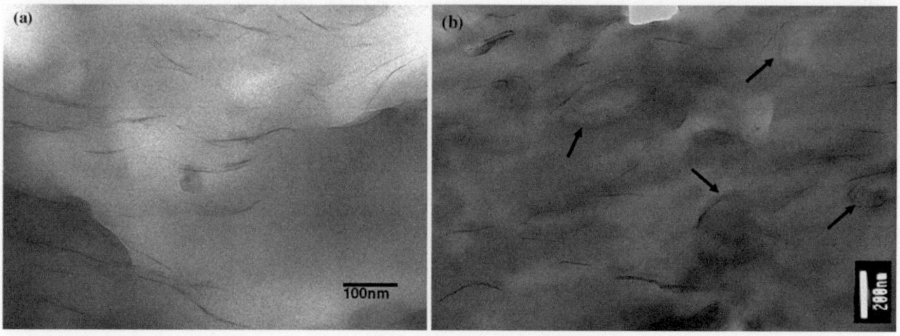

Fig. 4.5 TEM image of PS/ABS/clay nanocomposite (PS:ABS = 1:9, clay = 4 wt % Clay-111); **a** without stain and **b** stained with osmium tetroxide. Arrows point to some of the rubber boundaries. Reproduced with permission from [11]. Copyright 2011, Wiley Periodicals LLC

is examined through collections of low-energy-loss images that reveal information about the various phases of a polymer blend. Low-loss electron energy-loss spectroscopy (ELSS) exhibits heightened sensitivity to alterations in the chemical makeup of carbonaceous compounds due to its reliance on the molecules' electronic structures [13]. With energy filters integrated into TEM tools, EFTEM imaging only requires low beam exposure to attain resolutions of 1 nm and beyond, thus eliminating the necessity for staining. Additionally, the contrast between polymeric phases can often be removed at specific energy losses, resulting in a crossover area within the spectrum that enhances the background, facilitating the observation of non-polymeric components.

Figure 4.6a presents a collection of bright-field and low-loss EFTEM images depicting a natural rubber (NR)/poly(styrene-butyl acrylate) [P(S-BA)] blend with a 70/30 weight ratio, accompanied by the EELS spectra for the blend's individual components. In the bright-field image, one phase's domains stand out, appearing brighter against a darker matrix. The EFTEM images displayed a contrast inversion around 50 eV, attributed to overlapping EELS spectra of the blend components within this energy-loss range. While the spectra were qualitatively similar, their quantitative differences facilitated the identification of the two polymer phases: below 50 eV energy losses, the P(S-BA) domains appeared brighter (including in the bright-field images), in contrast to the dark NR phase. However, both phases showed increased brightness within the 50–90 eV energy loss range. The image taken at 90 eV revealed contrast variations within the P(S-BA) droplets and across the matrix. Most of the rubber was coalesced, yet some discrete rubber particles were identifiable, as noted by arrows.

Figure 4.6b displays a bright-field image alongside an EFTEM series of the NR/P(S-BA)/clay nanocomposite, showcasing a 70/30 weight ratio of the blend components and a nanoclay loading of 5 phr. The images highlight marked morphological and structural contrasts when compared to the unfilled polymer blend. While the clay is well-dispersed throughout the film, the bright-field image does not enable identification of the specific blend component where the clay is situated. Contrast inversion images in the ~40–50 eV range indicate that the clay particles align parallel to the film's surface. Furthermore, loss images within the 50–90 eV range reveal non-uniform clay distribution, with concentrations found in the darker P(S-BA) areas. Alongside exfoliated clay platelets, agglomerates with thicknesses reaching several tens of nanometers are also noted. The shape of the dispersed P(S-BA) domains in the nanocomposite is different from that in the unfilled blend. In the unfilled blend, P(S-BA) domains are distinctly separated, exhibiting an aspect ratio close to one, whereas in the nanocomposite, they comprise clay-containing agglomerates with an aspect ratio of 2–5.

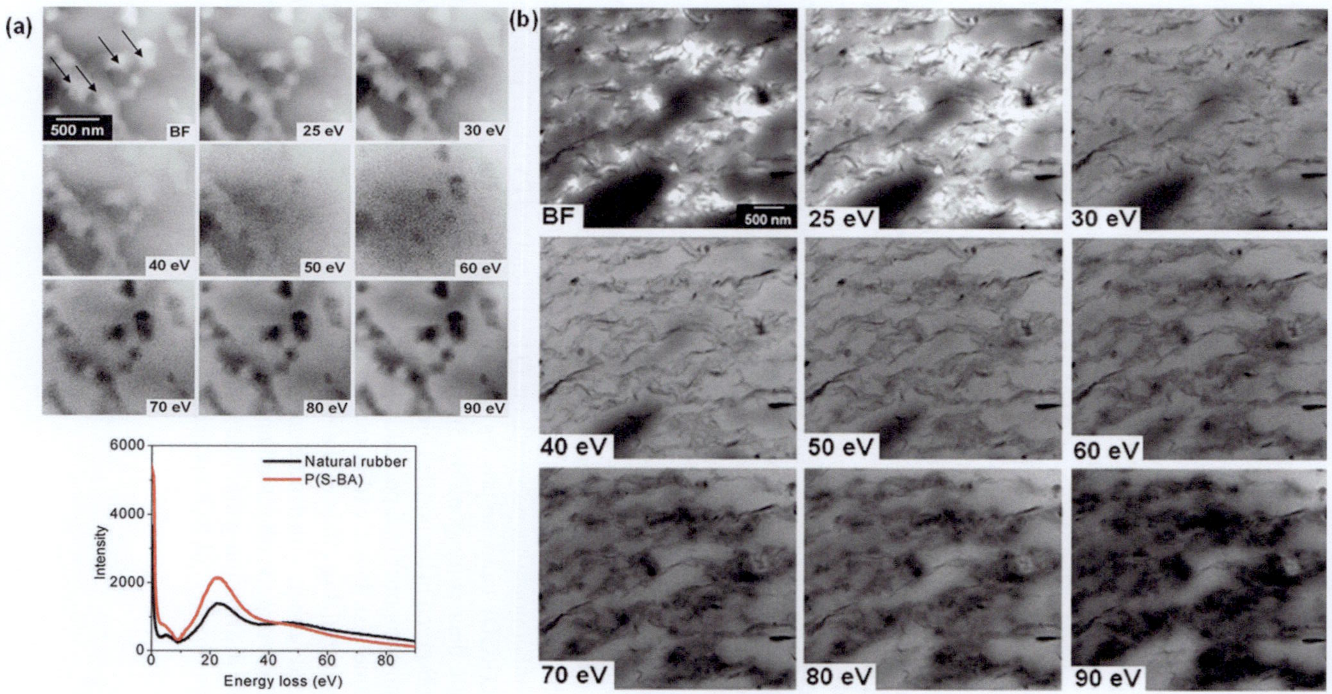

Fig. 4.6 **a** (Top left) Bright-field and low-loss EFTEM images (25–90 eV energy loss) of a thin cut from a NR/P(S-BA) blend with a weight ratio of 7:3. EELS spectra of NR and P(S-BA). Arrows indicate rubber particles. **b** (Top left) Bright-field and low-loss EFTEM (25–90 eV energy loss) images of a thin cut from the clay nanocomposite of a NR/P(S-BA) blend with a weight ratio of 7:3. Reproduced with permission from [10]. Copyright 2010, American Chemical Society

4.4 Atomic Force Microscopy

Atomic-force microscopy (AFM) is an advanced characterization technique capable of imaging nearly any surface, such as polymer blends. AFM measures and localises various forces, including adhesion strength, magnetic forces, and mechanical properties. It utilises a sharp tip about 10–20 nm in diameter mounted on a cantilever, constructed from Si or Si_3N_4 [1]. The weak interaction force between the tip and the specimen surface causes a slight deflection in the cantilever. An optical lever system measures this angle of deflection, producing an electrical signal that corresponds to the movement of the cantilever. AFM operates in different modes, primarily distinguished by the degree of contact between the tip and the specimen. In contact mode, the tip remains in continuous contact with the specimen. In tapping or intermittent contact mode, the tip oscillates and intermittently touches the specimen. In non-contact mode, the tip does not make contact with the specimen at all [2].

AFM produces a genuine 3D surface profile of specimens, unlike electron microscopy, which only yields a 2D representation. Additionally, samples analysed with AFM do not need any special preparation, such as metal coating, which can permanently alter or damage them. AFM can operate in ambient air, unlike electron microscopes that require an expensive vacuum environment. It also provides superior resolution compared to SEM, achieving true atomic resolution in both ultra-high vacuum and liquid conditions. High-resolution AFM delivers resolution similar to that of TEM. However, a limitation of AFM in comparison to SEM is its image size; SEM can capture areas measuring several square millimetres with a depth of field of millimeters. In contrast, AFM can image a maximum height in the micrometre range and has a maximum scanning area of about $150 \times 150\ \mu m^2$ [1].

Figure 4.7 shows AFM images for both an unfilled (70/30) poly(vinylidene fluoride) (PVDF)/thermoplastic polyurethane (TPU) blend and PVDF/TPU/reduced graphene oxide (RGO) nanocomposites. The 2D height images are depicted in Fig. 4.7a–d, while their corresponding 3D profiles are presented in Fig. 4.7a1–d1. The large holes observed in the unfilled blend (Fig. 4.7a, a1) indicate incompatibility between the blend components. Upon adding 0.1 wt% RGO, the holes were reduced in size and became distributed within the continuous phase (Fig. 4.7b, b1), highlighting the compatibilizing effect of the nanofiller. At a 0.3 wt% RGO loading, the hole size decreased further, with a more uniform distribution in the matrix (Fig. 4.7c, c1). However, increasing the filler loading beyond this resulted in larger holes and a non-uniform hole distribution within the matrix (Fig. 4.7d, d1). This indicates that the compatibilization effect of RGO peaked at 0.3 wt% loading [14].

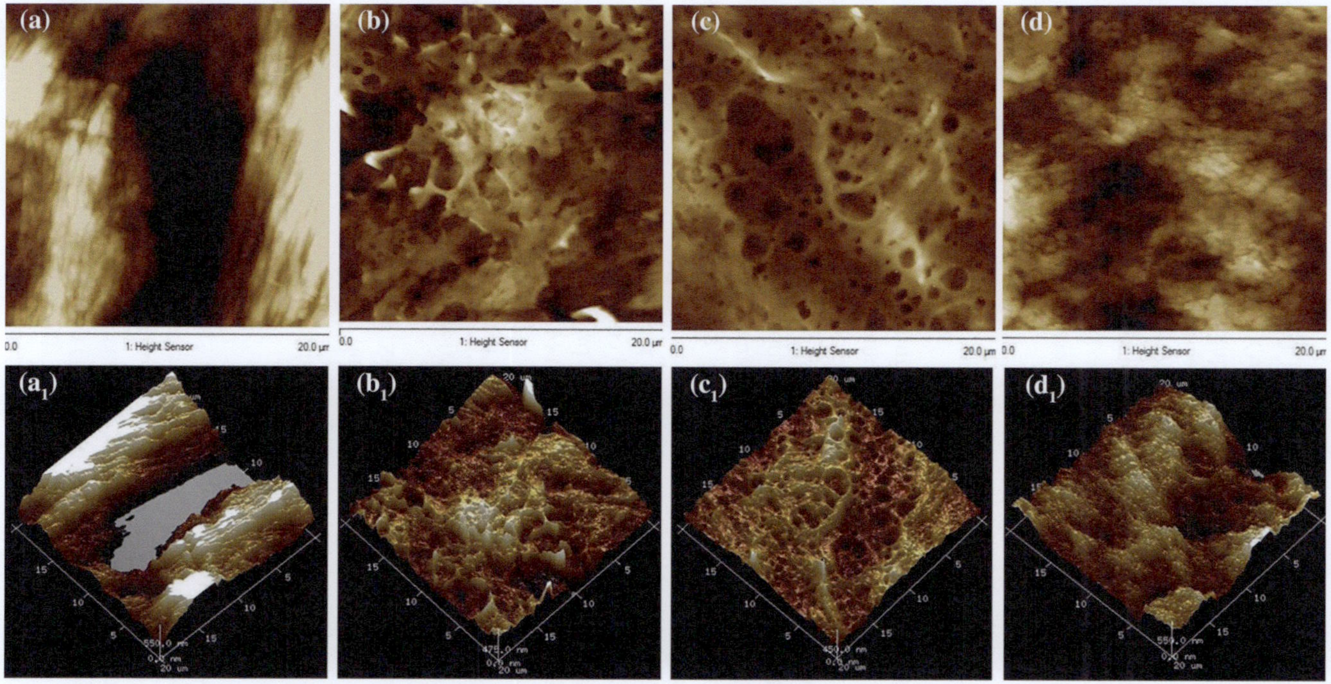

Fig. 4.7 AFM micrographs of incompatible and RGO compatibilized PVDF/TPU (70/30) blends after THF etching: **a** 7030-0 RGO, **b** 7030-0.1 RGO, **c** 7030-0.3 RGO, and **d** 7030-0.5 RGO; **a$_1$**, **b$_1$**, **c$_1$**, and **d$_1$** are the corresponding 3D profiles of **a**, **b**, **c**, and **d**, respectively. Reproduced with permission from [14]. Copyright 2018, Wiley Periodicals LLC

4.5 Wide-Angle X-Ray Diffraction

When X-rays pass through crystalline materials, they are diffracted, creating a diffraction pattern. Wide-angle X-ray diffraction (WAXD or XRD), sometimes referred to as wide-angle X-ray scattering (WAXS), is frequently employed to characterise crystalline materials. This method is based on the constructive interference of monochromatic X-rays that strike a crystalline sample. X-rays are generated by a cathode ray tube, filtered to produce monochromatic radiation, collimated, and aimed at the sample. As the incident rays interact with the sample, they generate a diffraction pattern. The condition for constructive interference is described by Bragg's law, equation (4.1):

$$2d \sin \theta = n\lambda \tag{4.1}$$

In this equation, n represents an integer, λ signifies the wavelength of the radiation, d denotes the spacing between the atomic planes, and θ (the Bragg angle) is the angle between the incident rays and the atomic planes of the sample. Consequently, Bragg's law connects the wavelength of incoming X-rays to the diffraction angle and the lattice spacing found in the crystalline sample. The sample is analysed across a series of 2θ values, allowing for the determination of d spacings. Since each set of lattice planes possesses a distinct d spacing, analysing diffraction peaks helps in identifying the lattice structure of the sample.

WAXD is the most fundamental and direct method for determining crystallinity (χ_c). This is based on the principle that the total coherent scattering by N atoms does not depend on the state of aggregation. This principle results in the fundamental equation for determining crystallinity equation (4.2):

$$\chi_c = \frac{\int_0^\infty s^2 I_c(s) ds}{\int_o^\infty s^2 I(s) ds} \tag{4.2}$$

where I represents the total scattered intensity, I_c denotes the scattered intensity attributed to the crystals (Bragg reflections), and s is defined by equation (4.3):

$$s = \frac{2 \sin \theta}{\lambda} \tag{4.3}$$

where λ represents the wavelength. However, this technique for assessing crystallinity presents three main challenges. The first is the need to subtract incoherent scattering from the overall scattering. The second is that the measurement of scattered intensity (I) can only be conducted within a restricted range of s. The third challenge is that the scattering associated with the crystal planes (I_c) contains diffuse components due to thermal vibrations and paracrystalline defects. Therefore, separating the crystalline contributions from the amorphous ones in the total scattering is quite complex.

Various practical methods exist for assessing crystallinity. The peak resolution technique follows the method outlined by Hermans and Weidinger [15]. A 2θ scan is conducted, capturing both the crystalline peaks and the amorphous halo. The diffractogram for the semi-crystalline polymer features distinct Bragg reflections with a total intensity I_c, along with an amorphous halo that possesses a total scattered intensity I_a. Crystallinity is calculated using equation (4.4):

$$\chi_c = \frac{I_c}{I_c + KI_a} \tag{4.4}$$

where K is a calibration factor that can be set to unity for comparison. However, to determine it absolutely, the K factor must be established using a definitive method, such as the Ruland method or through density measurements [3].

The differential intensity measurement method is appropriate for polymers that show notable reflections during scanning. In these instances, distinguishing between the crystalline and amorphous components becomes challenging. To address this, an amorphous standard and a "highly crystalline standard" are prepared, and scans are conducted for both standard samples and the sample with unknown crystallinity. The total scattered intensity across the measured 2θ range is normalised for all three samples. Differences in scattered intensities are noted: between the unknown sample and the amorphous sample ($I_u - I_a$), as well as between the crystalline and amorphous samples ($I_c - I_a$), recorded at various increments of 2θ. Crystallinity is assessed using the expression [3]:

$$\chi_C = \frac{\sum_{2\theta}(I_u - I_a)}{\sum_{2\theta}(I_c - I_a)} \tag{4.5}$$

The Ruland [16] method is a fundamental technique for assessing crystallinity. It emphasises that measuring the total intensity of distinct Bragg peaks is simpler than evaluating the entire crystalline intensity, which encompasses diffuse crystalline peaks as well. Equation (4.6) represents the basic equation introduced by Ruland:

$$\chi_c = \left[\frac{\int_{s_0}^{s_1} s^2 I_c ds}{\int_{s_0}^{s_1} s^2 I ds}\right] \times K\left(s_0, s_1, D, \overline{f^2}\right) \tag{4.6}$$

where K is a correction factor, which is a function of the selected s-range ($s_0 - s_1$), D is the disorder function, and $\overline{f^2}$ is the mean square atomic scattering factor for the polymer. The K-factor is given by equation (4.7):

$$K = \frac{\int_{s_0}^{s_1} s^2 \overline{f^2} ds}{\int_{s_0}^{s_1} s^2 \overline{f^2} D^2 ds} \tag{4.7}$$

where D is dependent on the nature of the disorder present in the crystals. For first-order defects, such as thermal vibrations, D is given by equation (4.8):

$$D^2 = \exp\left(-ks^2\right) \tag{4.8}$$

where k is a constant. For disorder of the second kind, i.e., paracrystalline defects, D is given in equation (4.9):

$$D^2 = \frac{2\exp\left(-as^2\right)}{1 + \exp\left(-as^2\right)} \tag{4.9}$$

where a is a constant.

A graph of s^2I against s^2 is created, with adjustments made for incoherent and air scattering. The crystalline component is distinguished from the amorphous one by drawing a smooth line at the base of the peaks. As stated by Ruland, this technique identifies ordered regions greater than 2–3 nm as crystalline. The subsequent steps involve fitting; measurements across various s-ranges should produce uniform crystallinity values. The parameter K is adjustable and must remain consistent across all s-ranges [4].

XRD is utilised to investigate the crystal structures of polymer blend nanocomposites and to assess how nanofillers impact crystalline morphology. Figure 4.8 shows the XRD patterns for PVDF, ethylene–vinyl acetate copolymer (EVA), a (50/50 volume ratio) PVA/EVA blend, and PVA/EVA/CNT nanocomposites with 0.56 vol% CNT content, all prepared through various mixing sequences. Pure PVDF displayed four diffraction peaks at $2\theta = 18.4°$, $18.6°$, $20.2°$, and $26.9°$, corresponding to the (100), (020), (110), and (021) reflections of the α-crystalline phase of PVDF. Pure EVA also showed a diffraction peak at $2\theta = 21.5°$. The PVDF/EVA blend and the nanocomposites exhibited the primary diffraction peaks of the PVDF phase alongside the diffraction peak of the EVA phase. The incorporation of CNTs did not affect the XRD profile of the blend, regardless of the mixing sequence [17].

XRD is extensively utilised to examine the morphology of layered silicate nanocomposites by observing the peak associated with the basal reflection from clay particles. In intercalated nanocomposites, the layered silicate's repetitive multilayer structure is preserved, facilitating the measurement of interlayer spacing with XRD. This spacing expands due to the intercalation of polymer chains into the clay galleries, resulting in a shift of the X-ray diffraction peak to lower angles compared to the original clay. Conversely, in exfoliated structures where the multilayer arrangement of layered silicates is not preserved, XRD diffractograms of exfoliated nanocomposites display no peak [8].

Figure 4.9a presents the XRD plots for Cloisite 20A (C20A) and the nanocomposites made from polystyrene (PS) and polypropylene (PP) with C20A. In the 19PS/79PP/2C20A composition, the intensity of the key C20A peak decreased, resulting in a broad peak observed at $2\theta = 2.6°$ ($d_{001} = 3.53$ nm), indicating intercalation. When the clay concentration was raised to 5 wt%, the position of the C20A peak remained mostly the same, while its intensity saw a notable increase compared to the nanocomposite containing 2 wt% nanoclay.

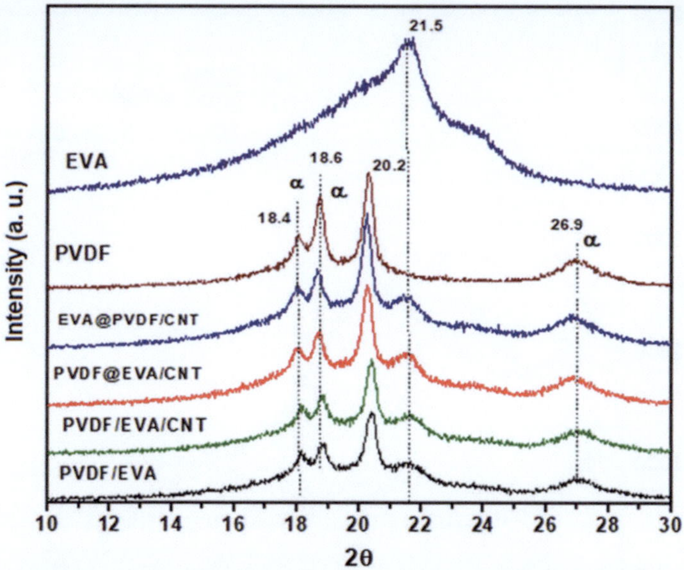

Fig. 4.8 X-ray diffraction (XRD) patterns of PVDF, EVA, PVDF/EVA blend and the CNT-based composites as a function of the mixing procedure. Reproduced with permission from [17]. Copyright 2021, Wiley Periodicals LLC

This observation is attributed to the weak interactions between C20A and the matrices, along with the fact that increasing the clay content resulted in greater parallel stacking of the silicate layers. Figure 4.9b shows the XRD plots for C20A and PS/maleic anhydride grafted PP (PP-g-MA)/C20A nanocomposites. The XRD plots for the nanocomposites lacked the characteristic peak of C20A, suggesting a significant degree of intercalation or exfoliation. The enhanced dispersion quality of C20 in the PS/PP-g-MA blend compared to the PS/PP blend is attributed to the presence of polar MA grafted groups in the PP chains [18].

4.6 Small-Angle X-Ray Scattering

The small-angle X-ray scattering (SAXS) technique is an effective characterization method for assessing the average size, shape, and internal structure of particles that range from 1 to 100 nm. In the context of polymer nanocomposites, SAXS is frequently utilised to study their structure, whether through the observation of peak shifts towards lower scattering angles or by analysing the power-law dependence [1].

SAXS analyses diffraction patterns for angles (2θ) ranging from 0.1° to 10°, typically focusing on values below 1°. The structural parameters averaged by SAXS are gathered from all particles in the sample and their various orientations. This approach differs from microscopy, which examines only a small portion of the sample

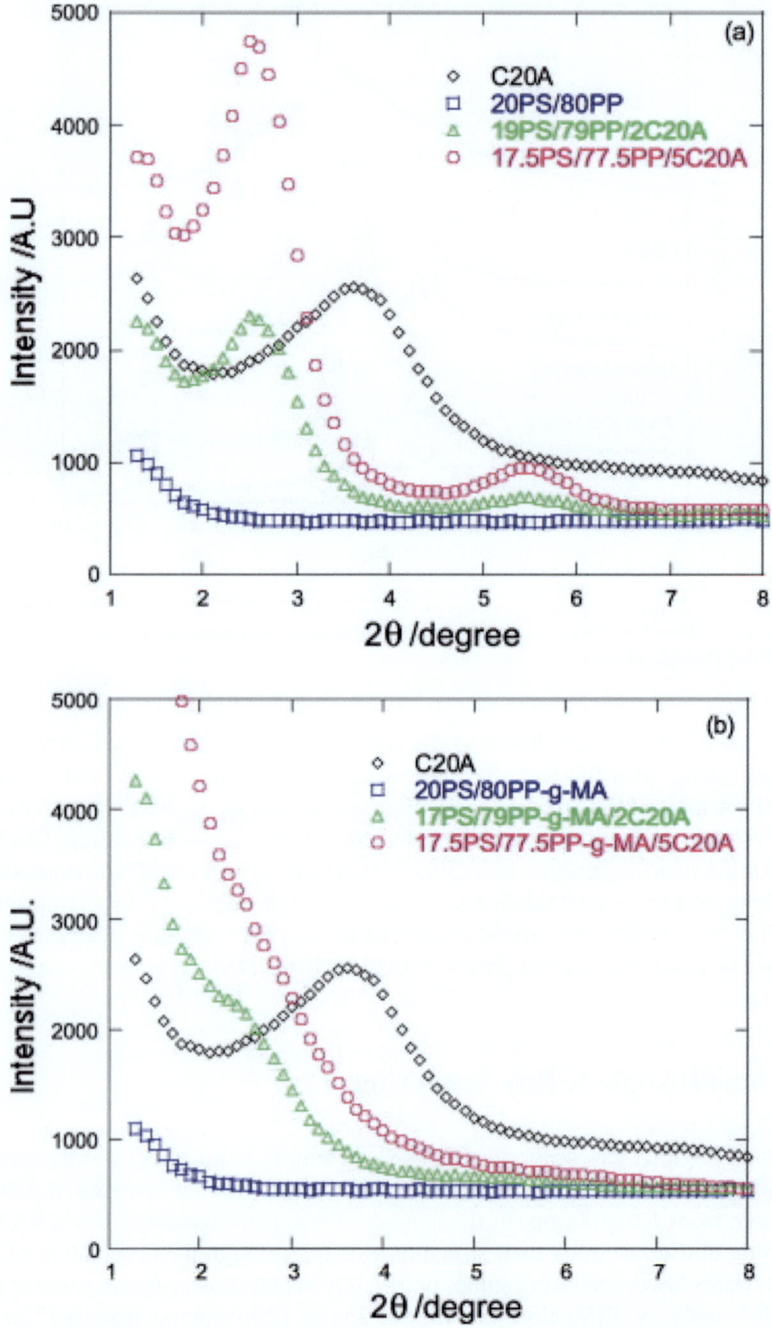

Fig. 4.9 XRD patterns of **a** unmodified and C20A modified (20/80) PS/PP blends and **b** unmodified and C20A modified (20/80) PS/PP-g-MA blends. Reproduced with permission from [18]. Copyright 2004, Elsevier Science Ltd.

after magnification, as SAXS investigates the entire irradiated sample. However, SAXS does not reveal details of individual particles unless they are sufficiently prominent within the specimen to be deemed representative.

The signal strength of SAXS is directly related to the square of a particle's volume. As a result, detecting small particles becomes challenging when larger particles are present. This also implies that SAXS can effectively identify the formation or growth of larger particles [19].

One technique for obtaining scattering curves uses the distance distribution function of electrons, p(r). This function shows the likelihood of locating a pair of electron densities at a distance r and is based on geometric considerations. The scattering curve is represented by the scattering intensity function I(q), as described in equation (4.10):

$$I(q) = 4\pi \int_0^\infty p(r) \frac{\sin qr}{qr} dr \tag{4.10}$$

where λ is the wavelength of the radiation and q is the scattering vector, which is given by equation (4.11):

$$q = \frac{4\pi}{\lambda} \sin\theta \tag{4.11}$$

where 2θ is the scattering angle [20].

The dimension of q is the inverse of length, which is why a scattering pattern is referred to as "structure in reciprocal space" [19].

4.6.1 Data Measurement, Processing and Reduction

Figure 4.10 displays a standard SAXS setup. Different types of detectors, including a 2D area detector and at least two beam monitors, are used in SAXS. The 2D area detector captures the angle-resolved SAXS signal that has been elastically scattered by the sample. Depending on the detector type, the scattered signal collected on the 2D detector goes through several standard 2D data correction processes, such as dark current subtraction, image distortion and linearity correction, and flat field correction. In the following step, the scattering angle 2θ or the scattering vector q corresponding to each pixel's position on the SAXS image is calculated using the relevant coordinate system. This calculation utilises experimental parameters like the sample-to-detector distance and the detector's pixel size. For isotropic images along the azimuthal direction, such as nanoparticle solutions, polar coordinates are usually applied. The next step involves azimuthally averaging the 2D images to create a one-dimensional (1D) scattering curve. In scenarios where the images are not isotropic

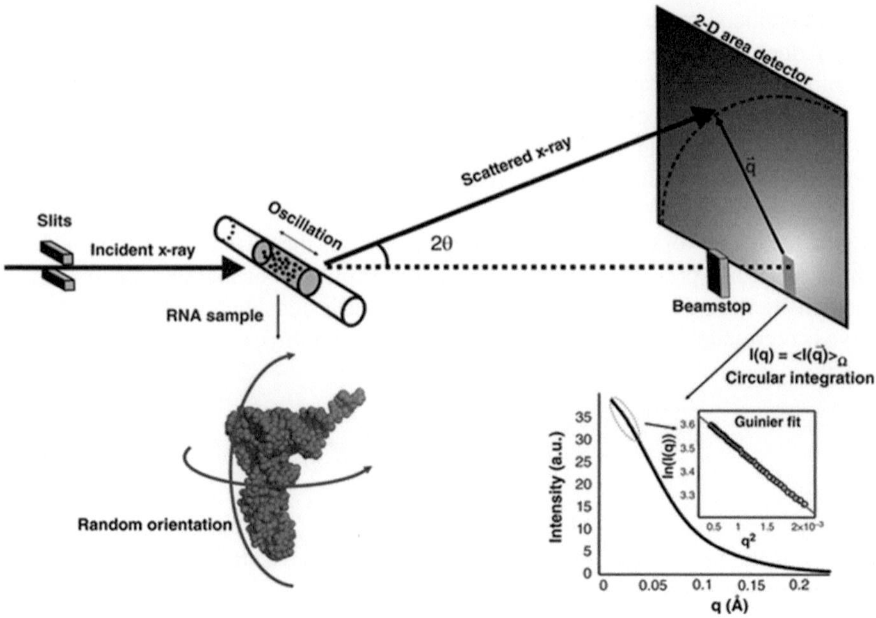

Fig. 4.10 Schematic of a SAXS experiment. The top panel represents a typical SAXS setup at the G1 station of Cornell's High Energy Synchrotron Source (CHESS). An incident X-ray beam is collimated using two (or more) sets of slits, then passes through a quartz capillary that contains a plug of diluted sample solution. The solutes are schematically shown as black dots in the upper panel; they are magnified in the lower left panel. To minimize radiation damage, the sample is oscillated during the exposure. X-Rays scattered by the sample are captured by a 2D area detector. The isotropic scattering pattern arises from the random orientation of the RNA in solution. A beamstop blocks the direct beam to avoid detector damage. The scattering intensity is integrated at each angle to produce a 1-dimensional Intensity versus q curve. As an example, the scattering profile of tRNA, the structure shown in the lower left panel, is plotted in the lower right panel. As described in the text, its radius of gyration can be readily computed from a linear fit to the low q (or Guinier) region of this curve plotted as ln(I(q)) versus q^2 in the inset. Reproduced with permission from [21]. Copyright 2016, Wiley Periodicals LLC

azimuthally, different coordinate systems are applied based on the nanoparticles' orientation [22].

4.6.2 Background Subtraction

The intensity of scattered radiation I contains a background signal due to air, the beamline optics, including the X-ray windows, and the sample cell. This background signal is removed by subtracting the intensity of radiation scattered by the empty cell (I_{cell}) from that scattered by the sample in the same container ($I_{sample/cell}$). The intensity after correction (I_{sample}) is given by equation (4.12):

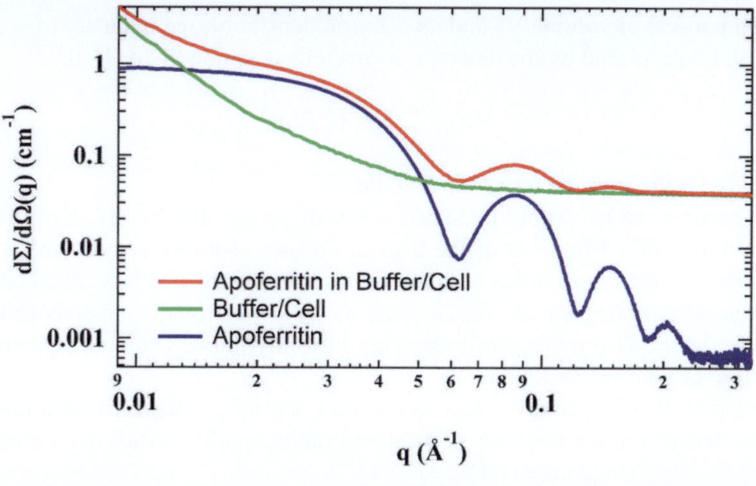

Fig. 4.11 SAXS data of 10 mg/mL apoferritin on an absolute intensity scale. Apoferritin is a protein cage with a diameter of 12 nm. Apoferritin in buffer in a quartz capillary (red), buffer in the same quartz capillary (green), and background subtracted data (blue). Reproduced with permission from [22]. Copyright 2016, American Chemical Society

$$I_{sample} = I_{sample/cell} - I_{cell} \tag{4.12}$$

For particles dispersed in a solvent, the scattering contribution from the particles ($I_{particle}$) is calculated using equation (4.13):

$$I_{particle} = I_{particle/solvent} - \varphi_{solvent} I_{solvent} \tag{4.13}$$

where $\varphi_{solvent}$ is the volume fraction of the solvent and $I_{particle/solvent}$ and $I_{solvent}$ are the scattering intensities of solution and the solvent, respectively, after making necessary corrections for the empty cell scattering in both cases [22]. Figure 4.11 presents a typical example of background subtraction.

4.6.3 Form Factor and Structure Factor

The scattering from a single particle, made up of numerous sub-particles, arises from the interference of waves that reach the detector from each sub-particle within it. The square of the sum of all wave amplitudes at the detector produces an interference pattern. The oscillation mode of this pattern, indicative of the particle's shape, is referred to as the "form factor" [19]. This form factor yields insights into the mean structural properties, such as size and shape, of the individual sub-particles. In contrast, the structure factor highlights the positional correlation among the sub-particles [22].

For a particle of volume V_1 and an electron density ρ_1, the intensity of scattered radiation $I_1(q)$, as read by the detector, is given by equation (4.14) [19]:

$$I_1(q) = I_0 \rho_1^2 V_1^2 P(q) \tag{4.14}$$

where P(q) is the form factor of the particle.

For particles to be visible in SAXS when dispersed in a matrix, their electron density must differ from that of the matrix. Greater visibility occurs with a larger difference in electron densities between the two materials, which is called contrast. Consequently, altering the electron density of the solvent (matrix) can make certain particles visible. This method of improving visibility through solvent adjustment is referred to as contrast variation.

For a particle of volume V_1 and an electron density ρ_1, dispersed in a matrix of electron density ρ_2, the intensity of scattered radiation after background correction is given by $\Delta I_1(q)$, equation (4.15):

$$\Delta I_1(q) = I_0 (\Delta\rho)^2 V_1^2 P(q) \tag{4.15}$$

where $\Delta\rho$ is given by equation (4.16):

$$\Delta\rho = \rho_1 - \rho_2 \tag{4.16}$$

For N identical particles, the scattered intensity is given by equation (4.17):

$$\Delta I(q) = N \Delta I_1(q) S(q) \tag{4.17}$$

where S(q) is the structure factor. For a dilute solution, the value of S(q) can be approximately taken as one [19]. Therefore, the scattering intensity can be treated as a product of the form factor and the structure factor [22].

From equations (4.15) and (4.17), one can conclude that the SAXS signal increases significantly with particle volume. For spherical particles, since particle volume is proportional to the cube of the radius, the SAXS signal is proportional to the sixth power of the radius. Furthermore, from equation (4.15), it can be stated that the SAXS signal depends on the square of the contrast and is thus independent of the sign of the contrast. Consequently, voids in a matrix material produce the same intensity as matrix particles in a void matrix.

4.6.4 *Effect of Polydispersity*

The scattering curves of polydisperse samples of N particles is given by equation (4.18):

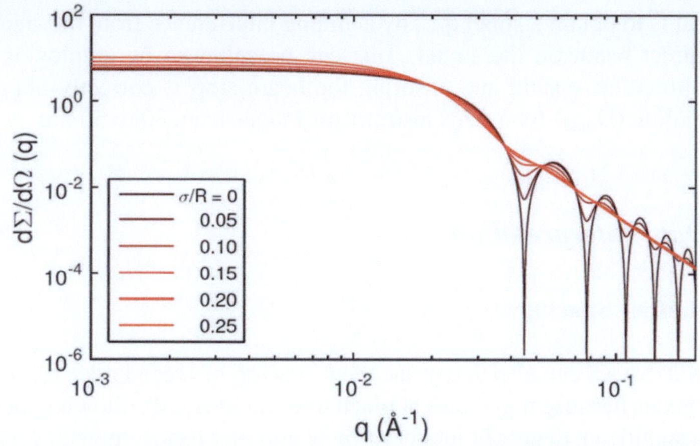

Fig. 4.12 Form factor scatterings of spherical particles with varying degree of polydispersity. The radius is 10 nm, volume fraction is 0.01, and $\Delta\rho/re = 0.04$ Å^{-3}. A rule of thumb is that when the particles have a size polydispersity of ≥15%, the form factor oscillations smear out. Reproduced with permission from [22]. Copyright 2016, American Chemical Society

$$\Delta I(q) = I_0 \sum_{i=1}^{N} (\Delta\rho)_i^2 V_i^2 P_i(q) \tag{4.18}$$

where $P_i(q)$, V_i and $\Delta\rho$ are the form factor, volume, and contrast, respectively, of the ith particle. This summation yields an averaged form factor that lacks sharp minima [19]. As polydispersity size (σ/R, with σ being the root-mean-square deviation of the size distribution and R the average radius) increases, the oscillations of the form factor become smeared, as shown in Fig. 4.12 [22].

4.6.5 Resolution in SAXS

Similar to microscopy, SAXS can only detect a limited range of sizes. The lower limit, q_{min}, is influenced by the primary beam and the collimation system's quality. The upper limit, q_{max}, results from the signal decreasing to the noise level. The scattering profile gathered from q_{min} to q_{max} helps in resolving particle features between D_{min} and D_{max}. These limits, D_{min} and D_{max}, are defined by equations (4.19) and (4.20):

$$D_{\min} \approx \frac{\pi}{q_{\max}} \tag{4.19}$$

$$D_{\max} \approx \frac{\pi}{q_{\min}} \tag{4.20}$$

The goal is to obtain a small q_{min} by avoiding interference from the significantly stronger direct beam on the signal. This can be achieved by employing a high-quality collimation system and ensuring the beam stop is correctly aligned. The usual resolution (D_{max}) for SAXS instruments ranges from 30 to 50 nm.

4.6.6 Data Interpretation

4.6.6.1 Lattice Spacing

Like WAXD, SAXS can also determine the d-spacing between lattice planes. SAXS is advantageous because it operates at much lower angles (2θ), allowing the identification of a significant degree of intercalation in polymer nanocomposites. This intercalation may be misinterpreted as exfoliation in WAXD. The d-spacing is calculated from the peak position, as shown in equation (4.21):

$$d_{Bragg} = \frac{2\pi}{q_{peak}} \tag{4.21}$$

4.6.6.2 Information from Intensity

Equations (4.7) and (4.9) can be summarised as equation (4.22):

$$\Delta I(q) = K \cdot P(q) \cdot S(q) \tag{4.22}$$

In equations (4.15) and (4.17), all constant terms have been consolidated into K. Consequently, the intensity of scattered radiation, following background correction, relies not only on the form factor and the structure factor but also on K, which depends on factors like particle volume, contrast, and concentration. The form factor delivers insights into the particle's shape, size, and internal density distribution, while the structure factor pertains to inter-particle interactions, including inter-particle distance and order degree. Typically, the form factor is more significant than the structure factor, and to access it, experiments should be carried out in the dilute regime where the scattering particles do not interact with one another.

Radius of Gyration

Guinier Approximation: The Guinier approximation demonstrates that asymptotic behavior manifests in the small q region, allowing any form factor function to be approximated as a Gaussian function at small angles. This approach offers a straightforward and precise method for measuring particle size using SAXS. The technique necessitates scattering data devoid of a structure factor, a requirement often met in dilute solutions. Guinier states that SAXS intensity at small angles ($q < 1/R_G$) is expressed by equation (4.23):

$$\Delta I(q) = a_0 e^{-\left(\frac{R_G^2 q^2}{3}\right)} \tag{4.23}$$

Here, R_G denotes the particle's radius of gyration, while a_0 signifies the extrapolated zero-angle intensity. Equation (4.23) does not provide insights into the particles' shape or internal structure. Nevertheless, if the shape is known or assumed, one can calculate the particle dimensions from R_G. For example, the mean radius R of spherical particles with uniform density can be obtained using equation (4.24):

$$R = \sqrt{5/3} R_G \tag{4.24}$$

In the so-called Guinier plot, the logarithm of intensity is plotted against q^2, as shown in the inset of the plot in Fig. 4.10. The slope of the straight-line plot is $R_G^2/3$ and the y-intercept is a_0 [22, 23]. If the scattering particles are monodisperse spheres, a_0 is given by equation (4.25):

$$a_0 = (\Delta\rho)^2 N_P \frac{V_{Sp}^2}{V} \tag{4.25}$$

where N_p is the number of particles, V is the sample volume and V_{sp} is the particle volume given by equation (4.26):

$$V_{sp} = \frac{4}{3}\pi R^3 \tag{4.26}$$

For a polydisperse system, the obtained size parameter is the average of squares $\langle R_G^2 \rangle$ and only in sufficiently monodisperse cases, $\langle R_G \rangle = \sqrt{\langle R_G^2 \rangle}$. If the data points on a Guinier plot do not form a straight line, this signals polydispersity. In such cases, Guinier's approximation is not applicable for determining particle size distribution, and inversion techniques must be employed instead.

For lamellar particles, such as clay platelets, of thickness T, Guinier's approximation is given by equation (4.27):

$$q^2 I(q) = a_0 e^{-(R_T^2 q^2)} \tag{4.27}$$

where R_T is the thickness radius of gyration and is given by equation (4.28):

$$R_G^2 = \frac{T^2}{12} \tag{4.28}$$

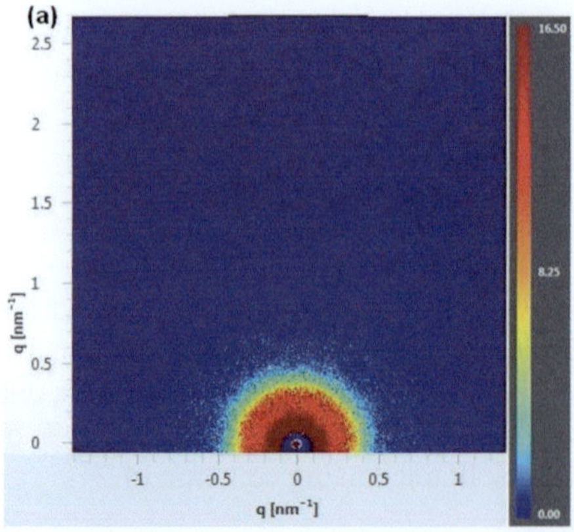

Fig. 4.13 **a** 2D SAXS scattering for the matrix. **b** 2D SAXS scattering for nanocomposite 2. Reproduced with permission from [24]. Copyright 2022, Elsevier Science Ltd.

Guinier's plot also helps determine the resolution limit, q_{min}, since at very small angles, the data points fall below the well-fitted line due to a decrease in intensity caused by the beam stop [19].

Yazid et al. [24] carried out SAXS studies to determine the radius of gyration of nanofiller particles in a blend of nano-boron carbide reinforced (50/10/40) natural rubber (NR), liquid natural rubber (LNR), and high-density polyethylene (HDPE). These nanocomposites were labelled as Nanocomposite x, where x represents the nanofiller content (wt%). Figure 4.13 illustrates the 2D SAXS scattering patterns for the matrix (Fig. 4.13a) and Nanocomposite 2 (Fig. 4.13b). Scattering patterns were recorded for all nanocomposites, with the excess scattering signal in them compared to the unfilled blend attributed to the nanofiller's presence. Both the unfilled blend and nanocomposites exhibited a "ring pattern" in their scattering patterns, indicating a lack of preferential orientation of the nanoparticles within the samples. Therefore, the samples were considered isotropic. The 2D scattering profiles were transformed into 1D scattering profiles, and data subtraction was performed to eliminate scattering contributions from the matrix. The resulting 1D scattering profiles are displayed in Fig. 4.14. An increase in filler loading led to a proportional rise in scattering intensity. The radius of gyration of the filler was computed from the profiles for all samples, with results shown in Table 4.1. It was found that the radius of gyration remained nearly constant, showing no effect from filler loading, which suggests that the filler was well-dispersed without agglomeration at increased filler loadings.

Particle Structure

Porod approximation: The Porod region provides information about the local structure of particles. The so-called fractal dimension of the scattering objects can be obtained from the plot of log(I) versus log(q). At high q, the following relationship

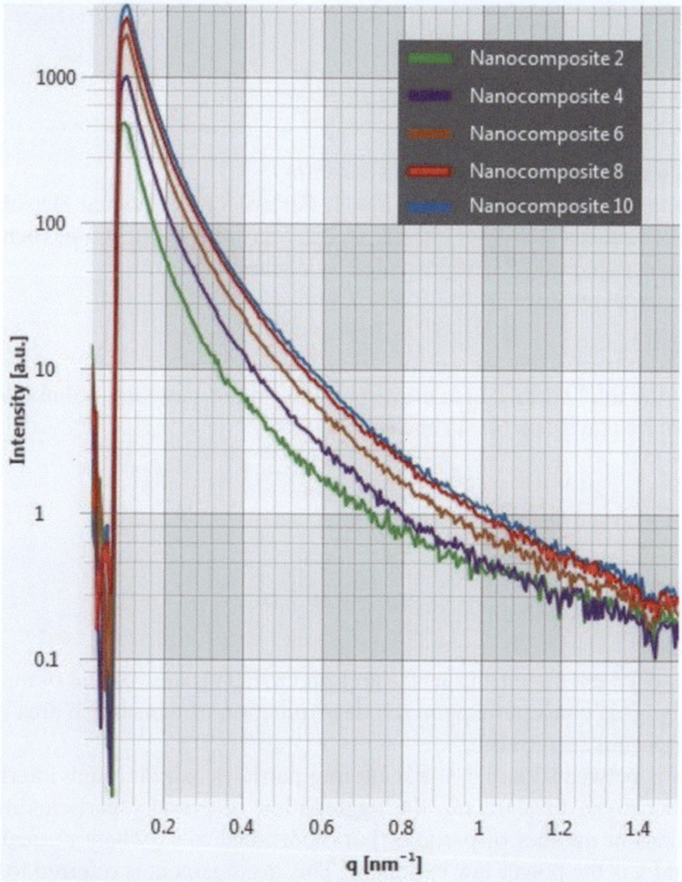

Fig. 4.14 1D SAXS scattering profiles. Reproduced with permission from [24]. Copyright 2022, Elsevier Science Ltd.

Table 4.1 Radius of gyration of fillers

Sample	Radius of gyration, R_t(nm)
Nanocomposite 2	4.7
Nanocomposite 4	4.5
Nanocomposite 6	4.8
Nanocomposite 8	4.6
Nanocomposite 10	4.9

Reproduced with permission from [24]. Copyright 2022, Elsevier Science Ltd.

is obeyed equation (4.29):

$$I(q) = \frac{P}{q^n} \tag{4.29}$$

where P is a constant called the Porod constant.

Scattering by rigid rods results in n = 1. For randomly oriented 3D solid objects with smooth surfaces and sharp interfaces, the value of n is 4 [23]. For such particles, Porod demonstrated that P is represented by equation (4.30):

$$P = 2\pi(\Delta\rho)^2 S/V \tag{4.30}$$

where S is the total surface area corresponding to the sampling volume V [25, 26] and is given by equations (4.31) and (4.32) [22]:

$$S = N_P \overline{S_P} \tag{4.31}$$

$$\frac{P}{a_0} = \frac{2\pi \overline{S_P}}{\overline{V_P^2}} \tag{4.32}$$

where $\overline{S_p}$ and $\overline{V_P^2}$ are the mean surface area and mean square volume of the particles, respectively. This concept helps in the determination of the surface area of particle systems or porous materials.

When n is between 3 and 4, the scattering particles exhibit rough interfaces with fractal dimensions. In a fractal, the aggregation of primary particles means that the total mass or number of particles is proportional to r^{ν}, where r equals the pair distance and ν is the power law exponent. This arrangement is referred to as a mass fractal. Therefore, the average density at a pair distance r is proportional to $r^{\nu-3}$. The power law scaling varies according to the fractal type. For polymer coils among the scattering particles, the exponent corresponds to the excluded volume parameter. In a 3D network of particles, ν equals 3, while for a Gaussian chain in a dilute environment, ν equals 2. Fully swollen coils and collapsed coils have ν values of 5/3 and 3, respectively. This scaling leads to a power law recognised as a Porod exponent; scattering by a fractal relates proportionally to $q^{-\nu}$ [23, 27].

It is well-known that the Porod exponents, or power law exponents, for a one-dimensional object like an infinitely thin rod, are −1, for a two-dimensional object such as an infinitely thin disk, are −2, and for a three-dimensional object like a sphere, are −4, respectively [25]. Additionally, the Fourier transform of a delta function—which represents a zero-dimensional (0D) object—results in a constant, indicating that its Porod exponent is 0. These findings are pertinent only to systems with distinct interfaces and do not apply to fractals, which have been addressed in the previous paragraph.

Let us consider a rod. When viewed from a short distance (in real space), which corresponds to the high q region (in reciprocal space), the rod appears as a 3D object

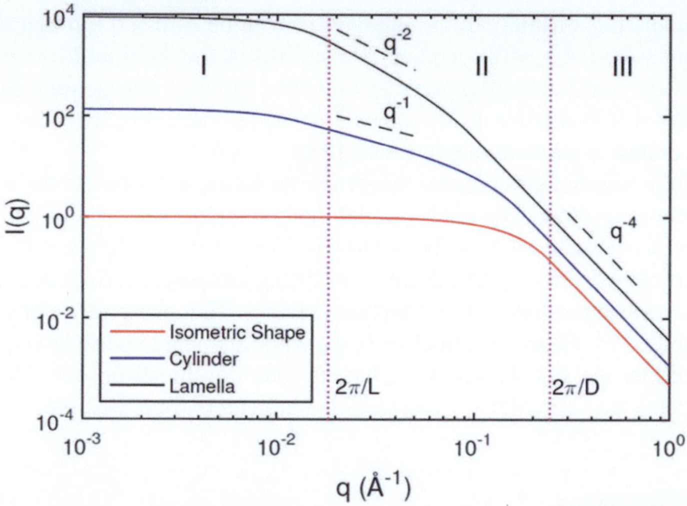

Fig. 4.15 Calculated form factor scatterings of objects with various dimensions and sharp interfaces. The R_g of the isometric particle is 1 nm. The cylinder has a diameter of 2.8 nm (R_g of 1 nm) and length 34.5 nm (overall R_g of 10 nm). The thickness and width of the lamella are 3.4 nm (R_g of 1 nm) and 34.5 nm, respectively (overall R_g of 10 nm). These data were calculated from the Guinier–Porod model. Reproduced with permission from [22]. Copyright 2016, American Chemical Society

(region III of Fig. 4.15). However, when the distance exceeds its diameter D but is less than its length L, which corresponds to the intermediate q range, the rod appears as a 1D object (region II of Fig. 4.15). When viewed at a distance greater than L, it appears as a 0D object, resulting in a flat intensity region at low q values (region I of Fig. 4.15). Thus, the shape of the scattering curve for a rod [28], and consequently the Porod exponent, varies depending on the q range used for observation: $I(q) \propto q^{-4}$ when $q > 2\pi/D$, $I(q) \propto q^{-1}$ when $2\pi/L < q < 2\pi/D$, and $I(q) \propto q^0$ when $q < 2\pi/L$ (Fig. 4.15). The scattering of the rod is comparable to that of a disk with thickness T and diameter D. Its form factor shows $I(q) \propto q^{-2}$ when $2\pi/D < q < 2\pi/T$ (region II of Fig. 4.15) and $I(q) \propto q^{-4}$ when $q > 2\pi/T$ (region III of Fig. 4.15). A sphere has only one Porod region, $q > 2\pi/D$, with an exponent of −4. Oriented systems may exhibit different behavior [22, 29].

When assessing single-particle structure, the structure factor is often significant. To mitigate its influence, diluting the sample is required, though this is not always practical for several reasons:

1. To achieve adequate particle scattering above the background, particularly for low-contrast samples, a minimum concentration is required.
2. A change in concentration may lead to a change in structure, as in the case of surfactants in solution, and this could be the objective of the study.

Interparticle forces contribute to the formation of the structure factor. The extent of these interactions depends on particle concentration and their mutual forces [19].

The characteristics of interparticle interactions can be assessed through the SAXS intensity measured at $q = 0$, specifically from I(0). A reduction in I(0) compared to the signal from non-interacting particles indicates repulsive interactions. In contrast, an increase in I(0) relative to the signal from non-interacting particles points to attraction or aggregation among particles [21].

Often, it is beneficial to remove the structural factor, as it hinders the analysis of single-particle structure. The conventional method includes interpreting the particle structure from the scattering profile at large angles, after excluding small-angle data points. A more effective approach involves fitting a theoretical S(q) in conjunction with the pair-distance distribution function (PDDF). The inter-particle contributions are represented by fitting an effective S(q), while the single-particle contributions are captured by the PDDF, with S(q) having only a minimal impact. This method also facilitates the extraction of interparticle parameters from S(q) [30].

Degree of Orientation

Non-spherical particles may show a favoured orientation when subjected to shear or strain. The azimuthal profiles of randomly oriented particles and spherical ones remain constant in intensity along a circular path, with the primary beam centred. However, introducing a preferred orientation in the sample causes the intensity profile to oscillate, peaking at specific azimuth angles. The amplitude of these oscillations is useful for determining the orientation. Two methods are utilised and explored in the subsequent sections.

Herman's Orientation Factor

Herman's orientation parameter is given by equation (4.33):

$$f = \frac{3\langle\cos^2\varphi\rangle - 1}{2} \tag{4.33}$$

The term $\langle\cos^2\varphi\rangle$ is given by equation (4.34):

$$\langle\cos^2\varphi\rangle = \frac{\int_0^{\pi/2} I(\varphi)\cos^2\varphi \sin\varphi d\varphi}{\int_0^{\pi/2} I(\varphi) \sin\varphi d\varphi} \tag{4.34}$$

where $I(\varphi)$ indicates the intensity distribution around the azimuthal angle φ at maximum scattering intensity, as illustrated in Fig. 4.16. The intensity values considered in calculating Herman's orientation parameter are measured azimuthally in an anticlockwise direction from 0° to 90°. When $f = 1$, the intensity peak is perfectly aligned at $\varphi = 0$. If $f = -1/2$, the intensity peaks entirely towards $\varphi = 90°$. Confusion occurs when $f = 0$, as this results in an intensity profile that is either constant, indicating no orientation, or perfectly peaked at $\varphi = 54.7356°$ [31].

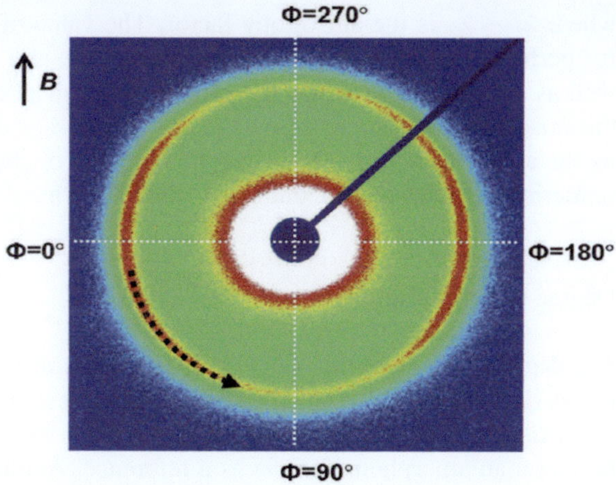

Fig. 4.16 The determination of Herman's orientation factor (f) using two-dimensional SAXS patterns. The integrated intensity corresponding to Φ = 0°–90° is used to obtain the orientation parameter. Reproduced with permission from [31]. Copyright 2019, Elsevier Science Ltd.

Anisotropy Factor

Cinader and Burghardt [32] analysed the anisotropy within a scattering pattern by calculating the anisotropy factor from an azimuthal scan. A unit vector u, representing a point on the azimuthal scan, was defined based on the azimuthal angle φ, as shown in equations (4.35) and (4.36):

$$u_1 = \cos\varphi \tag{4.35}$$

$$u_2 = \sin\varphi \tag{4.36}$$

A simple representation of anisotropy was provided by the weighted average of the second moment tensor of *u*, equation (4.37) [33]:

$$\langle \mathbf{uu} \rangle = \begin{bmatrix} \langle u_1u_1 \rangle & \langle u_1u_2 \rangle \\ \langle u_1u_2 \rangle & \langle u_2u_2 \rangle \end{bmatrix} = \begin{bmatrix} \langle \cos^2\varphi \rangle & \langle \cos\varphi \sin\varphi \rangle \\ \langle \cos\varphi \sin\varphi \rangle & \langle \sin^2\varphi \rangle \end{bmatrix} \tag{4.37}$$

where $\langle \ldots \rangle$ represents an average weighted by the azimuthal intensity distribution I(φ). For example, $\langle \cos^2\varphi \rangle$ is given by equation (4.38):

$$\langle \cos^2\varphi \rangle = \frac{\int_0^{2\pi} \cos^2\varphi I(\varphi) d\varphi}{\int_0^{2\pi} I(\varphi) d\varphi} \tag{4.38}$$

The difference in the eigen values of $\langle \mathbf{uu} \rangle$ is taken in equation (4.39):

$$\lambda_1 - \lambda_2 = \sqrt{\left[(\langle u_1u_1 \rangle - \langle u_2u_2 \rangle)^2 + 4\langle u_1u_2 \rangle^2 \right]} \tag{4.39}$$

where $\lambda_1 - \lambda_2$ is the anisotropy factor. The value of the anisotropic factor is one for perfect orientation, irrespective of direction, and zero for isotropic systems as well as for systems in which the orientations cancel each other. Possible values of the anisotropy lie between zero and one. The average direction of alignment is given by the eigen vectors of $\langle \mathbf{uu} \rangle$. Thus, this method of characterizing the orientation of scattering particles has no ambiguity, unlike Herman's orientation parameter.

Degree of Crystallinity

The degree of crystallinity (DOC) characterises materials that can be either amorphous (DOC = 0) or crystalline (DOC = 1). It estimates the volume of the sample that exists in a crystalline state. However, since this measurement is relative, it is based on an amorphous sample as a reference. Additionally, it is crucial to understand that the term "amorphous" in this context refers to both the lack of crystallinity and the undesirable crystal forms. Consequently, even amorphous reference samples may show Bragg reflections in their scattering curves.

Two scattering experiments are conducted to estimate DOC. The first measures the intensity of scattering from the reference sample [$\Delta I_{ref}(q)$], while the second measures the scattering intensity of the sample under investigation [$\Delta I_{sample}(q)$]. DOC is then calculated using equation (4.40):

$$DOC = \left[\int_{q_{min}}^{q_{max}} q^2 \{ \Delta I_{sample}(q) - \Delta I_{ref}(q) \} \right] / \int_{q_{min}}^{q_{max}} q^2 \Delta I_{sample}(q) dq \tag{4.40}$$

Before calculating DOC, it is essential to perform background subtraction; otherwise, any remaining baseline will cause a considerable rise in the integrands due to multiplication with q^2.

If an amorphous sample isn't available, one can use an assumed smooth background curve, though this will yield less accurate results. Steenstrup proposed an effective method for determining the background curve [34].

While XRD is commonly employed to examine the crystallinity of polymeric materials, SAXS is infrequently applied for this purpose due to challenges in background determination [19].

4.7 Nuclear Magnetic Resonance

Nuclear Magnetic Resonance (NMR) is a non-destructive technique for characterising different polymeric materials, including composites. The underlying principle of NMR is that atomic nuclei exhibit magnetic properties that yield chemical information. Subatomic particles like protons, neutrons, and electrons possess spin. In stable isotopes such as ^{12}C, ^{16}O, and ^{32}S, spins are paired, resulting in a nucleus with no overall spin. Conversely, elements like ^{1}H, ^{13}C, ^{31}P, ^{15}N, and ^{19}F have unpaired spins, giving their nuclei overall spin. The overall spin of a nucleus is determined by the following rules: (a) A nucleus with an even count of protons and neutrons exhibits no spin. (b) A nucleus where the total number of protons and neutrons is odd has a half-integer spin, such as ½, 3/2, or 5/2. (c) A nucleus with an odd number of both protons and neutrons has an integer spin, such as 1, 2, or 3 [1].

NMR operates through the interaction of the nuclear magnetic moment with a magnetic field, thus being influenced by localized field variations resulting from molecular movements, structural changes in molecules and crystals, and shifts in chemical composition. Proton-1 h is commonly utilised for characterizing polymer composites. The information acquired from NMR primarily relates to the relaxation behaviours of the nuclear magnetic moments, recorded using pulsed NMR techniques. The energy interchange between nuclear moments and the surrounding lattice is outlined by the spin–lattice relaxation time, T_1, whereas the interaction among nuclear magnetic moments is identified by the spin–spin relaxation time, T_2. Given that these relaxation times are considerably impacted by molecular movements and structural alterations, they offer both qualitative and quantitative insights into the dynamic environment surrounding the nuclei.

NMR is commonly used to confirm the functionalization of polymers and nanofillers in polymer blend nanocomposites. In this regard, Singh et al. [35] employed ^{1}H NMR to verify the functionalization of PMMA in PVDF/functionalized-PMMA blend nanocomposites. PMMA was functionalized with 2-amino-5-nitrobenzoic acid. Figure 4.17 displays the NMR spectrum of the functionalized PMMA. The sharp peak at a δ value of 0.823 ppm indicates the protons of the methyl side group. The peak at δ 1.799 ppm corresponds to the protons of the CH_2 group in the polymer chain backbone. The peak for the aliphatic proton of the amine group is observed at δ 3.585 ppm. This peak's presence at a slightly elevated value is due to the C=O group directly attached to the NH_2 group, confirming the chemical interaction between the amine and carbonyl groups of PMMA. Additionally, three aromatic proton signals appear at 7.134, 7.829, and 8.210 ppm for Hd, He, and Hf (as noted in the inset of Fig. 4.17) from the aromatic moiety, clearly demonstrating the functionalization of PMMA. The shift in these δ values is ascribed to the influence of highly electronegative groups near these protons.

Wang et al. [36] detail the use of ^{1}H NMR to confirm nanoparticle functionalization, where silica nanoparticles were modified with epoxide groups (Epoxy-SiO_2) and a combination of epoxide groups with long PMMA chains (Epoxy-MSiO_2). In the ^{1}H NMR studies (Fig. 4.18), the presence of peaks at 2.96 and 2.89 ppm for epoxide groups, along with a methoxyl group peak at 3.63 ppm, verified the success of the functionalization process.

Brito et al. [37] employed proton NMR relaxometry to examine the molecular dynamics of PLA/starch/organoclay and PLA/starch/PVA/organoclay nanocomposites. The spin relaxation time (T_1H) plots for starch, PLA, and the PLA/starch blend as a function of frequency are illustrated in Fig. 4.19a. The T_1H of the blend was higher than that of pure starch, indicating that the presence of PLA restricts the

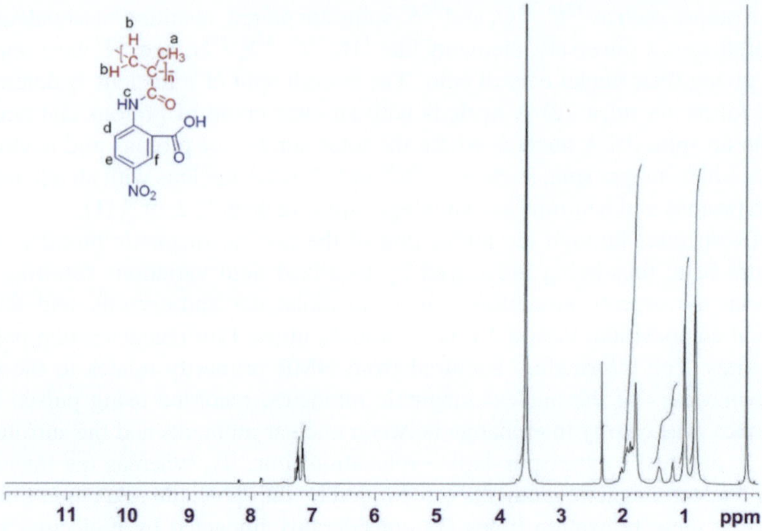

Fig. 4.17 NMR spectrum of functionalized-PMMA. Reproduced with permission from [35]. Copyright 2018, Wiley Periodicals LLC

Fig. 4.18 ^{1}H NMR spectra of the representative products: (II) Epoxy-SiO_2, (III) Epoxy-$MSiO_2$ and (V) PMMA-COOH. Reproduced with permission from [36]. Copyright 2017, American Chemical Society

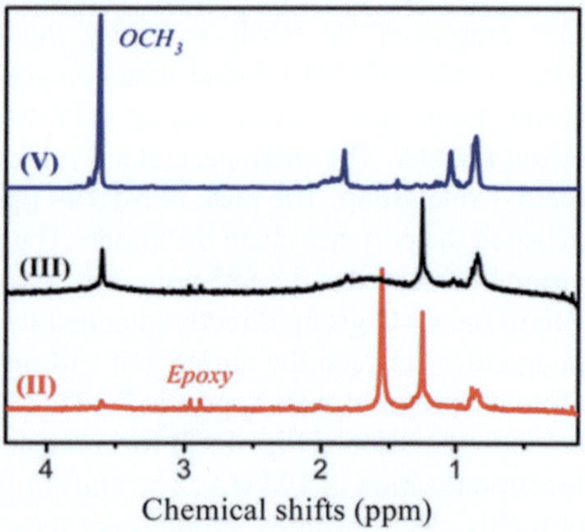

molecular motion of starch at lower frequencies, leading to the dominance of PLA molecular dynamics. Starch's molecular motion was only detected at higher frequencies, given that the blend's T_1H was lower than that of neat PLA. These findings are attributed to the encapsulation of starch within the PLA matrix.

Figure 4.19b shows the relaxation times for the PLA/starch/PVA blend and its nanocomposite containing 1 wt% organoclay Viscogel B8. Two categories of proton populations were distinguished based on the T_1H values of the individual components within the PLA/starch blends. The comparison of T_1H values to those of

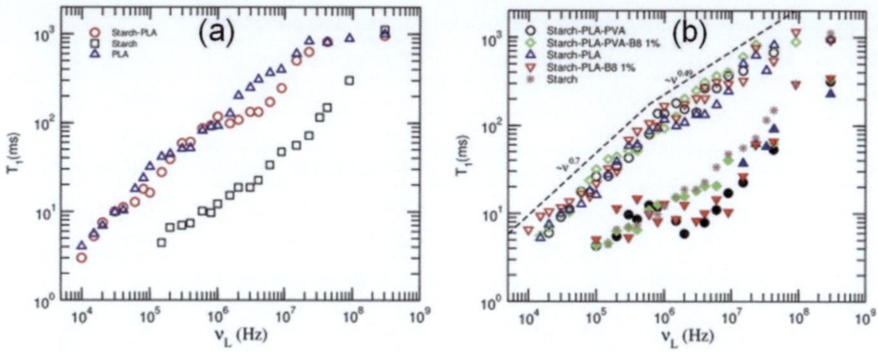

Fig. 4.19 **a** Proton spin–lattice relaxation time T_1 of polymers starch, PLA and starch-PLA blend measured as a function of the frequency. **b** Proton spin–lattice relaxation time T_1H of starch-PLA blend with organoclay (Viscogel B8) and PVA measured as a function of frequency. The dotted line is the best fit of the renormalized Rouse model. Reproduced with permission from [37]. Copyright 2013, Elsevier Science Ltd.

the PLA/starch blend revealed variations in molecular dynamics resulting from the incorporation of clay and PVA.

The T_1H of the starch-PVA-PLA blend showed a trend similar to that of the PLA/starch/clay nanocomposite. This indicated that the PVA and clay at 1% loading demonstrated comparable compatibility with the PLA/starch blend.

4.8 Infrared Spectroscopy

Infrared (IR) spectroscopy serves as an effective method for material analysis. A sample's IR spectrum is generated by passing a beam of IR light through it. The IR spectrum is categorised into three primary sections: the far-IR (<400 cm^{-1}), mid-IR (4000–400 cm^{-1}), and near-IR (14,000–4000 cm^{-1}). The mid-IR range is mainly utilised in IR spectroscopy because most compounds display distinctive absorption or emission characteristics within this spectrum. This range can further be split into the functional group region (4000–1400 cm^{-1}) and the fingerprint region (400–1400 cm^{-1}). Two different molecules will exhibit similar functional group regions if they possess analogous functional groups, while their fingerprint regions are generally quite distinct [38].

IR spectroscopy is based on the vibrations of atoms within a molecule, categorised into two types: stretching and bending. Stretching refers to alterations in bond length, which can occur either symmetrically or asymmetrically, while bending vibrations result in changes to the bond angle. As IR radiation passes through a sample, it gets absorbed by the molecules if its frequency aligns with a specific bond vibration. This absorption of IR light energy at particular values allows for the identification of

Table 4.2 Details of mixing contents for blended polymer nanocomposites

Polymer blend (gm)	Nanoparticle (gm)	Type of nanoparticles	Sample code
0.200	0.000	–	JNH1
–	0.008	ZnO	JNH2
–	–	V-doped ZnO	JNH3
–	–	S-doped ZnO	JNH4
–	–	Dy-doped ZnO	JNH5

– +

different functional groups in a molecule, making IR spectroscopy a valuable method for determining chemical structures [39].

Singh et al. [35] utilised IR spectroscopy to examine PVDF/functionalized PMMA blend nanocomposites, maintaining a weight ratio of 70/30. The PMMA was treated with 2-amino-5-nitrobenzoic acid. Various nanofillers-including vanadium-, sulfur-, and dysprosium-doped zinc oxide—were assessed. The compositions are detailed in Table 4.2, with their Fourier Transform IR (FTIR) plots displayed in Fig. 4.20. Characteristic peaks for the C–H and C=O bonds in PMMA appeared at approximately 3000, 2995, and 1730 cm^{-1}. Notably, both PMMA and 2-amino-5-nitrobenzoic acid do not contain an aliphatic C–N bond. Nonetheless, functionalizing PMMA introduced an aliphatic C–N bond (–CO–NH–Ar), confirmed by an absorption band at around 1176 cm^{-1}. The presence of a secondary amine group in the modified PMMA was indicated by a unique peak in the FTIR spectrum, absent of the amine bending peak. The amine band shift was attributed to two neighbouring electronegative groups. The absorption bands at 489, 750, 762, 796, 838, and 976 cm^{-1}, along with those at 510 and 840 cm^{-1} in the samples JNH1 and JNH2, were allocated to the α-phase and β-phase of PVDF, respectively. However, most α-phase peaks diminished with the introduction of doped ZnO nanoparticles, leaving only a single peak at 482 cm^{-1} across all samples with doped Zno. The strong absorption bands at 510 and 840 cm^{-1} in the spectra of JNH3, JNH4, and JNH5 indicated a heightened presence of the β-phase, suggesting that altering the electronic properties of the nanofiller through doping could effectively modify the phases of PVDF and other polymers.

4.9 Rheology

The rheology and morphology of polymeric materials are closely interconnected. Researchers have employed rheology as an indirect means to examine the morphological features of polymer blends and nanocomposites [40]. Nevertheless, there has been limited research on using rheology to evaluate the morphology of polymer blend nanocomposites. It has been reported that the localisation of nanofillers at

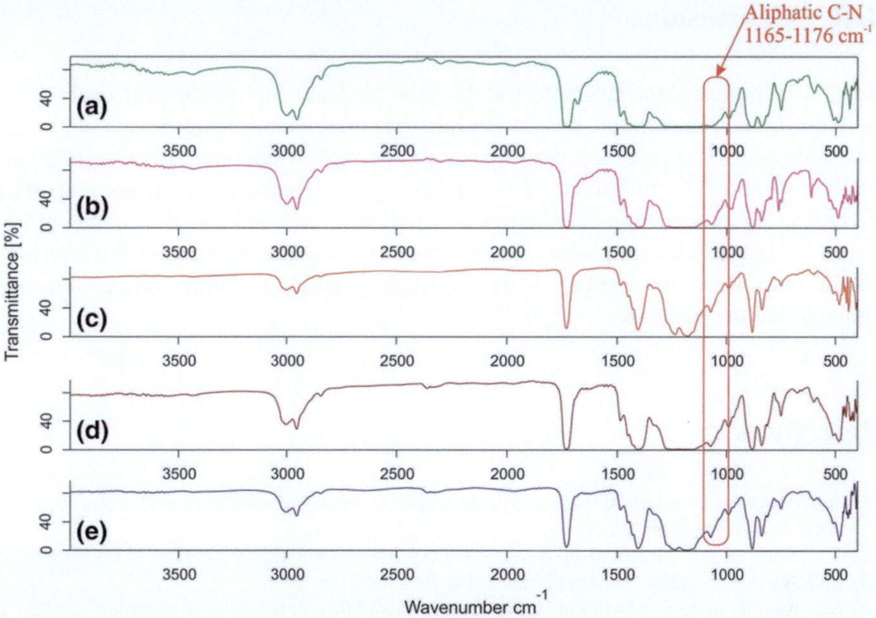

Fig. 4.20 FTIR spectra of **a** JNH1, **b** JNH2, **c** JNH3, **d** JNH4, and **e** JNH5. Reproduced with permission from [35]. Copyright 2018, Wiley Periodicals LLC

the interfaces promotes the development of a network of filler particles, resulting in gel-like behavior during frequency sweep rheological analyses, as evidenced by a plateau in the storage modulus plot and an increase in the complex viscosity plot in the low-frequency range [41, 42], as illustrated in Fig. 4.21.

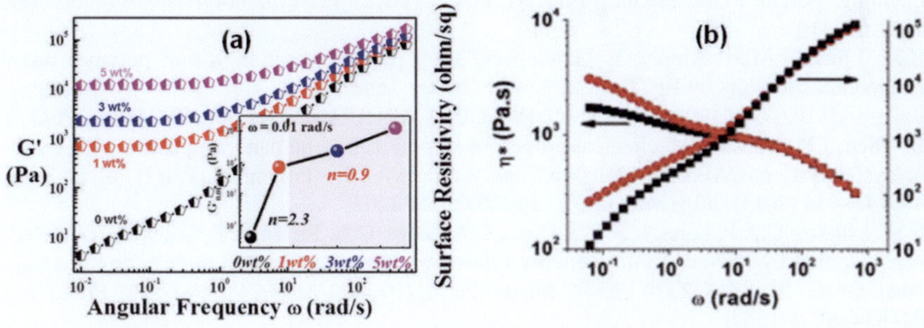

Fig. 4.21 **a** Rheology curves of blends with and without CNTs. Reproduced with permission from [43]. Copyright 2018, American Chemical Society. **b** The effect of interfacial localization of nanosilica on complex viscosity and storage modulus. Reproduced with permission from [42]. Copyright 2016, Elsevier Science Ltd.

4.10 Conclusions

Several techniques are available for the structural and morphological analysis of polymer blend nanocomposites. It is crucial to understand blend morphology, the positioning of filler particles, their dispersion quality—both qualitatively and quantitatively—and the interfacial adhesion between the blend components and filler particles. This knowledge is essential for quality control and enhancing the performance of blend nanocomposites. Therefore, structural and morphological analyses should employ complementary methods, since no single technique can completely characterize the structure.

References

1. S.S. Ray, *Clay-Containing Polymer Nanocomposites: From Fundamentals to Real Applications* (Elsevier, Amsterdam, 2013)
2. L.C. Sawyer, D.T. Grubb, G.F. Meyers, *Polymer Microscopy* (Springer, New York, 2008)
3. U.W. Gedde, *Polymer Physics* (Chapman & Hall, London, 1995)
4. S.S. Ray, R. Banerjee, *Sustainable Polylactide-Based Blends* (Elsevier, Amsterdam, 2022)
5. A. Auliawan, E.M. Woo, Nanocomposites based on vermiculite clay and ternary blend of poly(L-lactic acid), poly(methyl methacrylate), and poly(ethylene oxide). Polym. Compos. **32**, 1916–1926 (2011). https://doi.org/10.1002/PC.21194
6. A. Kelarakis, K. Yoon, Optical transparency in a polymer blend induced by clay nanofillers. Eur. Polym. J. **44**, 3941–3945 (2008). https://doi.org/10.1016/J.EURPOLYMJ.2008.08.030
7. T. Périé, A.C. Brosse, S. Tencé-Girault, L. Leibler, Co-continuous nanostructured nanocomposites by reactive blending of carbon nanotube masterbatches. Polymer (Guildf.) **53**, 984–992 (2012). https://doi.org/10.1016/J.POLYMER.2012.01.007
8. S.S. Ray, R. Banerjee, *Sustainable Polylactide-Based Composites* (Elsevier, Amsterdam, 2023)
9. M. Hemmati, H. Shariatpanahi, A. Fereidoon, J. Aalaie, M.G. Ahangari, Study of a polymer blend/nanoclay nanocomposite of polyethylene and polyamide 6 prepared by mechanical blending. Polym. Plast. Technol. Eng. **51**, 80–85 (2012). https://doi.org/10.1080/03602559.2011.617410
10. E.M. Linares, M.M. Rippel, F. Galembeck, Clay platelet partition within polymer blend nanocomposite films by EFTEM. ACS Appl. Mater. Interfaces **2**, 3648–3653 (2010). https://doi.org/10.1021/AM100812M/ASSET/IMAGES/LARGE/AM-2010-00812M_0005.JPEG
11. B. Chen, J.R.G. Evans, Mechanical properties of polymer-blend nanocomposites with organoclays: polystyrene/ABS and high impact polystyrene/ABS. J. Polym. Sci. Part B Polym. Phys. **49**, 443–454 (2011). https://doi.org/10.1002/POLB.22202
12. E.M. Linares, C.A.P. Leite, L.F. Valadares, C.A. Silva, C.A. Rezende, F. Galembeck, Molecular mapping by low-energy-loss energy-filtered transmission electron microscopy imaging. Anal. Chem. **81**, 2317–2324 (2009). https://doi.org/10.1021/AC8024834/SUPPL_FILE/AC8024834_SI_001.PDF
13. S. Yakovlev, M. Libera, Dose-limited spectroscopic imaging of soft materials by low-loss EELS in the scanning transmission electron microscope. Micron **39**, 734–740 (2008). https://doi.org/10.1016/J.MICRON.2007.10.019
14. M. Bera, U. Saha, A. Bhardwaj, P.K. Maji, Reduced graphene oxide (RGO)-induced compatibilization and reinforcement of poly(vinylidene fluoride) (PVDF)–thermoplastic polyurethane (TPU) binary polymer blend. J. Appl. Polym. Sci. **136**, 47010 (2019). https://doi.org/10.1002/APP.47010

15. P.H. Hermans, A. Weidinger, Crystallinity of precipitated cellulose. J. Polym. Sci. **5**, 565–568 (1950)
16. W. Ruland, X-ray determination of crystallinity and diffuse disorder scattering. Acta Crystallogr. **14**, 1180–1185 (1961). https://doi.org/10.1107/s0365110x61003429
17. E.C. Lopes Pereira, B.G. Soares, A.A. Silva, G.M.O. Barra, Master batch approach for developing PVDF/EVA/CNT nanocomposites with co-continuous morphology and improved electrical conductivity. J. Appl. Polym. Sci. **138**, 51164 (2021). https://doi.org/10.1002/APP.51164
18. S. Sinha Ray, S. Pouliot, M. Bousmina, L.A. Utracki, Role of organically modified layered silicate as an active interfacial modifier in immiscible polystyrene/polypropylene blends. Polymer (Guildf.) **45**, 8403–8413 (2004). https://doi.org/10.1016/J.POLYMER.2004.10.009
19. H. Schnablegger, Y. Singh, *The SAXS Guide* (Anton Paar GmbH, Graz, 2013)
20. O. Kratky, A survey, in *Small Angle X-Ray Scattering*, ed. by O. Glatter, O. Kratky (Academic Press, Cambridge, 1982), pp. 3–12
21. Y. Chen, L. Pollack, SAXS studies of RNA: structures, dynamics, and interactions with partners. Wiley Interdiscip. Rev. RNA **7**, 512–526 (2016). https://doi.org/10.1002/wrna.1349
22. T. Li, A.J. Senesi, B. Lee, Small angle X-ray scattering for nanoparticle research. Chem. Rev. **116**, 11128–11180 (2016). https://doi.org/10.1021/acs.chemrev.5b00690
23. G. Ochbaum, R. Bitton, Using small-angle X-ray scattering (SAXS) to study the structure of self-assembling biomaterials. Self-Assembl. Biomater. Mol. Des. Charact. Appl. Biol. Med. (Elsevier) 291–304 (2018). https://doi.org/10.1016/B978-0-08-102015-9.00015-0
24. H. Yazid, U.A. Anwar, A.S. Zaubidah, M. Nurulizzati, M. Sabtu, M.J. Andrianny et al., A combined method to probe the behaviour of the filler in polymer blend nanocomposites via X-ray diffraction and thermal measurement. Nano-Struct. Nano-Objects **32**, 100906 (2022). https://doi.org/10.1016/J.NANOSO.2022.100906
25. R.-J. Roe, *Methods of X-Ray and Neutron Scattering in Polymer Science* (Oxford University Press, New York, 2000)
26. P. Debye, H.R. Anderson, H. Brumberger, Scattering by an inhomogeneous solid. II. The correlation function and its application. J. Appl. Phys. **28**, 679–83 (1957). https://doi.org/10.1063/1.1722830
27. J.E. Martin, A.J. Hurd, Scattering from fractals. J. Appl. Crystallogr. **20**, 61–78 (1987). https://doi.org/10.1107/S0021889887087107
28. A. Loubat, L.M. Lacroix, A. Robert, M. Impéror-Clerc, R. Poteau, L. Maron et al., Ultrathin gold nanowires: soft-templating versus liquid phase synthesis, a quantitative study. J. Phys. Chem. C **119**, 4422–4430 (2015). https://doi.org/10.1021/acs.jpcc.5b00242
29. B. Hammouda, A new Guinier-Porod model. J. Appl. Crystallogr. **43**, 716–719 (2010). https://doi.org/10.1107/S0021889810015773
30. G. Fritz, A. Bergmann, O. Glatter, Evaluation of small-angle scattering data of charged particles using the generalized indirect Fourier transformation technique. J. Chem. Phys. **113**, 9733–9740 (2000). https://doi.org/10.1063/1.1321770
31. M.K. Kang, E.J. Cha, H.H. Song, Y.H. Na, Interlayer structure and magnetic field-induced orientation of modified nanoclays in polymer aqueous solution. Heliyon **5**, e02421 (2019). https://doi.org/10.1016/j.heliyon.2019.e02421
32. D.K. Cinader, W.R. Burghardt, X-ray scattering studies of orientation in channel flows of a thermotropic liquid-crystalline polymer. J. Polym. Sci. Part B Polym. Phys. **37**, 3411–3428 (1999). https://doi.org/10.1002/(SICI)1099-0488(19991215)37:24%3c3411::AID-POLB4%3e3.0.CO;2-S
33. D.K. Cinader, W.R. Burghardt, Mixed orientation state induced by expansion flow of a thermotropic liquid-crystalline polymer. Macromolecules **31**, 9099–9102 (1998). https://doi.org/10.1021/ma981139l
34. S. Steenstrup, A simple procedure for fitting a backgound to a certain class of measured spectra. J. Appl. Crystallogr. **14**, 226–229 (1981). https://doi.org/10.1107/s0021889881009242
35. N. Singh, H. Madhav, S. Yadav, G. Jaiswar, Impact of vanadium-, sulfur-, and dysprosium-doped zinc oxide nanoparticles on various properties of PVDF/functionalized-PMMA blend

nanocomposites: structural, optical, and morphological studies. J. Appl. Polym. Sci. **136**, 47116 (2019). https://doi.org/10.1002/APP.47116

36. H. Wang, Z. Fu, X. Zhao, Y. Li, J. Li, Reactive nanoparticles compatibilized immiscible polymer blends: synthesis of reactive SiO_2 with long poly(methyl methacrylate) chains and the in situ formation of janus SiO_2 nanoparticles anchored exclusively at the interface. ACS Appl. Mater. Interfaces **9**, 14358–14370 (2017). https://doi.org/10.1021/ACSAMI.7B01728/ASSET/IMAGES/LARGE/AM-2017-01728S_0010.JPEG
37. L.M. Brito, F.V. Chávez, M.I. Bruno Tavares, P.J.O. Sebastião, Molecular dynamic evaluation of starch-PLA blends nanocomposite with organoclay by proton NMR relaxometry. Polym. Test. **32**, 1181–1185 (2013). https://doi.org/10.1016/J.POLYMERTESTING.2013.07.002
38. H.B. Stuart, *Infrared Spectroscopy: Fundamentals and Applications* (Wiley, Hoboken, NJ, 2004)
39. B. Jaleh, P. Fakhri, Infrared and Fourier transform infrared spectroscopy for nanofillers and their nanocomposites. Spectrosc. Polym. Nanocompos. (Elsevier Inc.) 112–29 (2016). https://doi.org/10.1016/B978-0-323-40183-8.00005-7
40. R. Banerjee, S.S. Ray, Role of rheology in morphology development and advanced processing of thermoplastic polymer materials: a review. ACS Omega **8**, 27969–28001 (2023). https://doi.org/10.1021/ACSOMEGA.3C03310/ASSET/IMAGES/LARGE/AO3C03310_0020.JPEG
41. L. Bai, S. He, J.W. Fruehwirth, A. Stein, C.W. Macosko, X. Cheng, Localizing graphene at the interface of cocontinuous polymer blends: morphology, rheology, and conductivity of cocontinuous conductive polymer composites. J. Rheol. (N. Y. N. Y.) **61**, 575–587 (2017). https://doi.org/10.1122/1.4982702
42. E. Jalali Dil, N. Virgilio, B.D. Favis, The effect of the interfacial assembly of nano-silica in poly(lactic acid)/poly(butylene adipate-co-terephthalate) blends on morphology, rheology and mechanical properties. Eur. Polym. J. **85**, 635–646 (2016). https://doi.org/10.1016/J.EURPOLYMJ.2016.07.022
43. X. Zhao, H. Wang, Z. Fu, Y. Li, Enhanced interfacial adhesion by reactive carbon nanotubes: new route to high-performance immiscible polymer blend nanocomposites with simultaneously enhanced toughness, tensile strength, and electrical conductivity. ACS Appl. Mater. Interfaces **10**, 8411–8416 (2018). https://doi.org/10.1021/ACSAMI.8B01704/SUPPL_FILE/AM8B01704_SI_001.PDF

Chapter 5
Nanoparticles as Interface Modifiers

Abstract This chapter reports an overview of theories and key research related to the interfacial modification of immiscible polymer blends with the help of various nanofillers. This strategy has been found to significantly change the morphology and, hence, properties of nanoparticle-containing immiscible polymer blend composites.

5.1 Introduction

Interfacial localisation of nanoparticles (NPs) has been found to bring about significant changes in the morphology and properties of immiscible polymer blends. In fact, various nanofillers have been successfully employed as compatibilizers in the production of polymer blends. Nanofiller-compatibilized blends having NPs selectively localised at the blend interface have gained significant importance in both industry and academia. Due to the small size and rigidity of NPs, these blends have been found to exhibit stronger interfacial stability in comparison with blends compatibilized with graft or block copolymers [1–3]. Furthermore, the NP-compatibilized blends exhibit unique morphological characteristics, which are responsible for enhancement in various properties (electrical [3, 4], mechanical [4, 5], etc.), depending on the nature of the nanofiller used. This chapter provides an overview of theories and key research related to the interfacial modification of immiscible polymer blends with the help of various nanofillers.

5.2 Nanoclay

The use of nanoclay as a compatibilizer in immiscible polymer blends has been reported by several research groups [5–11]. In this context, Ray et al. [5] conducted extensive studies on Cloisite 20A (C20A) reinforced polystyrene (PS)/polypropylene (PP) blends. Blend nanocomposites containing PS/PP/C20A in weight ratios of 19/79/2 and 17.5/77.5/5 were prepared in a Thermohaake mixer by first melting the

S. S. Ray and R. Banerjee, *Polymer Blend Nanocomposites*, Springer Series in Materials Science 354, https://doi.org/10.1007/978-3-032-00309-6_5

Fig. 5.1 XRD patterns of unmodified and C20A modified (20/80) PS/PP blends. Reproduced with permission from [5]. Copyright 2004, Elsevier Science Ltd.

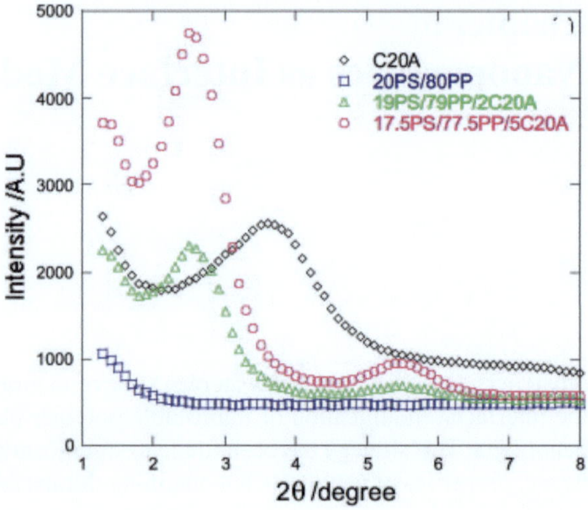

blend components and then mixing the melt with C20A for 10 min at 180 °C. The samples were compression molded for morphological analysis. Studies were also conducted on unfilled blends with a 20/80 PS/PP weight ratio to understand the effect of nanoclay addition. The (X-ray diffraction) XRD patterns of C20A and the various compositions are presented in Fig. 5.1. The 19PS/79PP/2C20A blend exhibited a reduced intensity of the characteristic peak of C20A. A broad peak was observed at $2\theta = 2.6°$ (d001 = 3.53 nm), indicating intercalation. For the nanocomposite with 5 wt% clay loading, this peak position was identical; however, the intensity was significantly higher, indicating lower interaction with the polymer chains and more parallel stacking of the clay platelets due to the higher filler loading.

The fact that chains of both blend components were intercalated in the clay galleries was confirmed in the XRD patterns of the nanocomposites (Fig. 5.2) of the individual blend components. Both PP/clay and PS/clay nanocomposites exhibited an intercalated structure. In the XRD plot of the PS/C20A nanocomposite with 5 wt% clay content, the characteristic peak of C20A was observed at $2\theta = 3.3°$ (d = 2.67 nm), indicative of a poorly intercalated structure. For the same filler loading, the PP/C20A nanocomposite showed a d-spacing of 2.95 nm ($2\theta = 3°$). Interestingly, the blend nanocomposite at 5 wt% filler loading had a d-spacing of 3.53 nm ($2\theta = 2.6°$), indicating that chains of both blend components were intercalated within the clay galleries. The researchers proposed that, due to the common intercalation, the silicate layers were pushed towards the interface, resulting in a situation where the silicate layers were shared by the two polymers. Thus, C20A served as an interfacial compatibilizer. The presence of intercalated silicate layers at the blend interface was confirmed in TEM studies (Fig. 5.3).

The compatibilization by C20A resulted in a reduction of interfacial tension, leading to a decrease in the size of the PS domains, as observed in scanning electron microscopy (SEM) micrographs. The addition of 5 wt% nanoclay was found

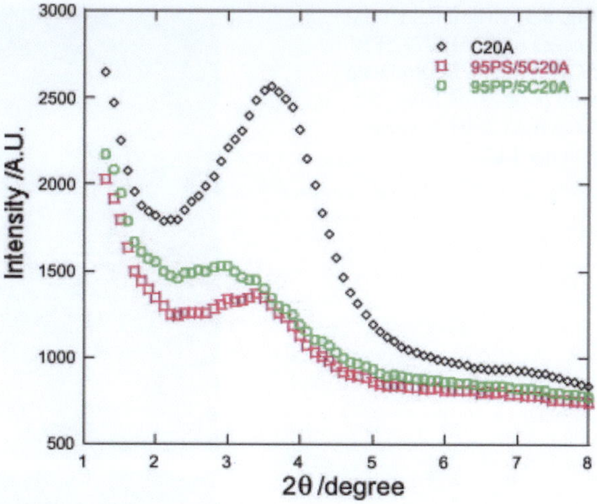

Fig. 5.2 XRD patterns of PS/5C20A and PP/5C20A hybrids. Reproduced with permission from [5]. Copyright 2004, Elsevier Science Ltd.

to decrease the PS domain size from 8 μm (domain size of the unfilled blend) to 1 μm. The compatibilization action of clay was confirmed by measuring the interfacial tension of the unfilled blend and the blend nanocomposites using the drop deformation method. Adding only 0.5 wt% nanoclay led to a reduction in interfacial tension from 5.1 to 3.4 mN/m, indicating the clay's interfacial activity localized at the interface, similar to classical compatibilization. The blend nanocomposites exhibited a significant increase in elongation at break compared to the unfilled blends. This was attributed to the fact that C20A not only acted as a filler but also as a compatibilizer, enhancing the adhesion between the phases. Thus, through their extensive studies, the researchers established the compatibilization effect of clay in nanoclay-reinforced polymer blends.

Similar findings were reported by Wang et al. [10] in their studies on PS/PP/organically modified montmorillonite (OMMT) nanocomposites. The lower particle size of PS in the blend nanocomposite, compared to the unfilled blend, was attributed to the intercalation of both PP and PS chains within the clay galleries, as presented schematically in Fig. 5.4. The researchers proposed that the layer structure resembled a nanobox (Fig. 5.4a). The confinement of the PP and PS chains within the nanobox resulted in the forced mixing of the polymers, leading to a homogeneous distribution of the silicate layers. However, since phase separation was evident on a micrometer scale, it was reasonable to assume that some parts of the polymer chains remained outside the clay galleries (Fig. 5.4b). The segments of the chains outside the clay galleries acted as a compatibilizer, similar to block copolymers. The interfacially localized clay platelets were stiffer than the polymer chains, resulting in high interfacial bending energy. This facilitated the formation of small domains and further promoted compatibilization.

Reduced domain size due to interfacial modification by nanoclay was also reported by Hong et al. [9] in their studies on poly(butylene terephthalate) (PBT)/polyethylene (PE)/OMMT nanocomposites. At low clay loadings (1–3 phr), thin clay tactoids were

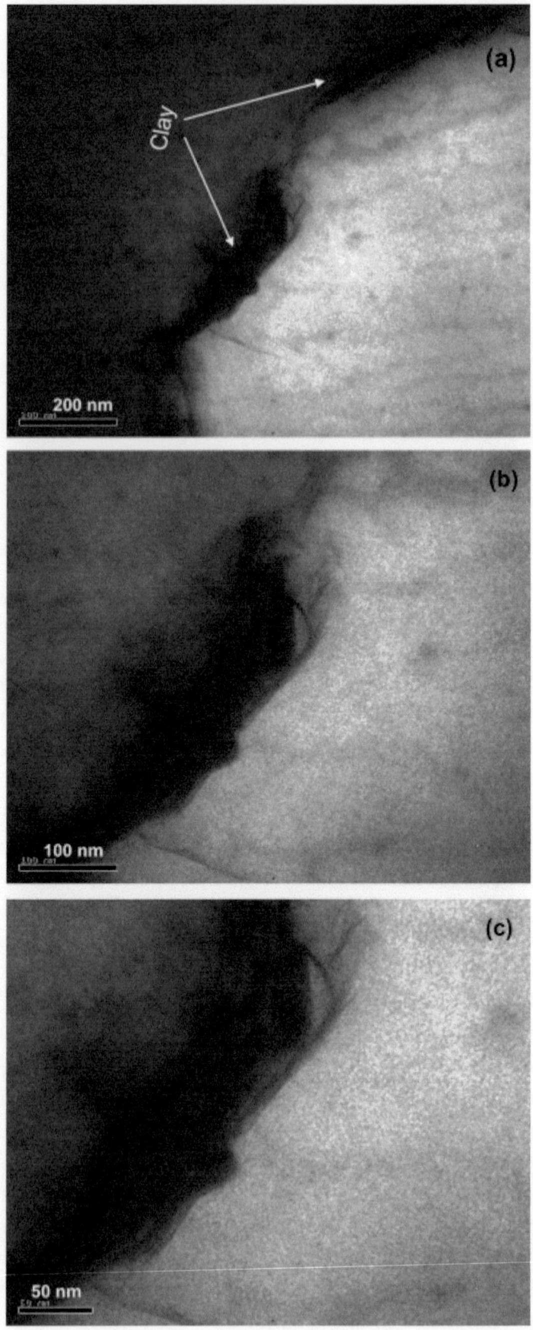

Fig. 5.3 Bright field TEM images of 17.5PS/77.5PP/5C20A blend. Reproduced with permission from [5]. Copyright 2004, Elsevier Science Ltd.

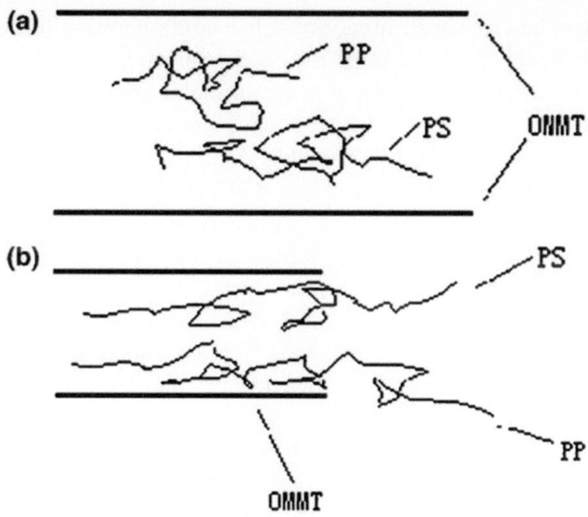

Fig. 5.4 Schematic representation of the intercalated structure in PP/PS/OMMT blends: **a** PP and PS confined in the same gallery of OMMT, and **b** parts of PP and PS molecules located outside the gallery serving as a compatibilizer. Reproduced with permission from [10]. Copyright 2003, Wiley Periodicals LLC

present at the interface of the blend components. As the clay loading increased, the additional clay incorporated into one of the blend components. The nanoclay at the interface caused changes in interfacial tension, which suppressed droplet coalescence and therefore reduced droplet size. The interfacial localization of nanoclay, which suppresses droplet coalescence and leads to a decrease in droplet size, has also been reported by other researchers [8–11].

It can be summarized that the interfacial localization of nanoclay in an immiscible polymer blend results in a compatibilization effect, which leads to a reduction in domain size. Intercalation of both blend components into the clay galleries is advantageous for this interfacial localization. Thus, nanoclay can act not only as a filler but also as an interface modifier and therefore a compatibilizer.

5.3 Carbon Nanotubes

The use of carbon nanotubes (CNTs) as modifiers in immiscible polymer blends has been reported by several research groups [4, 12–17]. Since the formation of a percolating network of interconnected CNTs at the blend interface results in enhanced electrical conductivity for ternary compositions, thereby improving their performance as conductive polymer composites, various strategies have been proposed to localize CNTs at the interface of blend components. One strategy devised by Chen et al. [16] involved trapping CNTs with graphene oxide (GO) at the interface of poly(L-lactide) (PLLA)/ethylene-co-vinyl acetate (EVA) blends. The studies were motivated by GO's proven ability to stabilize at the interface of an immiscible polymer blend, exhibiting a compatibilizing effect [18, 19], and by the fact that the simultaneous presence of GO and CNTs in a polymer matrix resulted in the trapping of CNTs on the GO sheet, hindering CNT agglomeration through π–π stacking interactions [20, 21]. These facts led the researchers to believe that the simultaneous introduction of GO and CNTs into a polymer blend might trap CNTs by GO at the blend interface, resulting in improved electrical conductivity. The composites were prepared by diluting GO/PLLA and CNT/PLLA masterbatches through melt processing with PLLA and EVA, maintaining a PLLA: EVA weight ratio of 60:40. The CNT content varied from 0.1 to 2 wt%, while the GO content remained at 0.3 wt%. The samples were designated as CNTx/GO 0.3/PLLA-EVA, where x indicated the CNT content. For comparison, studies were also conducted on the unfilled PLLA-EVA blend and the ternary compositions CNTx/PLLA-EVA and GO 0.3/PLLA-EVA. All samples were produced in an internal mixer at a melt temperature and rotor speed of 60 rpm. SEM studies (Fig. 5.5) after the selective removal of EVA revealed a co-continuous blend morphology. Compositions containing GO exhibited increased co-continuity. In the CNT 0.1/PLLA-EVA sample (Fig. 5.5a, b), only a few CNTs were located at the interface, unlike the GO 0.3/PLLA-EVA sample (Fig. 5.5c, d), where all GO particles were at the interface. In the CNT0.1/GO0.3/PLLA-EVA sample (Fig. 5.5e, f), both GO and CNT particles were found at the interface, with a significantly higher number of CNTs compared to CNT 0.1/PLLA-EVA. Furthermore, some CNTs were observed around GO sheets, forming loose aggregates conducive to the formation of a percolated network of CNTs at the interface, indicating enhanced electrical conductivity at a low CNT concentration.

TEM micrographs (Fig. 5.6) revealed similar findings. In the CNT-reinforced blend (Fig. 5.6a), most CNTs were located in the EVA phase. In the ternary composite (Fig. 5.6b), a significant number of CNTs were found trapped around the GO sheets, preferentially localized at the interface or on the EVA side of the interface, near the interface.

With the help of a schematic representation (Fig. 5.7), the researchers proposed possible mechanisms for the entrapment and interfacial localization of the CNTs. For the ternary composites (Fig. 5.7a), during melt processing, the CNTs tended to migrate from the PLLA phase to the EVA phase due to the significantly lower viscosity of EVA [22]. Once the CNTs reached the interface because of shear stress, they were quickly transferred to the EVA phase [23]. For the quaternary composites (Fig. 5.7b), due to the large aspect ratio and the smaller volume of a single CNT, the CNTs exhibited a higher transfer rate but lower interfacial stability compared to the GOs. During melt processing, the time required for the CNTs to migrate from the PLLA phase was shorter than the time needed by the GOs to reach the blend interface. However, once the GOs reached the interface, they could trap the CNTs from the EVA phase to the interface through π–π stacking interaction (possibility I). Alternatively, it could have also occurred (possibility II) that the GOs trapped CNTs in the PLLA phase, and both filler particles (CNTs trapped in GOs) migrated to the interface, or that the GOs delayed the diffusion rate of the CNTs. Since the GO/CNT interaction was strong and the formation of a GO-CNT complex was thermodynamically more favorable compared to the formation of CNT aggregates, it can be said that during the initial stage of melt processing, CNTs were absorbed on the GO sheets forming a GO-CNT complex, which exhibited a lower rate of transfer and greater interfacial

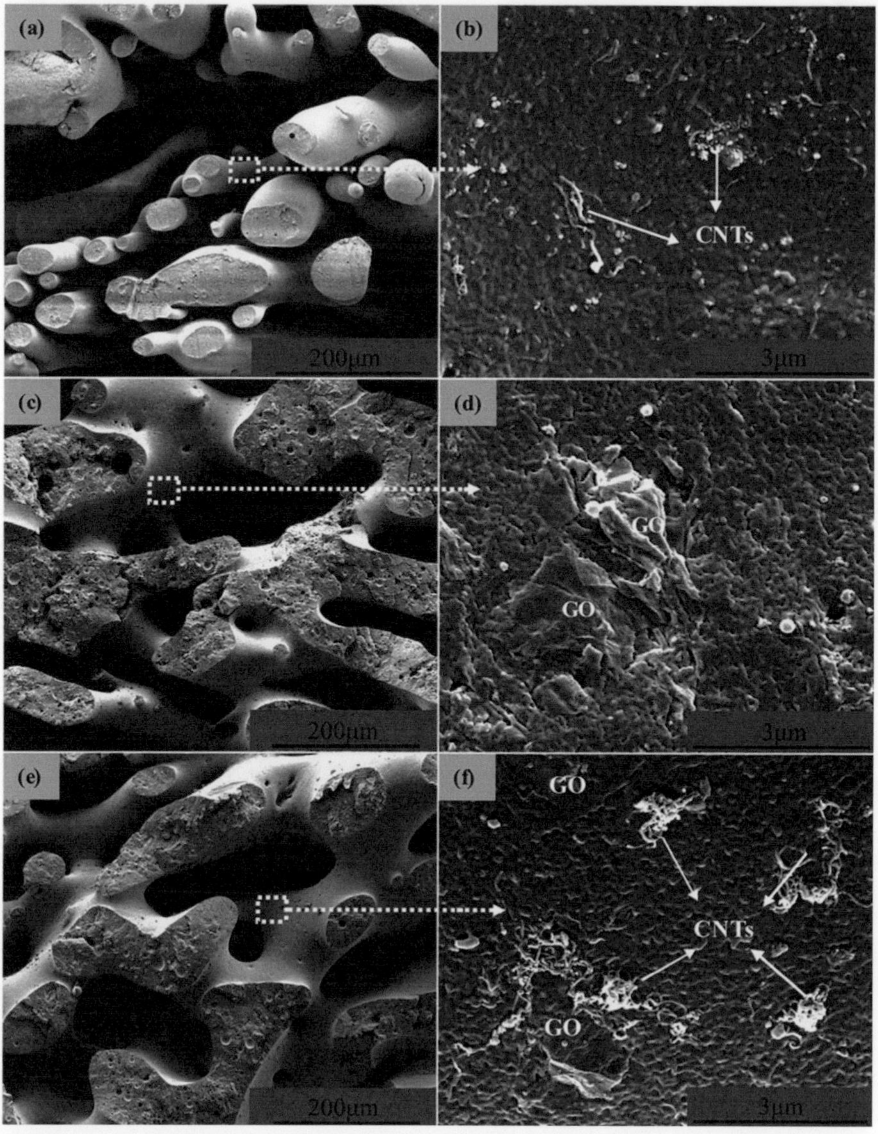

Fig. 5.5 SEM images show the morphologies of representative samples and the dispersion states of CNTs and GO. **a** and **b** CNT0.1/PLLA-EVA, **c** and **d** GO0.3/PLLA-EVA, and **e** and **f** CNT0.1/GO0.3/PLLA-EVA. (**b**), (**d**), and (**f**) images are obtained at higher magnification of (**a**), (**c**), and (**e**), respectively. Reproduced with permission from [16]. Copyright 2013, Royal Society of Chemistry

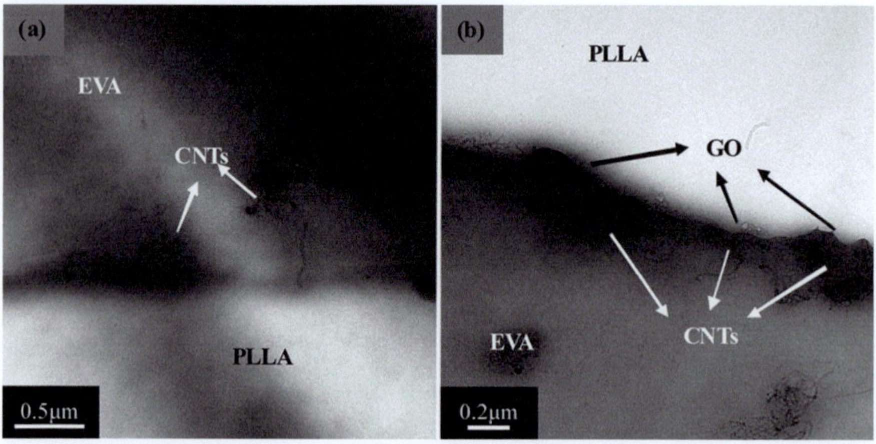

Fig. 5.6 TEM images show the localization of CNTs and GO in the different composites. **a** CNT0.1/PLLA-EVA and **b** CNT0.1/GO0.3/PLLA-EVA. Reproduced with permission from [16]. Copyright 2013, Royal Society of Chemistry

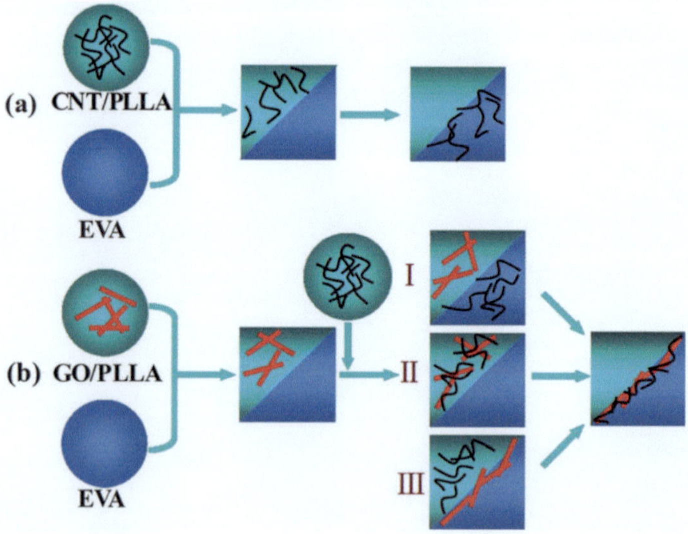

Fig. 5.7 Schematic representations show the migration of CNTs in **a** CNT/PLLA-EVA and **b** CNT/GO/PLLA-EVA composites. Reproduced with permission from [16]. Copyright 2013, Royal Society of Chemistry

stability compared to the CNTs. Another possibility (possibility III) is that the GOs, already present at the interface, exhibited a shielding effect, preventing the CNTs from transferring to the EVA phase. However, since the TEM micrographs showed the localization of the CNTs at the interface or in the EVA phase close to the interface, the trapping mechanism corresponded to the first two possibilities.

The quaternary composites demonstrated a significantly lower percolation threshold (0.06 wt%) compared to the ternary composites that contained only CNTs (0.24 wt%). The electrical resistivity of the quaternary compositions was seven orders of magnitude lower, even when the CNT concentration was as low as 1 wt%.

The researchers thus established that CNTs can be trapped at the blend interface with GOs, which facilitates the formation of a percolating network, resulting in a remarkable improvement in electrical properties.

Zhao et al. [4] used a different strategy for the interfacial localization of CNTs. The researchers synthesized CNTs with reactive epoxide groups and long poly(methyl methacrylate) (PMMA) tails. These reactive CNTs were melt-processed with PLLA and poly(vinylidene fluoride) (PVDF) in a 1:1 ratio. Processing took place in a Haake Polylab QC at 190 °C and 50 rpm for 10 min. The CNT content of the blend nanocomposites was 1 wt%. The blend nanocomposites prepared in this way were subjected to morphological analysis. For comparison, studies were also conducted on pristine OH-CNT reinforced PLLA/PVDF blends with the same composition, which were prepared similarly. TEM images of the blend nanocomposites are presented in Fig. 5.8. The blend reinforced with OH-CNTs had the filler particles localized in the PLLA phase due to the higher affinity of the OH-CNTs for PLLA. In contrast, the blend reinforced with the reactive CNTs had the filler particles localized at the interface. The reactive CNTs acted as a thermodynamic compatibilizer, bridging the two phases of the blend through physical entanglements between the grafted PMMA tails and the PVDF chains, as well as through chemical reactions between the epoxide groups and the carboxyl end groups of PLLA, as schematically presented in Fig. 5.9. This compatibilization by CNTs led to a stable morphology that showed no coarsening during static thermal annealing. Furthermore, there was a significant improvement in mechanical properties, including ductility, tensile strength, thermal properties, and electrical conductivity. The researchers established that chemically modified CNTs can be used as reactive compatibilizers, resulting in a significant enhancement of the properties of immiscible blends through interface modification.

The strategy of interfacial localization of CNTs through their interactions with both blend components was also utilized by Chen et al. [12] in their studies on carboxyl group-functionalized CNT-reinforced co-continuous PS/PMMA blends containing 40 wt% PS. The nanocomposites were prepared in a torque rheometer by mixing the blend components at 200 °C and 30 rpm rotor speed for 1 min, followed by the rapid addition of the nanofiller and mixing all components for an additional 9 min at 100 rpm rotor speed. TEM studies revealed that the CNTs were present at the blend interface, and the extent of this localization depended on the carboxyl content on the CNT surfaces. The interfacial localization was attributed to the balance of π–π interactions between the PS and the CNTs, and the dipole–dipole interactions between the PMMA and the carboxyl groups on the CNT surfaces. Complete localization was achieved when the carboxyl content was approximately 0.73 wt%, which corresponded to a balance between π–π interactions and dipole–dipole interactions.

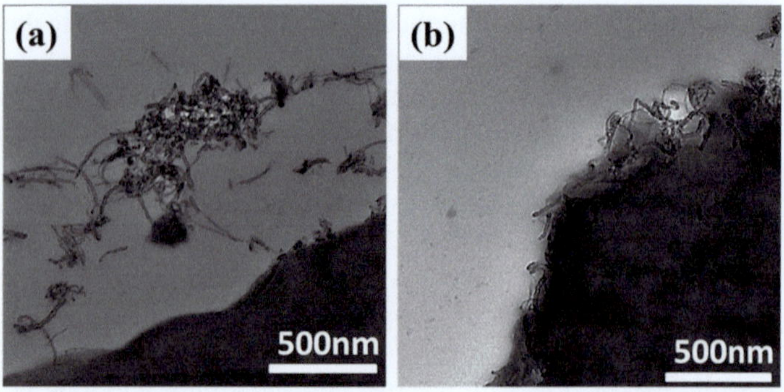

Fig. 5.8 TEM images of **a** pristine hydroxy-CNT incorporated PLLA/PVDF blends and **b** reactive CNTs with PMMA tails compatibilized PLLA/PVDF blends with 1 wt% CNTs (dark part represents PVDF and gray part represent PLLA). Reproduced with permission from [4]. Copyright 2018, American Chemical Society

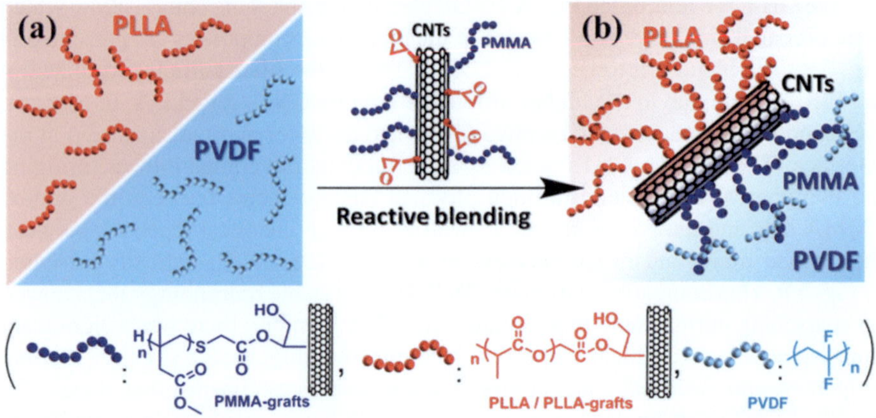

Fig. 5.9 Schematic diagram of the reactive CNTs with long PMMA tails for the compatibilizer of immiscible PLLA/PVDF blends **a** blends without CNTs with weak interface; **b** CNTs compatibilized blends with strengthened interface. Reproduced with permission from [4]. Copyright 2018, American Chemical Society

The interfacially localized CNTs formed a percolating network, resulting in a significant improvement in electrical conductivity. The formation of a percolating network of CNTs at the blend interface due to π–π interactions was also reported by Cohen et al. [13].

The strategy for achieving interfacial localization of CNTs through interaction with both blend components was also applied by Kar et al. in their studies on CNT-reinforced PVDF/acrylonitrile-butadiene-styrene (ABS) blends containing 50 wt% PVDF [14]. However, the researchers utilized a unique approach by wrapping the CNTs in a mutually miscible polymer, PMMA. The blend nanocomposites were prepared using melt processing with all components in a twin-screw micro-compounder at 230 °C and a screw speed of 60 rpm for 20 min. The miscibility of the PMMA coating on the CNTs with both blend components resulted in interfacial localization of the filler particles and a compatibilization effect, unlike uncoated CNT-reinforced blends where the CNTs were localized in the thermodynamically favored PVDF phase. Such a tailored interface led to a significant improvement in mechanical properties and electrical conductivity. Additionally, the PMMA-wrapped CNT-reinforced blends exhibited shielding of electromagnetic radiation primarily through an absorption mechanism, in contrast to blends reinforced with uncoated CNTs, where the dominant mechanism was reflection. This was attributed to the high dielectric constant of the PMMA coating on the CNTs.

It can be summarized that interfacial localization of CNTs in immiscible polymer blends can be achieved by inducing interactions between the filler particles and both blend components. This interaction may involve a chemical reaction, as seen in reactive compatibilization, or it may be physical, such as π–π interactions and dipole–dipole interactions. Alternatively, another component, such as GO, may be used to trap the CNTs at the blend interface. Whatever strategy is employed for interfacial localization, the outcome is the modification of the blend interface, resulting in a significant improvement in mechanical properties and electrical conductivity.

5.4 Graphene Oxide

The use of graphene oxide as an interface modifier in immiscible polymer blends has been reported by researchers [3, 24, 25]. Different strategies have been employed for interface modification using graphene oxide. In this context, Kultravut et al. [25] grafted poly(glycidyl methacrylate) (PGMA) onto reduced graphene oxide (rGO) to produce PGMA-grafted rGO (rGO-PGMA). First, graphene (GE) was ground and oxidized using the Brodie method to produce graphene oxide (GO). Then, GO was grafted with α-bromoisobutyryl bromide (an ATRP initiator) to produce an ATRP initiator-grafted GO. Finally, GMA was polymerized onto rGO-PGMA. The hydroxyl content of the GO surface was reduced during polymerization through a partial reduction process. PLA/poly(trimethylene terephthalate) (PTT)/rGO-PGMA nanocomposites containing 60 wt% PLA and varying filler loadings were prepared using two-step melt processing. PTT was first mixed with rGO-PGMA for 2.5 min in a melt mixer at 240 °C and 120 rpm. PLA was then added to the premix, and melt processing was continued for another 2.5 min. For comparison, studies were also conducted on PLA/PTT/GE nanocomposites prepared using the same process. TEM micrographs (Fig. 5.10A) revealed that, unlike the PLA/PTT/GE nanocomposites, where GE agglomerates (indicated by white arrows) were found in the PTT phase [Fig. 5.10A (a)], in the PLA/PTT/rGO-PGMA nanocomposites, the filler particles were localized at the blend interface [Fig. 5.10A (b)]. Such selective localization of rGO-PGMA particles was observed at all filler loadings (Fig. 5.10B) and was attributed to the reaction of the epoxy group of GMA on the rGO-PGMA with the

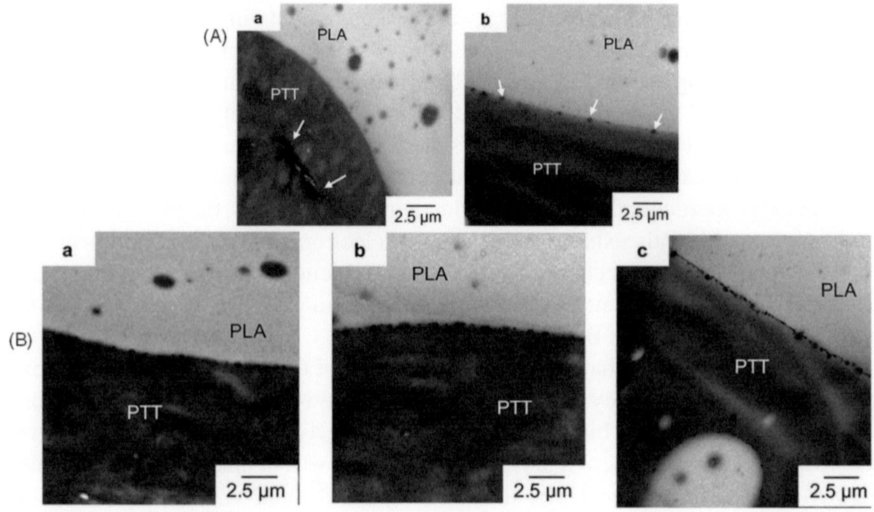

Fig. 5.10 **A** TEM micrographs of **a** PLA/PTT/GE (60/40/0.2) and **b** PLA/PTT/rGO-PGMA (60/40/0.2) composites. **B** TEM micrographs of the PLA/PTT/rGO-PGMA composite with the composition of **a** 60/40/0.3, **b** 60/40/0.4, and **c** 60/40/0.5. Reproduced with permission from [25]. Copyright 2021, American Chemical Society

end groups of both blend components. The interfacially localized filler particles created a conductive path, resulting in a significant improvement in the thermal and electrical conductivities of the PLA/PTT/rGO-PGMA nanocomposites compared to the PLA/PTT/GE nanocomposites. The researchers thus demonstrated that by chemically modifying GE and inducing chemical reactions of the modified filler particles with both blend components, it is possible to achieve interfacial localization of filler particles. Nanocomposites with such tailored interfaces exhibit superior thermal and electrical conductivities. The chemical modification of GO to achieve interfacial localization of filler particles due to their interaction with both blend components, leading to enhanced morphology and superior electrical conductivity, was also reported by Tan et al. [3].

Bai et al. [26] tailored the mixing sequence to achieve the interfacial localization of r-GO particles in PLA/PS blends with 48 vol% PLA content. The researchers reported that when r-GO was premixed with PLA and the premix was subsequently compounded with PS, the NPs were localized at the blend interface. Furthermore, the presence of the r-GO particles at the interface stabilized the co-continuous blend morphology and reduced coarsening during annealing. In fact, nanocomposites with filler loadings higher than 0.56% did not coarsen over time. However, this interfacial localization of filler particles was not achieved in nanocomposites prepared by mixing PS with r-GO, followed by blending the premix with PS in the subsequent step. The filler particles remained in the PS phase even after annealing, and the blend morphology was not stabilized. After 1 h of annealing, the electrical conductivity of the (PLA/r-GO)/PS nanocomposite reached 1.97×10^{-5} S/cm at 180 °C, which was over 2 orders of magnitude higher than that of the (PS/r-GO)/PLA nanocomposite and the PLA/r-GO nanocomposite with the same filler loading. Furthermore, the percolation threshold of the (PLA/r-GO)/PS nanocomposites was 50 times lower than that of the PLA/r-GO nanocomposites. All these observations clearly indicated that the interfacial localization of r-GOs facilitated the formation of a percolating network of conductive NPs.

Thus, it may be summarized that graphene oxide can serve as an interface modifier for immiscible polymer blends, leading to a compatibilization effect that results in a more stable blend morphology and improved electrical conductivity at low filler loadings. Interfacial localization of the nanoparticles can be achieved through their chemical modification to induce reactions or interactions with both blend components. Alternatively, by adjusting the mixing sequence, the filler particles can also be trapped at the interface. The effect of mixing sequence on the migration and localization of nanoparticles has been discussed in Chap. 2.

5.5 Other Nanofillers

Researchers have also modified the interfaces of immiscible polymer blends using various nanofillers, including carbon black (CB) [27], boehmite [28], polyhedral oligomeric-silsesquioxane (POSS) [29, 30], silica [31–33], and Janus NPs [34, 35]. Dil et al. [31] achieved interfacial localization of nanosilica particles in PLA/PBAT blends by tailoring the mixing sequence and investigating the effect of the interfacially localized silica particles on the blend morphology. The blend nanocomposites were prepared in an internal batch mixer at 50 rpm and 180 °C for 10 min. Two mixing strategies were employed:

Pr1 The addition of filler particles to a PLA/PBAT melt (one-step mixing).

Pr2 Premixing the filler with PLA, followed by blending the pre-mix with PBAT (two-step process).

AFM studies (Fig. 5.11) revealed that the filler particles in the blend nanocomposites prepared by Pr1 were selectively localized in the PBAT phase, while those in the blend composites prepared by Pr2 were found at the interface.

Furthermore, as observed in Fig. 5.12, the interfacial localization of the filler particles resulted in a finer blend morphology, with a decrease in the average size of the PBAT domains in the PLA/PBAT (80/20) blend from 1.3 ± 0.2 μm to 0.8 ± 0.1 μm and a narrowing of the domain size distribution.

Both the unfilled and filled PLA/PBAT (50/50) blends displayed a co-continuous morphology. However, unlike the unfilled blend, the filled blend with 3 wt% interfacially localized silica showed a remarkably stable morphology that did not coarsen during annealing, resulting in no deterioration of mechanical properties. The researchers demonstrated that by adjusting the mixing sequence, it is possible

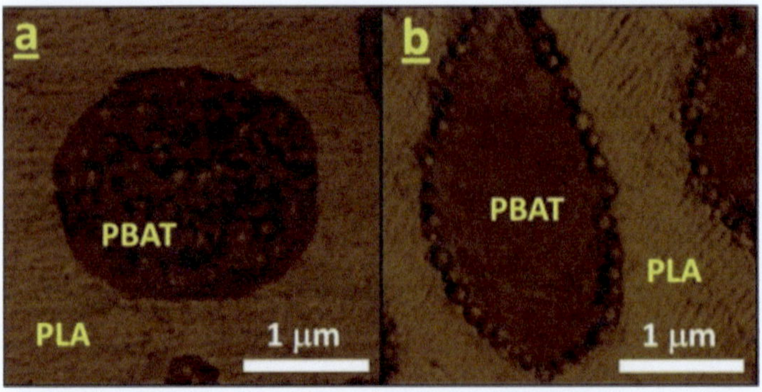

Fig. 5.11 Different localizations of nano-silica particles in PLA/PBAT (70/30) blends: **a** AFM image of the blend with 3 wt% of nano-silica prepared using Pr1, **b** AFM image of the blend with 3 wt% of nano-silica prepared using Pr2. The white scale bars indicate 1 μm. Reproduced with permission from [31]. Copyright 2016, Elsevier Science Ltd.

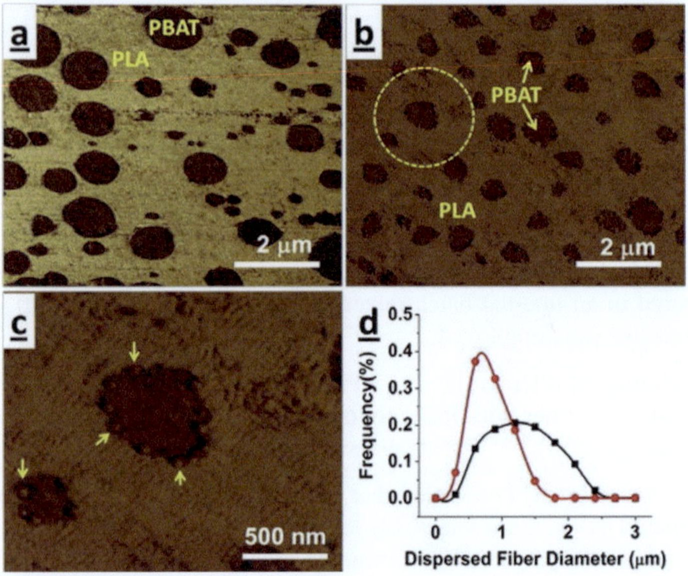

Fig. 5.12 **a** AFM image of the neat PLA/PBAT(80/20) blend, **b** AFM image of PLA/PBAT(80/20) blend with 3 wt% of nano-silica localized at the interface, **c** higher magnification image of the marked area in (**b**), **d** dispersed PBAT fiber diameter distribution for PLA/PBAT(80/20) without (■) and with (●)3 wt% nano-silica. The arrows in (**c**) point to selected nano-silica particles at the interface. Reproduced with permission from [31]. Copyright 2016, Elsevier Science Ltd.

to interfacially localize silica particles in immiscible polymer blends. These interfacially localized filler particles aid in refining the blend morphology and providing stability against coarsening during annealing.

Li et al. [28] advanced research on interface modification with nanofillers by demonstrating that the interfacial localization of filler particles might not be sufficient for a compatibilization effect. The researchers prepared nanocomposites of PLLA/poly(1,4-butylene succinate) (PLLA/PBSU) blends reinforced with pristine boehmite nanorods (p-BNRs) and those modified with an epoxide group (m-BNRs) through melt processing in a micro-compounder at 190 °C and a screw speed of 50 rpm for 10 min. The PLA: PBSU weight ratio of the compositions was 7:3, and the nanofiller content varied from 0 to 5 wt%. Compression-molded sheets of the melt-blended samples were used for further studies. TEM micrographs of the PLLA/PBSU with 3 wt% p-BNRs and m-BNRs are presented in Fig. 5.13. In nanocomposites with both types of filler, PBSU was found dispersed as 100 nm sized domains of irregular shape. Furthermore, particles of both p-BNR and m-BNR were selectively localized at the blend interface. Thus, the morphological characteristics of both types of nanocomposites were apparently similar, and their properties were expected to be the same. However, the mechanical properties differed significantly.

The PLLA/PBSU blend nanocomposites containing m-BNRs exhibited higher values of elongation at break, notched impact strength, and yield strength compared to the unfilled PLLA/PBSU blends. In contrast, the p-BNR nanocomposites displayed significantly lower mechanical properties, despite the selective interfacial localization of all p-BNRs. The results indicated that the p-BNRs at the interface had a barren surface that did not interact with the blend components. On the other hand, the epoxide groups on m-BNRs reacted with the end carboxylic acid groups of both blend components, leading to the polymer chains being chemically bonded to the surface of the filler particles. The chemically bonded polymer chains on the interfacially localized nanorods could entangle with the molecular chains of both phases, thus further enhancing interfacial adhesion.

The difference in the interaction of the two types of nanofiller with the blend components was confirmed in FTIR studies on the blend nanocomposites containing 5 wt% nanofiller, conducted after etching out the polymers with a mixture of petroleum ether and chloroform. Since the polymers were dissolved, the FTIR spectra of the remaining materials were expected to resemble those of the respective nanofillers. This was indeed observed in the case of the p-BNR reinforced nanocomposite (Fig. 5.14). However, the remnants from the nanocomposite with m-BNRs exhibited a markedly different FTIR spectrum. A new absorption peak appeared around 1730 cm^{-1} (Fig. 5.14a), characteristic of carboxyl groups. This peak could be curve-fitted into two absorptions at 1714 and 1758 cm^{-1} (Fig. 5.14b), corresponding to the carboxyl groups of PBSU and PLLA, respectively. This indicated that both PBSU and PLLA chains were grafted onto the m-BNR rods, which was supported by the absorption peaks in the region of 1200–1000 cm^{-1} (Fig. 5.14c). Furthermore, the absorption peaks at 910 cm^{-1}, corresponding to the epoxy groups of m-BNRs, disappeared in the spectral curve of the nanocomposite. This meant that

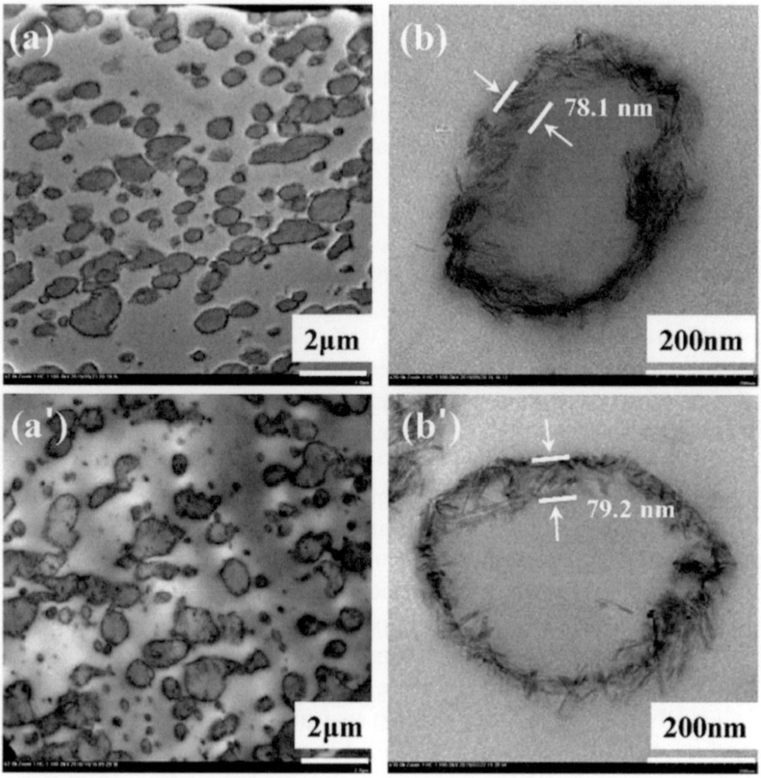

Fig. 5.13 TEM images for the PLLA/PBSU (70/30) blend sample with 3 wt% p-BNRs (**a**) (**b**) and 3 wt% m-BNRs (**a′**) (**b′**) with different magnifications. Reproduced with permission from [28]. Copyright 2020, Elsevier Science Ltd.

the epoxy groups of m-BNRs had reacted with the carboxylic end groups of PLLA and PBSU.

The researchers, therefore, through their in-depth studies, clearly established that the interfacial localization of filler particles in immiscible polymer blends is insufficient for interface modification through interaction with both blend components. The presence of functional groups with compatibility or reactivity towards both blend components facilitates this interaction. Only adequate interaction with both blend components results in significant enhancement of mechanical properties through interface modification.

Another way to achieve blend interface modification with nanofiller particles is through the molecular entanglements of the chains grafted onto the interfacially localized NPs and the molecular chains of the blend components. This was proposed by Wang et al. [33] in their studies on nanosilica reinforced PVDF/PLLA blends. Studies were conducted on blend nanocomposites with a 50:50 ratio of the blend

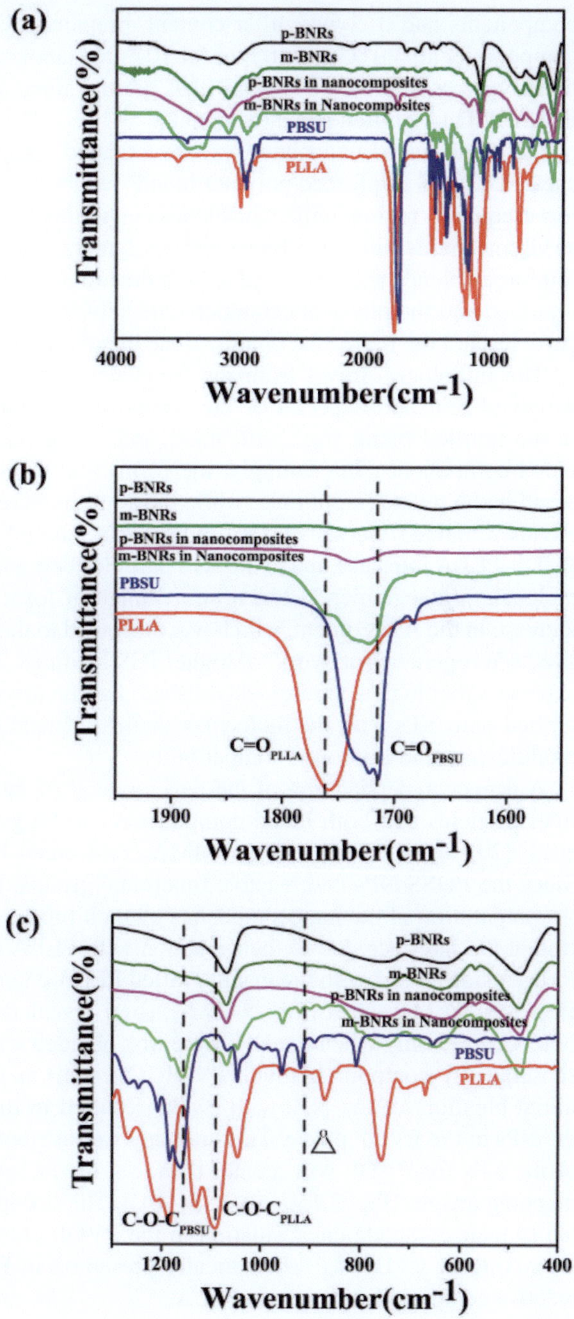

Fig. 5.14 FTIR spectra curves of the p-BNRs, m-BNRs, p-BNRs in nanocomposites, m-BNRs in nanocomposites, neat PBSU and neat PLLA in the wavenumber region of 4000–400 cm^{-1} (**a**), 2000–1500 cm^{-1} (**b**), and 1300–400 cm^{-1} (**c**). Reproduced with permission from [28]. Copyright 2020, Elsevier Science Ltd.

components and 0–5 wt% filler content, prepared by melt processing in a micro-compounder at 190 °C and 50 rpm for 10 min. Nanocomposites were prepared using three types of silica: bare silica, PLLA grafted silica (designated as GS), and Janus PLLA/PMMA grafted silica (designated as JGS). The linear viscoelastic properties of the blend nanocomposites were investigated using small amplitude oscillatory shear. The JGS reinforced polymer blends exhibited solid-like flow behavior in the low-frequency region, unlike the blends containing bare silica or GS, which behaved as viscous fluids, similar to linear homopolymers. The solid-like behavior of the JGS reinforced blends was attributed to both the ordered arrangement of JGS at the blend interface and the physical entanglements between the grafted long tails of the JGS particles and the molecular chains of the blend components.

The inferences drawn from the frequency sweep tests were supported by the weighted relaxation spectra of the compositions. The shape relaxation of PVDF in the unfilled blend was quite small, indicating poor compatibility between the blend components. Interestingly, the form relaxation peak of the JGS reinforced blend had a greater amplitude, while those of the bare silica and the GS reinforced blends remained the same. This indicated a strengthened relaxation effect at the PVDF-PLLA interface due to the JGS particles being interfacially anchored through molecular entanglements. The reinforcement of form relaxation increased with an increase in the JGS content, which was attributed to the formation of a more penetrative "heterogeneous network" at higher JGS loadings. The researchers, through their extensive rheological studies, established that the physical entanglements of chains grafted onto NPs with the molecular chains of blend components lead to interface modification and improved compatibility.

A deeper understanding of the development of interfacial interactions between filler particles and both blend components can be gained from the studies by Fu et al. [29] on POSS(epoxy)$_x$-g-PMMA$_y$ reinforced PVDF/PLLA (50/50) blends. Since the POSS NPs had reactive functional groups, it was anticipated that during the preparation of the nanocomposites by melt processing, the NPs would be localized at the interface due to their reaction with PLLA and the entanglements of the PMMA side chains with the in situ grafted PLLA chains and the blend components. However, TEM studies [Fig. 5.15(a$_1$), (b$_1$)] revealed that effective compatibilization occurred only when the NPs were first blended with PLLA and the premix was subsequently compounded with PVDF. One-step mixing or two-step mixing with initial blending of the NPs with PVDF resulted in the preferential localization of the NPs in the PVDF phase. This anomaly was attributed to the fact that the affinity of the NPs for PVDF was greater than that for PLLA, as confirmed by elemental mapping images [Fig. 5.15(a$_2$–a$_4$), (b$_2$–b$_4$)]. The strong affinity of the NPs for PVDF led to their complete encapsulation in the PVDF phase during one-step mixing or premixing in PVDF, as schematically presented in Fig. 5.16, thereby preventing reactive compatibilization with PLLA.

The researchers proposed a mechanism for the compatibilization effect observed when the NPs were first premixed with PLLA and the premix compounded with PVDF. During the melt processing of POSS(epoxy)$_x$-g-PMMA$_y$ with PLLA, the PLLA was grafted onto the POSS core in situ through the ring opening reaction of the

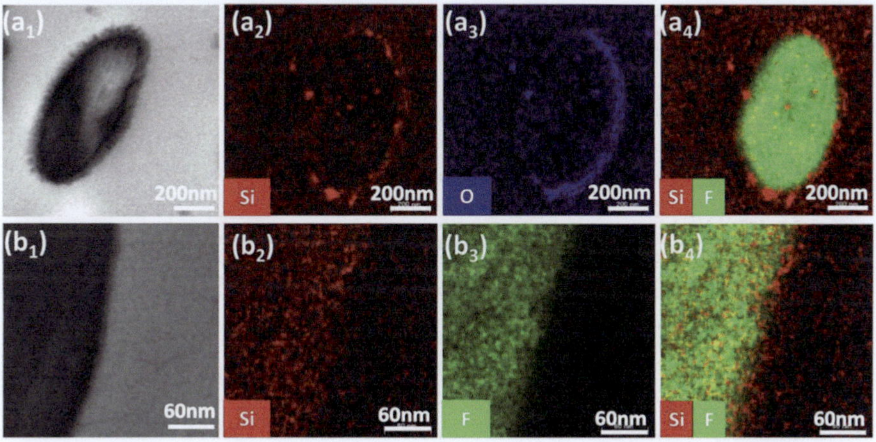

Fig. 5.15 **a** TEM and Elemental Mapping Images (EMI) of compatibilized PVDF/PLLA (50/50) blends by PLLA-premixed sequence; (**a**$_2$)–(**a**$_4$) EMI on (**a**) blends showing the signals of silicon, oxygen and combined fluorine silicon element, respectively; **b** TEM and Elemental Mapping Images (EMI) of compatibilized PVDF/PLLA (50/50) by one-step mixed strategy; (**b**$_2$)–(**b**$_4$) EMI on (**b**) blends showing the signals of silicon, fluorine and combined fluorine silicon element, respectively. The compatibilizer is 3 wt% POSS(epoxy)$_4$-g-PMMA$_4$ in both cases. Reproduced with permission from [29]. Copyright 2017, Elsevier Science Ltd.

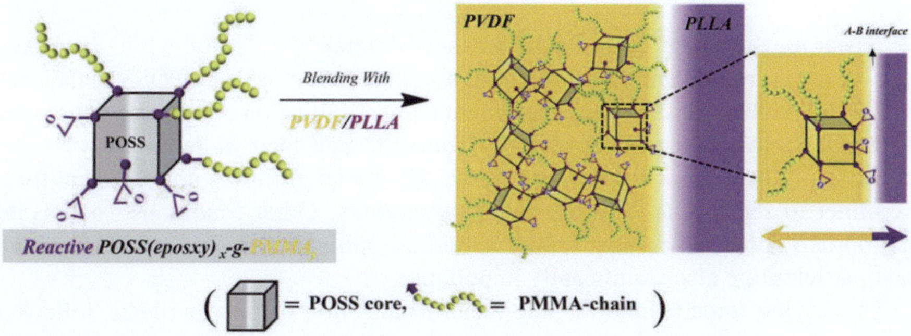

Fig. 5.16 Schematic diagram of the modified POSS encapsulated in the PVDF phase. Reproduced with permission from [29]. Copyright 2017, Elsevier Science Ltd.

COOH groups of PLLA with the epoxide groups of POSS. The subsequent blending with PVDF resulted in the assembly of the grafted POSS at the interface of the PVDF/PLLA blend. This effect stemmed from thermodynamic forces arising from the interactions between the PMMA tails and PVDF molecules and the entanglements of the grafted PLLA chains with the PLLA molecules. A Janus-like structure having the POSS at the core and the grafted PLLA and PMMA at the outer shell was produced in situ, resulting in effective compatibilization at the interface, as presented in Fig. 5.17. Since the PMMA chains grafted onto the POSS core contained about

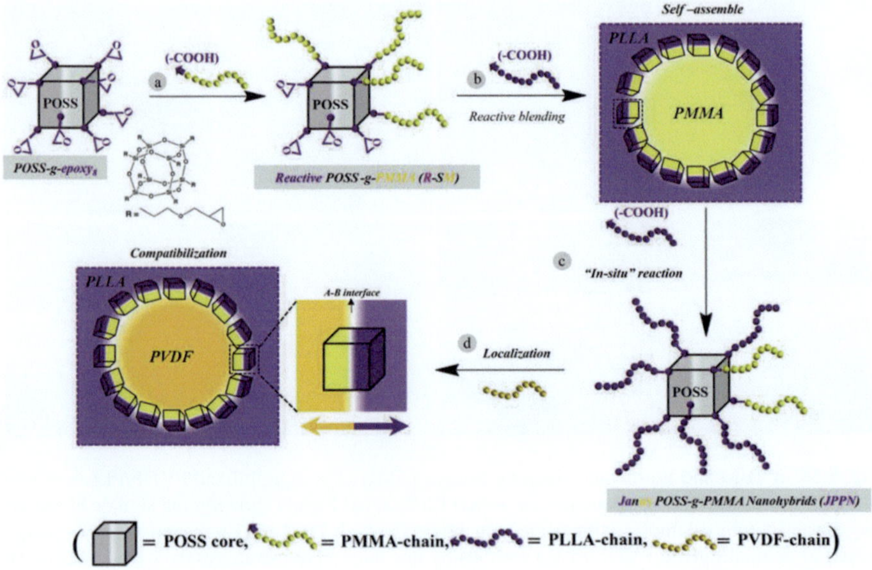

Fig. 5.17 Schematic diagram of the in-situ formation of the Janus-like structure by the reactive compatibilization of POSS(epoxy)$_x$-g-PMMA$_y$ for PLLA/PVDF blends. Reproduced with permission from [29]. Copyright 2017, Elsevier Science Ltd.

40 repeat units and the chain length was approximately 5.92 nm, while the POSS core size was around 1 nm, the formation of a Janus structure with POSS as the core was highly viable. Furthermore, maximum compatibilization effects were observed with POSS having three PMMA tails. Thus, through their in-depth analysis, the researchers established that having reactive functional groups is not sufficient for a nanofiller to serve as an effective interface modifier. Other conditions, such as an appropriate choice of mixing sequence and an optimally tailored structure of the nanoparticles, are also significantly important.

The driving force behind the interfacial localization of filler particles, followed by interface modification, can also be attributed to hydrogen bonding. This was demonstrated by Zhang et al. [27] in their studies on CB-reinforced thermoplastic polyurethane (TPU)/polyamide copolymer (COPA) with 20 wt% COPA content, for which different filler loadings were prepared by melt mixing. Four different mixing sequences were employed. TEM studies (Fig. 5.18) revealed that, regardless of the mixing sequence, most filler particles were localized at the interface; however, the extent of CB localization was significantly less in the composites prepared by adding COPA to the premixed TPU/CB. The interfacial localization of CB was attributed to hydrogen bonding between CB and both blend components, as confirmed by FTIR and DMA studies (Fig. 5.19).

From the FTIR spectral curves, the hydrogen bonding indices (R)—defined as the ratio of the bonded C=O at 1702 cm^{-1} to that of the free C=O at 1724 cm^{-1}—of TPU in the neat polymer, the unfilled blends, and the filled blends were calculated.

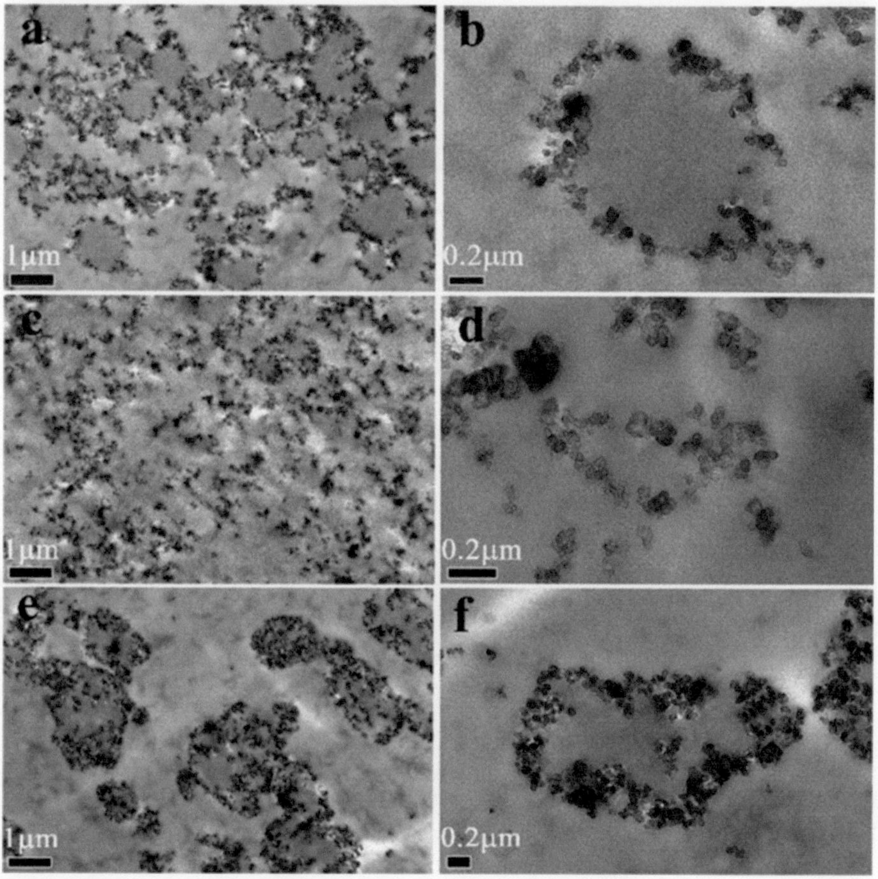

Fig. 5.18 TEM micrographs of the composites prepared by different mixing sequences: **a** and **b** (TPU/COPA)/CB, **c** and **d** (TPU/CB)/COPA and **e** and **f** (COPA/CB)/TPU. Reproduced with permission from [27]. Copyright 2017, Royal Society of Chemistry

As shown in Fig. 5.19B (a), R decreased after blending with 20 wt% COPA due to hydrogen bonding between TPU and COPA. However, the incorporation of CB particles led to an increase in R once more. The greater the CB loading, the higher the extent of the increase. This indicated that the CB particles were hydrogen bonded to TPU. Since the COPA peaks in Fig. 5.19A (a) were weak, the hydrogen bonding of CB with COPA was evaluated by dissolving the TPU/COPA/9CB sample in a formic acid/DMF (1:2) mixture, isolating the CB particles through vacuum infiltration, washing with the solvent mixture and THF, drying, and then subjecting the dried CB sample to FTIR analysis. The characteristic peaks of both TPU (at 1724, 1702, and 1411 cm^{-1}) and COPA (at 2918, 2850, and 1633 cm^{-1}) appeared in the IR spectrum. This clearly indicated that both TPU and COPA were hydrogen bonded to CB.

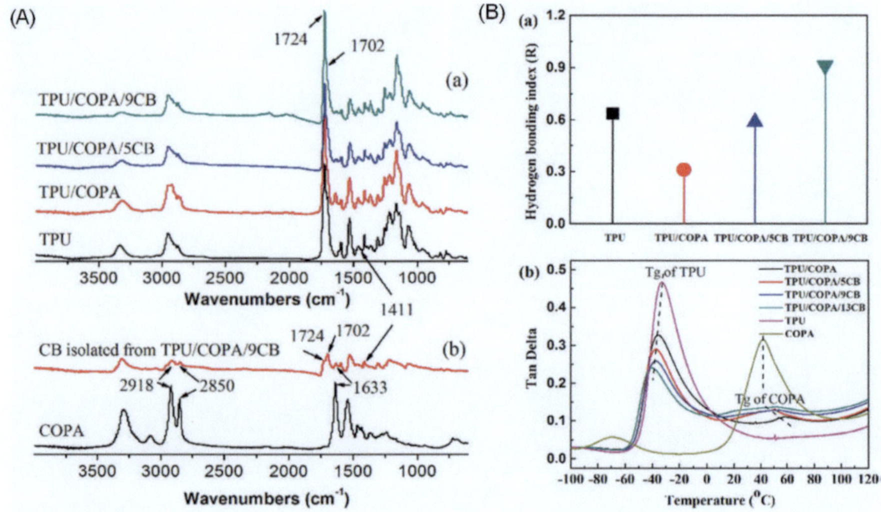

Fig. 5.19 **A** (**a**) FTIR spectra of TPU, TPU/COPA blend, TPU/COPA/5CB and TPU/COPA/9CB composites, (**b**) FTIR spectra of CB particles isolated from the TPU/COPA/9CB composite and COPA. **B** (**a**) Hydrogen bonding indexes (R) for TPU, TPU/COPA blend, TPU/COPA/5CB and TPU/COPA/9CB composites, and (**b**) DMA curves of TPU, COPA, TPU/COPA blend, TPU/COPA/5CB, TPU/COPA/9CB and TPU/COPA/13CB composites. Reproduced with permission from [27]. Copyright 2017, Royal Society of Chemistry

The tan δ curves from DMA temperature sweep studies on the neat polymers, unfilled blends, and filled blends are presented in Fig. 5.19B (b). Compared to the neat polymers, the glass transition temperature of TPU in the unfilled blend was lower, while that of COPA was higher, indicating an interaction between the two blend components. The incorporation of CB resulted in a decrease in the T_gs of both blend components compared to those of the unfilled blend, confirming that the CB interacted with both TPU and COPA, providing evidence in favor of hydrogen bonding. Thus, the researchers demonstrated that the driving force behind the interfacial localization of filler particles and the subsequent interface modification may be the hydrogen bonding of filler particles with both blend components.

Modifying the interface of immiscible polymer nanocomposites can also be accomplished by altering the shape of the nanoparticles, as seen with Janus particles (JPs). Researchers studying JP-reinforced polymer blends have found that the shape anisotropy (Janus balance) significantly affects nanoparticle interfacial activity [36] and can aid in minimizing the size of the dispersed domains [37].

According to You and Yu [34], the shape asymmetry of snowman-like JPs significantly affects the development of co-continuous morphology in polymer blends during melt processing. Blending with only 0.9 vol% snowman-like JPs nearly halved the concentration required for the onset of co-continuity, which was considerably lower than the apparent interfacial jamming concentration. Additionally, the JPs localized at the interface demonstrated a superior ability to stabilize the co-continuous blend morphology during annealing, with no redistribution of the JPs taking place. The interfacial activity of the snowman-like JPs is attributed to the fact that increased shape asymmetry of the JPs leads to a reduction in the radius of jamming curvature at the interface. The researchers suggested that by synthesizing JPs with tailored shape symmetry, it would be possible to achieve a high level of interfacial modification, which would promote the development of a stable co-continuous blend morphology.

5.6 Conclusions

In summary, nanofillers function as interface modifiers for immiscible polymer blends. This modification occurs when nanoparticles are selectively positioned at the interface of the blend, allowing them to engage with both components—either through reactions (reactive compatibilization) or other means, like hydrogen bonding or physical entanglements with macromolecular chains. The nanoparticles' molecular structure, including the presence of functional groups and grafted chains, is crucial for these interactions. Additionally, the processing conditions, particularly the mixing protocol, are essential. Effective compatibilization with nanoparticles leads to considerable enhancements in the properties of polymer blend composites. The effects of filler migration and localization on the characteristics of polymer blend nanocomposites are explored in Chap. 6.

References

1. A.M. Zolali, B.D. Favis, Compatibilization and toughening of co-continuous ternary blends via partially wet droplets at the interface. Polymer (Guildf) **114**, 277–288 (2017). https://doi.org/10.1016/J.POLYMER.2017.02.093
2. N. Virgilio, C. Marc-Aurèle, B.D. Favis, Novel self-assembling close-packed droplet array at the interface in ternary polymer blends. Macromolecules **42**, 3405–3416 (2009). https://doi.org/10.1021/MA802544Q/ASSET/IMAGES/LARGE/MA-2008-02544Q_0015.JPEG
3. Y. Tan, L. Fang, J. Xiao, Y. Song, Q. Zheng, Grafting of copolymers onto graphene by miniemulsion polymerization for conductive polymer composites: improved electrical conductivity and compatibility induced by interfacial distribution of graphene. Polym. Chem. **4**, 2939–2944 (2013). https://doi.org/10.1039/C3PY00164D
4. X. Zhao, H. Wang, Z. Fu, Y. Li, Enhanced interfacial adhesion by reactive carbon nanotubes: new route to high-performance immiscible polymer blend nanocomposites with simultaneously enhanced toughness, tensile strength, and electrical conductivity. ACS Appl. Mater. Interfaces **10**, 8411–8416 (2018). https://doi.org/10.1021/ACSAMI.8B01704/SUPPL_FILE/AM8B01704_SI_001.PDF
5. S.S. Ray, S. Pouliot, M. Bousmina, L.A. Utracki, Role of organically modified layered silicate as an active interfacial modifier in immiscible polystyrene/polypropylene blends. Polymer (Guildf) **45**, 8403–8413 (2004). https://doi.org/10.1016/J.POLYMER.2004.10.009
6. H. Eslami, M.R. Kamal, Elongational rheology of biodegradable poly(lactic acid)/poly[(butylene succinate)-co-adipate] binary blends and poly(lactic acid)/poly[(butylene succinate)-co-adipate]/clay ternary nanocomposites. J. Appl. Polym. Sci. **127**, 2290–2306 (2013). https://doi.org/10.1002/APP.37928

7. R.M. Salih, R.H. Mohammed, Influence of nanoclay as a compatibilizer and a reinforcement for polymer blends used as insulators. Prog. Rubber Plast. Recycl. Technol. **38**, 227–246 (2022). https://doi.org/10.1177/14777606221105298
8. A. Taguet, B. Otazaghine, M. Batistella, Kinetic and thermodynamic parameters guiding the localization of regioselectively modified kaolin platelets into a PS/PA6 co-continuous blend. Polymer (Guildf) **191**, 122277 (2020). https://doi.org/10.1016/J.POLYMER.2020.122277
9. J.S. Hong, H. Namkung, K.H. Ahn, S.J. Lee, C. Kim, The role of organically modified layered silicate in the breakup and coalescence of droplets in PBT/PE blends. Polymer (Guildf) **47**, 3967–3975 (2006). https://doi.org/10.1016/J.POLYMER.2006.03.077
10. Y. Wang, Q. Zhang, Q. Fu, Compatibilization of immiscible poly(propylene)/polystyrene blends using clay. Macromol. Rapid Commun. **24**, 231–235 (2003). https://doi.org/10.1002/MARC.200390026
11. M. Nofar, M.C. Heuzey, P.J. Carreau, M.R. Kamal, Effects of nanoclay and its localization on the morphology stabilization of PLA/PBAT blends under shear flow. Polymer (Guildf) **98**, 353–364 (2016). https://doi.org/10.1016/J.POLYMER.2016.06.044
12. J. Chen, X. Cui, Y. Zhu, W. Jiang, K. Sui, Design of superior conductive polymer composite with precisely controlling carbon nanotubes at the interface of a co-continuous polymer blend via a balance of π–π interactions and dipole-dipole interactions. Carbon N Y **114**, 441–448 (2017)
13. E. Cohen, L. Zonder, A. Ophir, S. Kenig, S. McCarthy, C. Barry et al., Hierarchical structures composed of confined carbon nanotubes in cocontinuous ternary polymer blends. Macromolecules **46**, 1851–1859 (2013). https://doi.org/10.1021/MA301903N/SUPPL_FILE/MA301903N_SI_001.PDF
14. G.P. Kar, S. Biswas, S. Bose, Simultaneous enhancement in mechanical strength, electrical conductivity, and electromagnetic shielding properties in PVDF–ABS blends containing PMMA wrapped multiwall carbon nanotubes. Phys. Chem. Chem. Phys. **17**, 14856–14865 (2015)
15. L. Azubuike, U. Sundararaj, Interface Strengthening of PS/aPA polymer blend nanocomposites via in situ compatibilization: enhancement of electrical and rheological properties. Materials **14**, 4813 (2021). https://doi.org/10.3390/MA14174813
16. J. Chen, Y. Shen, J.H. Yang, N. Zhang, T. Huang, Y. Wang et al., Trapping carbon nanotubes at the interface of a polymer blend through adding graphene oxide: a facile strategy to reduce electrical resistivity. J Mater Chem C **1**, 7808–7811 (2013). https://doi.org/10.1039/C3TC31702A
17. R. Jiang, T. Liu, Z. Xu, C.B. Park, L. Zhao, Improving the continuous microcellular extrusion foaming ability with supercritical CO_2 of thermoplastic polyether ester elastomer through in-situ fibrillation of polytetrafluoroethylene. Polymers (Basel) **11**, 1983 (2019). https://doi.org/10.3390/polym11121983
18. Y. Cao, J. Zhang, J. Feng, P. Wu, Compatibilization of immiscible polymer blends using graphene oxide sheets. ACS Nano **5**, 5920–5927 (2011). https://doi.org/10.1021/NN201717A/SUPPL_FILE/NN201717A_SI_001.PDF
19. J. Yang, C. Feng, J. Dai, N. Zhang, T. Huang, Y. Wang, Compatibilization of immiscible nylon 6/poly(vinylidene fluoride) blends using graphene oxides. Polym. Int. **62**, 1085–1093 (2013). https://doi.org/10.1002/PI.4396
20. M. Chen, T. Tao, L. Zhang, W. Gao, C. Li, Highly conductive and stretchable polymer composites based on graphene/MWCNT network. Chem. Commun. **49**, 1612–1614 (2013). https://doi.org/10.1039/C2CC38290C
21. Y. Li, T. Yang, T. Yu, L. Zheng, K. Liao, Synergistic effect of hybrid carbon nantube–graphene oxide as a nanofiller in enhancing the mechanical properties of PVA composites. J. Mater. Chem. **21**, 10844–10851 (2011). https://doi.org/10.1039/C1JM11359C
22. Y.Y. Shi, J.H. Yang, T. Huang, N. Zhang, C. Chen, Y. Wang, Selective localization of carbon nanotubes at the interface of Poly(L-lactide)/ethylene-co-vinyl acetate resulting in lowered electrical resistivity. Compos. Part B Eng. **55**, 463–469 (2013). https://doi.org/10.1016/J.COMPOSITESB.2013.07.012

23. A. Göldel, G.R. Kasaliwal, P. Pötschke, G. Heinrich, The kinetics of CNT transfer between immiscible blend phases during melt mixing. Polymer (Guildf) **53**, 411–421 (2012). https://doi.org/10.1016/J.POLYMER.2011.11.039
24. H. Bai, S. Deng, D. Bai, Q. Zhang, Q. Fu, Recent Advances In Processing Of Stereocomplex-Type Polylactide. Macromol. Rapid Commun. **38**, 1700454 (2017). https://doi.org/10.1002/marc.201700454
25. K. Kultravut, K. Kuboyama, V. Sedlarik, M. Mrlík, J. Osička, P. Dröhsler et al., Localization of Poly(glycidyl methacrylate) grafted on reduced graphene oxide in poly(lactic acid)/poly(trimethylene terephthalate) blends for composites with enhanced electrical and thermal conductivities. ACS Appl. Nano Mater. **4**, 8511–8519 (2021). https://doi.org/10.1021/ACSANM.1C01843/SUPPL_FILE/AN1C01843_SI_001.PDF
26. L. Bai, S. He, J.W. Fruehwirth, A. Stein, C.W. Macosko, X. Cheng, Localizing graphene at the interface of cocontinuous polymer blends: Morphology, rheology, and conductivity of cocontinuous conductive polymer composites. J. Rheol. (N Y N Y) **61**, 575–587 (2017). https://doi.org/10.1122/1.4982702
27. Q. Zhang, J. Wang, J. Yu, Z.X. Guo, Improved electrical conductivity of TPU/carbon black composites by addition of COPA and selective localisation of carbon black at the interface of sea-island structured polymer blends. Soft Matter **13**, 3431–3439 (2017). https://doi.org/10.1039/C7SM00346C
28. X. Li, Z. Fu, X. Gu, H. Liu, H. Wang, Y. Li, Interfacially located nanoparticles: barren nanorods versus polymer grafted nanorods. Compos. Part B Eng. **198**, 108153 (2020). https://doi.org/10.1016/J.COMPOSITESB.2020.108153
29. Z. Fu, H. Wang, X. Zhao, S. Horiuchi, Y. Li, Immiscible polymer blends compatibilized with reactive hybrid nanoparticles: morphologies and properties. Polymer (Guildf) **132**, 353–361 (2017). https://doi.org/10.1016/J.POLYMER.2017.11.004
30. N.T. Kilic, B.N. Can, M. Kodal, G. Ozkoc, The potential use of Epoxy-POSS as a reactive hybrid compatibilizers for PLA/PBAT blends: "effect of PBAT molecular weight and POSS type." Polym. Eng. Sci. **60**, 398–413 (2020). https://doi.org/10.1002/PEN.25295
31. E.J. Dil, N. Virgilio, B.D. Favis, The effect of the interfacial assembly of nano-silica in poly(lactic acid)/poly(butylene adipate-co-terephthalate) blends on morphology, rheology and mechanical properties. Eur. Polym. J. **85**, 635–646 (2016). https://doi.org/10.1016/J.EURPOLYMJ.2016.07.022
32. X. Zhang, M. Sun, W. Chen, Synergistic effects of silica nanoparticles and reactive compatibilizer on the compatibilization of polystyrene/polyamide 6 blends. Polym. Eng. Sci. **57**, 1301–1310 (2017). https://doi.org/10.1002/PEN.24511
33. H. Wang, X. Yang, Z. Fu, X. Zhao, Y. Li, J. Li, Rheology of nanosilica-compatibilized immiscible polymer blends: formation of a "heterogeneous network" facilitated by interfacially anchored hybrid nanosilica. Macromolecules **50**, 9494–9506 (2017). https://doi.org/10.1021/ACS.MACROMOL.7B02143/ASSET/IMAGES/LARGE/MA-2017-021438_0011.JPEG
34. W. You, W. Yu, Onset reduction and stabilization of cocontinuous morphology in immiscible polymer blends by snowmanlike Janus nanoparticles. Langmuir **34**, 11092–11100 (2018). https://doi.org/10.1021/ACS.LANGMUIR.8B02503/SUPPL_FILE/LA8B02503_SI_001.PDF
35. F.I. Seyni, B.P. Grady, Janus particles as immiscible polymer blend compatibilizers: a review. Colloid Polym. Sci. **299**, 585–593 (2021). https://doi.org/10.1007/S00396-021-04820-X/METRICS
36. D. Wu, J.W. Chew, A. Honciuc, Polarity reversal in homologous series of surfactant-free Janus nanoparticles: toward the next generation of amphiphiles. Langmuir **32**, 6376–6386 (2016). https://doi.org/10.1021/ACS.LANGMUIR.6B01422/ASSET/IMAGES/LARGE/LA-2016-01422J_0007.JPEG
37. T. Parpaite, B. Otazaghine, A.S. Caro, A. Taguet, R. Sonnier, J.M. Lopez-Cuesta, Janus hybrid silica/polymer nanoparticles as effective compatibilizing agents for polystyrene/polyamide-6 melted blends. Polymer (Guildf) **90**, 34–44 (2016). https://doi.org/10.1016/J.POLYMER.2016.02.044

Chapter 6
Properties of Polymer Blend Nanocomposites

Abstract The properties of polymer blend nanocomposites are primarily influenced by the positioning of nanoparticles. This chapter examines how particle migration and localization, when various nanofillers are used individually or in combination, affect the properties of blend nanocomposites.

6.1 Migration-Induced Properties

6.1.1 Nanoclays

The effect of migration and selective localization of clay particles on the properties of nanoclay-reinforced polymer blends has been studied by several research groups [1–15]. The work of Nofar et al. [12] provides an in-depth understanding of this subject. The researchers conducted studies on the properties of C30B (CN) reinforced polylactide (PLA)-based blends with poly(butylene adipate-co-terephthalate) (PBAT) produced through melt mixing. An amorphous PLA and a semi-crystalline PLA, designated as A-PLA and SC-PLA, respectively, were used for the studies. The compositions were prepared by twin screw extrusion at 160 °C and 100 rpm screw speed, followed by injection molding (IM) at 200 °C and 1500 bar. The transmission electron microscopy (TEM) images of ternary composites with 74 wt% PLA, 25 wt% PBAT, and 1 wt% nanoclay contents, as well as those with 48.5 wt% PLA, 48.5 wt% PBAT, and 3 wt% nanoclay contents, are presented in Fig. 6.1a. Using calculations based on the wetting parameter, clay was predicted to be primarily located at the interface, with some presence in the PLA phase. This localization tendency was observed only when SC-PLA was used. When A-PLA was used, clay was distributed throughout the composition, regardless of the PLA and clay contents. The researchers proposed that the nanoclay kinetics within the SC-PLA/PBAT blend were much simpler, facilitating the migration of clay particles according to thermodynamic principles. However, in A-PLA/PBAT blends, the short mixing time in the twin screw extruder (TSE) was insufficient for this migration. This explanation was supported by TEM studies (Fig. 6.1b) conducted on an A-PLA/PBAT blend with 1

S. S. Ray and R. Banerjee, *Polymer Blend Nanocomposites*, Springer Series in Materials Science 354, https://doi.org/10.1007/978-3-032-00309-6_6

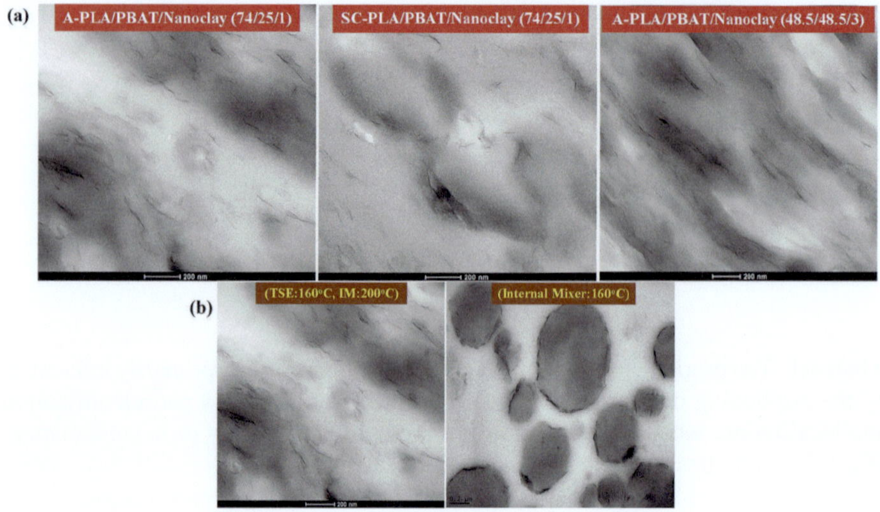

Fig. 6.1 **a** TEM images of PLA/PBAT/clay blend nanocomposites with A-PLA and SC-PLA as the major phase. **b** TEM images of A-PLA/PBAT/clay blend nanocomposites reflecting the nanoclay dispersion and localization when prepared through an internal mixer and a TSE. Reproduced with permission from [12]. Copyright 2019, Elsevier Science Ltd.

wt% nanoclay loading, produced in an internal mixer at 160 °C and 100 rpm, with a mixing time of 12 min. The increased mixing time led to the migration of filler particles to the thermodynamically favored region.

The incorporation of nanoclay into the blends did not result in a significant change in their strength and modulus. However, the addition of just 1 wt% nanoclay to A-PLA/PBAT led to a substantial increase in ductility from approximately 5% to about 220%. This remarkable improvement was likely due to the development of a more uniform and fine morphology resulting from the presence of nanoclay. The ductility of SC-PLA/PBAT also slightly improved with the incorporation of 1 wt% clay, rising from approximately 205% to roughly 213%. In contrast, in the blends with a 50/50 ratio, the addition of 3 wt% clay decreased the ductility of the blend. This was attributed to the more significant role of nanoclay as a solid particle reinforcement, which increased brittleness.

The difference in the extent of ductility improvement for the A-PLA/PBAT and SC-PLA/PBAT blends was attributed to the variation in the localization of nanoclays. In the SC-PLA/PBAT blend nanocomposites, the clay particles were localized at the interphase, whereas in the A-PLA/PBAT blend nanocomposites, they were dispersed throughout. The clay particles acted as points of stress concentration. Consequently, when stress was applied, the dispersed polymer droplets underwent deformation, in contrast to the clay particles at the interface, which could not deform. Furthermore, since the tensile test was conducted at room temperature in the solid state, the clay particles could not align themselves in the direction of the stress. Due to the resisting forces exerted by the clay particles, the droplets cracked at the interface of the SC-PLA/PBAT nanocomposites where the clay was localized. This led to a significantly lower ductility of the SC-PLA/PBAT nanocomposites compared to the A-PLA/PBAT nanocomposites. The researchers thus clearly demonstrated the influence of the localization of nanoclay on the mechanical properties of nanoclay-reinforced polymer blends. The influence of clay localization on the mechanical properties of nanoclay reinforced polymer blends has also been reported by other researchers [4, 5, 9, 16–18].

Wu et al. [11] investigated similar topics. The researchers conducted studies on nanoclay-reinforced polypropylene (PP)/polystyrene (PS) blends with a weight ratio of 70:30 for PP and PS. The nanocomposites were prepared by melt mixing in an eccentric rotor extruder and were designated as PP/PS/Mx, where x indicated the clay content in wt%. TEM studies (Fig. 6.2, right side) of the nanocomposites with clay loadings ranging from 1 to 7 wt% revealed that the clay particles were localized at the interface. Melt state dynamic rheological studies (Fig. 6.2, left side) showed a low-frequency plateau in the storage modulus (G′) curve when the clay loading exceeded 1 wt%. Furthermore, the storage modulus values increased as the clay loading rose to 5 wt% clay content. The low-frequency plateau was attributed to the formation of a network of clay particles at the interface. This network formation became more pronounced with an increase in clay concentration, leading to higher G′ values. At very high loadings, G′ values decreased, which was ascribed to the agglomeration of clay particles. A similar "solid-like" behavior (low-frequency G′ plateau) due to network formation by clay particles selectively localized at the interface has also been reported by other researchers [8, 19].

Although the enhanced barrier properties of polymer clay nanocomposites are well established [20–23], limited research has focused on the impact of nanoclay localization on the barrier properties of clay-reinforced immiscible polymer blends. In this context, Binis et al. [9] studied the gas permeability of PLA/natural rubber (NR)/Cloisite 30B (C30B) nanocomposites containing 10 wt% NR and 3 wt% clay. When all components were mixed simultaneously, the clay particles were located in the continuous PLA phase and at the interface, as seen in TEM studies. However, when C30B was premixed with PLA, and this mixture was blended with NR, the clay was selectively localized in the continuous PLA phase, resulting in the nanocomposite exhibiting oxygen and carbon dioxide permeability values of 0.68 barrer and 1.86 barrer, respectively. Thus, the researchers demonstrated that selective localization of nanoclay in the continuous phase can improve the gas barrier properties of nanoclay-reinforced immiscible polymer blends. Nevertheless, further research is needed in this area [24, 25].

The work of Ojijo et al. [13] highlights the effect of the localization of clay particles on the crystallization behavior of polymer blend nanocomposites. The researchers focused on PLA/poly(butylene succinate-co-adipate) (PBSA)/Cloisite 20A (C20A) nanocomposites with a PLA to PBSA weight ratio of 70:30. Nanocomposites containing 0.5, 1, 2, 6, and 9 wt% C20A loadings, designated as B0.5, B1, B2, B4, B6, and B9, respectively, were prepared by melt blending all components in an internal mixer at 185 °C and 60 rpm for 8 min. TEM studies revealed that

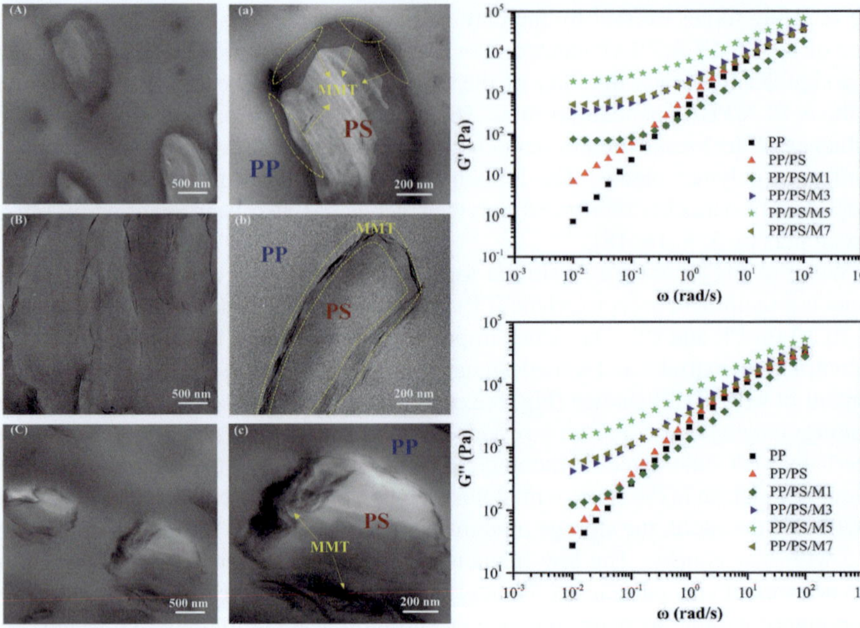

Fig. 6.2 Right side-TEM morphology of PP/PS/MMT blends with different MMT concentration. **A** and **a** PP/PS/M1, **B** and **b** PP/PS/M3, **C** and **c** PP/PS/ M7. **Left side**: (upper panel) storage modulus G′ and (bottom panel) loss modulus G″ of pure PP and PP/PS/MMT blends. Reproduced with permission from [11]. Copyright 2017, Elsevier Science Ltd.

in nanocomposites with very low clay content (regime 1, including B0.5 and B1), most clay was localized in the low viscosity PBSA dispersed droplets. In regime 2 (B2), which had a higher clay loading, some clay particles were present in the PLA phase, with many more at the interface. At even higher clay loadings (regime 3, including B4, B6, and B9), most clay migrated into the PLA phase. B6 exhibited a co-continuous morphology. The development of co-continuity was attributed to the increased presence of clay in PLA, which raised the PLA viscosity to the extent that the viscosity ratio of PBSA to PLA became equal to the volume fraction, leading to the onset of co-continuity. This co-continuous morphology broke down at a higher clay loading (B9). The breakdown was ascribed to a further increase in PLA viscosity, which altered the viscosity ratio, favoring deformation of the PBSA phase and disrupting co-continuity. The crystallization rate of the PLA phase (G) across the various compositions was examined. Except for B2 and B6, increasing clay loading resulted in a higher G, attributed to enhanced heterogeneous nucleation of crystals due to a greater concentration of clay particles in the PLA phase. Even though most clay was present in the PBSA phase in regime 1, a small amount of clay particles in the PLA phase increased with higher clay loading, resulting in an increased G. For B2, G decreased due to the increased presence of clay particles at the interface, which hindered crystal growth in the PLA phase and raised the activation energy for crystallization. In B6, the reduction in G was attributed to co-continuity, as the growth of PLA crystals was impeded by the co-continuous PBSA phase. The researchers thus demonstrated that the migration and localization of clay can significantly influence the crystallization behavior of nanoclay-reinforced immiscible polymer blends. The effect of nanoclay localization on the crystallization of polymer blend nanocomposites has also been discussed in other studies [15, 17, 26].

It can be summarized that the migration and localization of clay can affect the mechanical and barrier properties, as well as the crystallization behavior, of nanoclay-reinforced polymer blends. Therefore, tailoring morphology by controlling the localization of nanoparticles can also serve as a method to achieve the desired properties of polymer blend nanocomposites.

6.1.2 Carbon Nanotubes

The impact of the migration and selective localization of carbon nanotubes (CNTs) on the properties of CNT-reinforced polymer blends has been investigated by several research groups [27–46].

For example, Morshedi et al. [32] investigated the effect of CNT localization on the toughness and electrical properties of CNT-reinforced compatibilized PP/ethylene-octene copolymer (EOC) blends, which were prepared by melt mixing at 190 °C and a rotor speed of 90 rpm for 20 min. Three different feeding techniques were employed to produce the nanocomposite samples: (i) the direct method, in which all components [PP, EOC, CNTs, and maleic anhydride grafted polypropylene (PP-g-MA)] were fed into the mixer simultaneously; (ii) the sequential method, in which EOC, PP, and PP-g-MA were first fed into the mixing chamber, and after 2 min of mixing, the CNTs were added; and (iii) the masterbatch method, in which a concentrated masterbatch of PP, PP-g-MA, and CNTs was first produced, and then an appropriate quantity of this masterbatch was compounded with PP and EOC. In this method, the mixing times for producing the masterbatch and the final nanocomposite were 12 and 8 min, respectively. From the shift in the tan δ peak and the reduction of the damping area in dynamic mechanical thermal analysis (DMTA) studies, it was revealed that the nanocomposites prepared by the masterbatch method had most of the CNTs localized in the thermodynamically less preferred PP phase, although some could be located at the interface. In contrast, the nanocomposites prepared by the direct and sequential methods had the CNTs localized in the EOC phase and at the interface. Notched Izod impact strength values of the nanocomposites prepared by the sequential method were significantly higher than those of the nanocomposites prepared by the masterbatch method, which had the same composition. The nanocomposite S-PP8/EC2/N2, with 2phr CNT content and an 80:20 weight ratio of PP:EOC, prepared by the sequential method, exhibited a 56% higher impact strength than that of the unfilled blend and 17 times the impact strength of neat PP. In contrast, the same composition prepared by the masterbatch method (MB-PP8/EC2/N2) had an impact strength lower than that of the unfilled blend, though higher than that of neat PP. This was ascribed to the selective localization of the CNTs in the EOC particles and at the interface, which resulted in intensified interaction between the blend components and facilitated stress transfer and the superposition of stress fields. Such selective localization of the CNTs also led to the nanocomposites prepared by the sequential method exhibiting significantly higher electrical conductivity than those prepared by the direct method at the same filler loading (Fig. 6.3). The CNTs localized in the EOC particles and at the interface facilitated the formation of electrically conductive pathways due to the bridging of CNTs and the creation of an interconnected network of CNT particles. Such percolating networks were not formed by the poorly dispersed CNTs in the PP matrix of nanocomposites prepared by the masterbatch method.

The enhancement of the mechanical properties of blends due to selectively dispersed CNTs has also been reported by other research groups [42, 47]. By using selectively localized reactive CNTs as a compatibilizer in poly(vinylidene fluoride) (PVDF)/poly(L-lactide) (PLLA), Zhao et al. [42] achieved a combination of excellent mechanical properties along with thermal and electrical conductivities. The researchers employed CNTs with reactive epoxide groups and long poly(methyl methacrylate) (PMMA) tails. These reactive CNTs bridged the two phases of the blend through physical entanglements and chemical reactions, strengthening the interface, as schematically presented in Fig. 6.4A. The localization of the CNTs at the interface was confirmed by TEM studies (Fig. 6.4B). The blend nanocomposites were found to possess superior ductility, tensile strength, storage modulus, and electrical conductivity (Fig. 6.4C).

The effect of CNT migration and localization on the electrical properties of CNT-reinforced polymer blends was also reported by Salehiyan et al. [27]. The researchers conducted studies on CNT-reinforced PLA/PVDF (70:30 wt/wt ratio) blends at

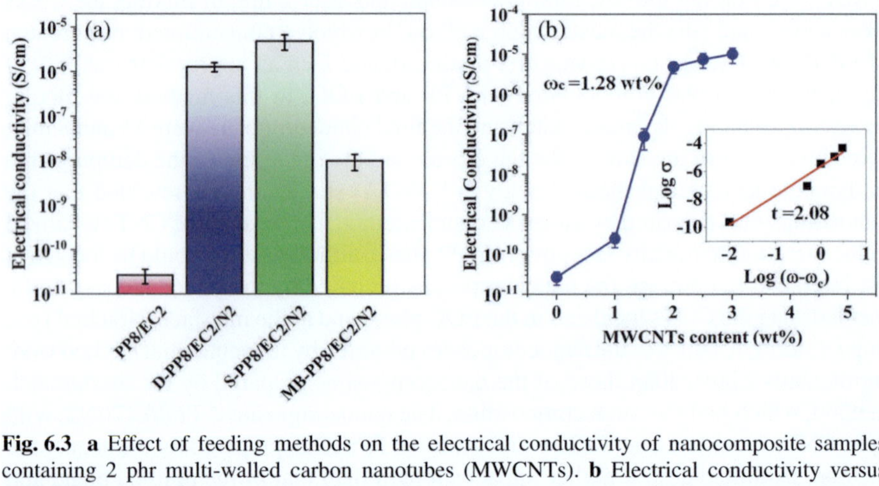

Fig. 6.3 **a** Effect of feeding methods on the electrical conductivity of nanocomposite samples containing 2 phr multi-walled carbon nanotubes (MWCNTs). **b** Electrical conductivity versus MWCNTs content for nanocomposite samples prepared by the sequential feeding method. D ⇒ direct method, S ⇒ sequential method MB ⇒ masterbatch method. Reproduced and modified with permission from [32]. Copyright 2022, Wiley Periodicals LLC

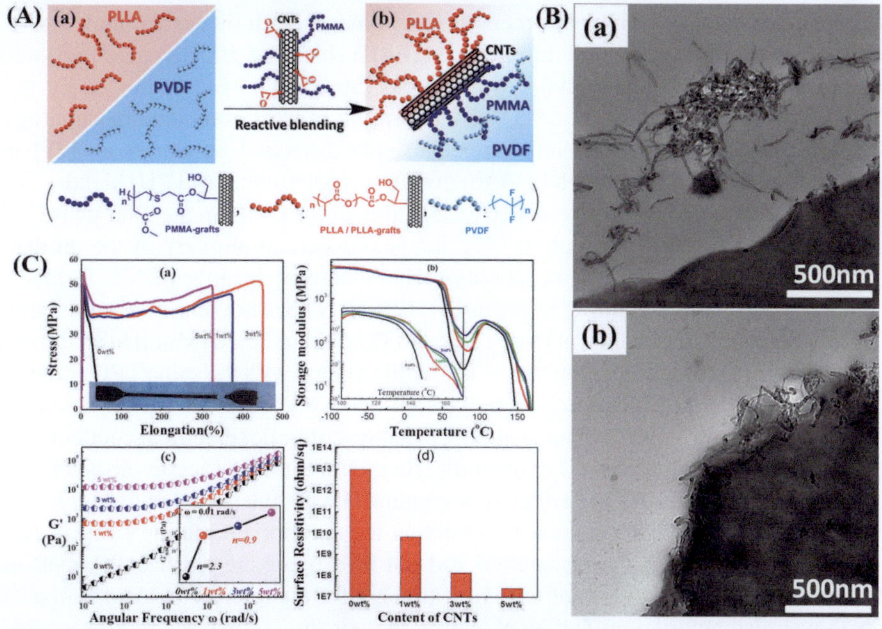

Fig. 6.4 **A** Schematic diagram of the reactive CNTs with long PMMA tails for the compatibilizer of immiscible PLLA/PVDF blends **a** blends without CNTs with weak interface; **b** CNTs compatibilized blends with strengthened interface. **B** TEM images of **a** pristine CNT incorporated PLLA/PVDF blends and **b** reactive CNTs with PMMA tails compatibilized PLLA/PVDF blends with 1 wt% CNTs (dark part represents PVDF and grey part represent PLLA). **C** **a** Strain–stress curves of the blends with and without CNTs; **b** DMA curves of the blends with and without CNTs; **c** Rheology curves of the blends with and without CNTs; **d** electrical conductivity of the blends with and without CNTs. Reproduced with permission from [42]. Copyright 2018, American Chemical Society

loading levels of 0.25, 0.75, and 1 wt%, which were prepared by melt processing in an extruder. Comparative studies were performed on nanocomposites created using two methods: (i) a two-step method, where the CNTs were pre-mixed with PVDF in the first step, followed by compounding the pre-mixture with PLA to facilitate the migration of the filler particles toward the interface, and (ii) a single-step extrusion method where all components were blended simultaneously. In melt-state rheological studies, the appearance of a plateau in the storage modulus (G′) curve in the low-frequency region indicated the formation of a network. A stronger network formation was noted in the nanocomposites prepared using the two-step method, attributed to the localization of CNTs at the interface. This also resulted in higher values of the dielectric constant. Furthermore, when the nanocomposites were annealed in the linear viscoelastic region, their G′ and dielectric properties increased significantly. This was attributed to the coarsening of the morphology, leading to greater CNT-CNT contacts at the interface. The researchers thus demonstrated that interfacial localization of CNTs enhances the formation of a network structure, which improves the dielectric constant of CNT-reinforced polymer blends. Such interfacial localization also improves electromagnetic interference (EMI) shielding effectiveness (SE) and electrical conductivity, as shown in other works [36, 39].

The combined effect of the selective localization of CNTs on blend morphology and the properties of blend nanocomposites was discussed by Shi et al. [47] in their studies on CNT-reinforced PLA/thermoplastic polyurethane (TPU) blends. The researchers observed that the selective localization of the CNTs in the TPU phase led to a morphological transformation from the island-sea morphology of the unfilled blend to a quasi-cocontinuous morphology. The selectively localized CNTs formed a percolating structure, resulting in super-toughening. The nanocomposite containing 2 wt% selectively dispersed CNTs and a PLA:TPU ratio of 70:30 exhibited an impact strength eight times greater than that of the unfilled blend with the same TPU content.

Different strategies have been employed to selectively localize CNTs at the interface of immiscible polymer blends, aiming to achieve superior electrical conductivity at a low percolation threshold. For example, Baudouin et al. [48] accomplished interfacial localization of CNTs (2 wt%) in polyamide (PA)/ethylene-acrylate copolymer blends with a 10 wt% PA content by selecting the appropriate mixing strategy and the PA used. However, the researchers did not investigate the electrical properties of the compositions. The use of a mixing strategy for the interfacial localization of CNTs was also reported by Zonder et al. [49] in their studies on CNT-reinforced PA/ polyethylene (PE) blends. The researchers achieved the lowest electrical resistivity (10^8 Ω cm) at a loading of only 0.75 wt% CNTs in a blend containing 75% PA, when the CNTs were pre-mixed in PE and subsequently compounded with PA. This was attributed to the migration of the CNTs from the PE phase to the interface, resulting in the formation of a percolating network. The composition exhibiting the lowest electrical resistivity also showed an increased low-frequency melt G', indicating a correlation between the two properties, as reported by Chen et al. [45], whose work has been discussed later in this section.

Another strategy for interfacial localization of CNTs is adjusting the mixing time. For example, Zhou et al. [41], in their studies on CNT-reinforced PMMA/PS blends, controlled the migration of CNTs by regulating the processing time of melt processing a PMMA-CNTs masterbatch with PS. Maximum interfacial localization of the CNTs was reached at a processing time of 2.5 min. The greatest enhancement in electrical conductivity and dielectric constant occurred with interfacial localization of CNTs for (PMMA-CNTs)/PS (50/50) with 1.06% CNT content, attributed to the interface promoting the double percolation effect. The fact that the percolation threshold was not low enough was linked to the migration of some CNTs to the thermodynamically favored PS phase, in addition to the attrition of the CNTs during processing, which resulted in a reduction of their aspect ratio. The successful creation of an interfacially localized percolating network of CNTs in a co-continuous blend by adjusting the processing time was also reported by Huang et al. [50], in their studies on CNT-reinforced PA/ poly(-caprolactone) (PCL) blends. The researchers noted a percolation threshold of 0.025 wt%.

A different strategy for the interfacial localization of CNTs was employed by Chen et al. [45] in their studies on CNT-reinforced PMMA/PS blends. The researchers used carboxyl group-functionalized CNTs. The interfacial localization of CNTs was attributed to the balance of π–π interactions between the PS and the CNT surfaces and the dipole–dipole interactions between PMMA and carboxyl groups on the CNT surfaces. An ultralow percolation threshold of 0.73 wt% CNT was achieved in the blend with 60 wt% PMMA content, which was attributed to the localization of CNTs at the interface of a co-continuous structure. Furthermore, the percolation threshold in electrical conductivity matched the percolation threshold in rheology.

Researchers have also employed a third blend component to achieve interfacial localization of CNTs. For example, Cohen et al. [51] achieved interfacial localization of CNTs in a co-continuous PA/PP blend by introducing a minor third blend component, pyridine-modified poly(ethyleneco-methacrylic acid) (designated as mEMMA1). The mEMMA1 formed strong π–π interactions with the CNTs, resulting in the localization of a percolated CNT network at the PA/PP interface, as evident in TEM studies. The ternary blend nanocomposites exhibited lower electrical resistivity than the co-continuous binary. The tuning of the localization of CNTs in co-continuous blends through the addition of a third component and its subsequent effect on electrical properties has also been reported by Gültner et al. [52].

A unique strategy to localize CNTs at the interface of co-continuous blends was employed by Kar et al. [36] in their studies on PVDF/ABS/CNT blend nanocomposites. The researchers wrapped the CNTs with PMMA, a polymer that is miscible with both blend components. The PMMA-wrapped CNTs were melt-processed with PVDF and ABS, taken in a 1:1 weight ratio. A customized interface with enhanced stress transfer was obtained in the blends containing PMMA-wrapped CNTs. The composite with 3 wt% PMMA-wrapped CNTs exhibited a 108% increase in tensile strength and a 48% increase in Young's modulus compared to the neat blends. This was significantly higher than the improvement (29% increase in tensile strength and 27% increase in Young's modulus) achieved in nanocomposite blends of the same composition but fabricated with pristine CNTs. The researchers attributed this improvement to the localization of the PMMA-wrapped CNTs at the interface, which, in the absence of the PMMA wrapping, were selectively localized in the thermodynamically preferred PVDF phase. Such interfacial localization of CNTs also led to an enhancement in electrical conductivity and EMI SE. The increase in electrical conductivity over nanocomposites with the same loading of pristine CNTs was attributed to a more interconnected network of CNTs at the continuous interface. The nanocomposite blends with pristine CNTs shielded EM waves primarily by reflection, while those containing PMMA-wrapped CNTs (3 wt%) shielded EM waves mainly by absorption. The researchers thus achieved a simultaneous improvement in mechanical properties and EMI SE by interfacially localizing CNTs in co-continuous blends.

Unlike most studies, which achieved the highest electrical conductivity by selectively localizing CNTs at the interface of co-continuous blends, there are reports of enhancing electrical conductivity by selectively localizing CNTs in one phase of blends with different morphologies. For example, Zhao et al. [53] obtained the highest electrical conductivity and dielectric constant in CNT-reinforced 50/50 PVDF/PS blends when the CNTs were selectively localized in the PVDF phase. This selective localization was achieved by melt processing a PVDF/CNT masterbatch with PS. The high electrical conductivity and dielectric constant were attributed to the formation of a conductive path through the PVDF phase.

Similarly, Zhang et al. [38] employed morphological regulation to create conductive networks in PLLA/PCL blends with droplet/matrix morphology. Stereocomplex crystallites (SCs) formed in the PLLA phase. The SCs produced during melt processing enhanced the viscosity and elasticity of the PLLA phase, causing the PLLA domains to shrink. Consequently, the previously dispersed PCL phase became continuous, as schematically illustrated in Fig. 6.5. This morphological regulation led to the localization of CNTs in the continuous PCL phase, resulting in a significant increase in the electrical conductivity and the EMI SE of the nanocomposites. The electrical conductivity of the PLLA/PCL/Multiwall CNT (MWCNT or MWNT) (PLLA/PCL 60/40) nanocomposites with 0.08 wt% CNT content increased from 2.1×10^{-12} Sm^{-1} without PDLA to 1.1×10^{-5} Sm^{-1} with 20 wt% PDLA in the PLLA phase. The percolation threshold of the PLLA/PCL/CNT (PLLA/PCL 60/40) nanocomposites was also reduced from 0.13 vol% to 0.017 vol% due to the presence of 20 wt% PDLA in the PLLA phase. The EMI SE of the (48PLLA + 12PDLA)/PCL/0.8CNT composite was nearly double that of the PLLA/PCL/0.8CNT composite. The researchers attributed this improvement in electrical properties and EMI SE to the simultaneous organization of conductive paths facilitated by the development of the continuous PCL phase.

Other strategies have been used to selectively localize CNTs in one of the blend components. For example, Li and Shimizu [54] fabricated a CNT-reinforced PVDF/PA6 blend featuring a unique microstructure with a co-continuous morphology, where the CNTs are selectively dispersed in the PA6 phase and nanometer-sized PA6 domains contain CNTs dispersed in the PVDF phase. This distinctive morphology was achieved with the aid of high shear. Details regarding the processing and morphological characteristics are discussed in Sect. 2.2.3.4 of Chap. 2. The researchers found that the formation of PA6 nanodomains in the PVDF phase enhanced the electrical conductivity and increased the ductility of the fabricated blend nanocomposites.

In yet another report, Gao et al. [55] fabricated anisotropic CNT-reinforced polycarbonate (PC)/PE blends with a 1:3 PC:PE ratio using an extrusion-hot stretching–quenching process. PC, the dispersed phase, formed fibrils, and the CNTs predominantly localized on the surface of the dispersed PC fibrils. Anisotropy in electrical properties was observed and attributed to the alignment of the fibrils in the stretching direction. Moreover, the resistivity in the perpendicular direction increased more rapidly with the stretch ratio.

Therefore, it can be summarized that the key to achieving CNT-reinforced immiscible polymer blends with superior electrical conductivity and mechanical properties is to establish an interconnected network of CNTs. A percolating structure at very low CNT loading is typically observed in co-continuous blends that have selectively localized CNTs at the interface. This selective interfacial localization of CNTs can be achieved through various strategies, such as adjusting the mixing protocol, mixing time, and functionalization of the CNTs. Electrical percolation typically corresponds

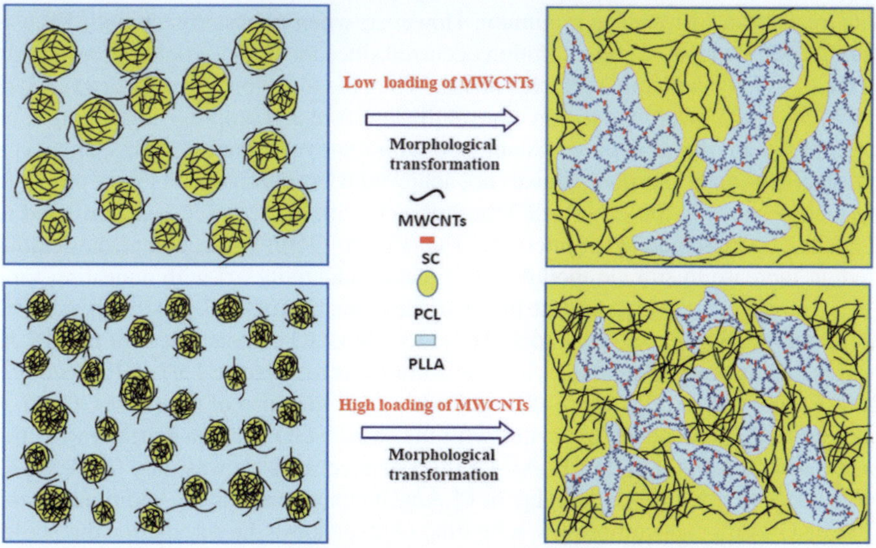

Fig. 6.5 Schematic illustrations of the morphological transformation and the MWCNT redistribution in PLLA/PCL/MWCNT (PLLA/PCL 60/40) nanocomposites with low and high loadings of MWCNTs with 20 wt% PDLA in the PLLA phase. Reproduced with permission from [38]. Copyright 2017, Royal Society of Chemistry

with rheological percolation. Electrical anisotropy can be achieved by fabricating CNT nanocomposite blends using the extrusion-hot stretching–quenching process, ensuring that the CNTs are preferentially positioned on the surface of the dispersed phase fibrils, which are aligned in the direction of stretching.

6.1.3 Graphene and Graphene Oxide

The effect of the migration and selective localization of graphene/graphene oxide (GO) on the properties of CNT-reinforced polymer blends has been studied by several research groups [30, 33, 56–62].

For example, Bai et al. [57], in their studies on reduced-GO (r-GO) reinforced PLA/PS blends with a 48:52 volume ratio of PLA and PS, observed that r-GO was trapped at the interface as multilayers during melt processing and annealing. However, this trapping was achieved only through sequential compounding, where a masterbatch of PLA and r-GO was first prepared and then processed with PS. When a masterbatch of PS and r-GO was melt compounded with PLA, no migration of r-GO to the interface or PLA was observed. The researchers attributed this observation to the fact that when r-GO was premixed with PLA and the masterbatch was subsequently compounded with PS, the filler particles began migrating towards the thermodynamically more favorable PS but were trapped at the interface, where they reached a free energy minimum. However, when a masterbatch with PS was compounded with PLA, no migration occurred since the filler particles were already in the thermodynamically preferred phase. When the filler content exceeded 0. 0.028 vol%, the r-GO was found to percolate at the interface and suppress coarsening. The co-continuous morphology was stabilized with relatively small domains. However, the stabilization of morphology was not achieved when compositions were prepared by melt processing the PS + r-GO masterbatch with PLA, which was attributed to the lack of percolation at the interface. The (PLA/r- GO)/PS nanocomposites exhibited an increase in storage modulus (G′) and conductivity (σ) with annealing time [Fig. 6.6A (a), (b)], which was attributed to the formation of the 3-D network of r-GO at the interface [Fig. 6.6A (c), (d)]. At 1.12 vol% r-GO content, the σ of the (PLA/r-GO)/PS nanocomposite after 1 h of annealing was found to be 1.97×10^{-5} S/cm at 180 °C, which was two orders of magnitude higher than that of the (PS/r-GO)/PLA and the PLA/r-GO nanocomposites with the same r-GO content. Furthermore, the percolation threshold of the (PLA/r-GO)/PS nanocomposites was only 0.028 vol%, which was 50 times lower than that of PLA/r-GO nanocomposites. Additionally, the (PLA/r-GO)/PS nanocomposites, with filler content above the percolation threshold, exhibited gel-like behavior in frequency sweep rheological studies, unlike the (PS/r-GO)/PLA nanocomposites, which exhibited liquid- like behavior (Fig. 6.6B).

Similarly, Kultravut et al. [60] achieved a combination of enhanced thermal and electrical conductivity through the interfacial localization of nanoparticles in polyglycidyl methacrylate (PGMA) grafted r-GO reinforced PLA/poly(trimethylene terephthalate) (PTT) blends. This interfacial localization resulted from the reaction between the epoxy group of the GMA in the functionalized r-GO and the end groups of the blend components.

Reports indicate that superior electrical conductivity can be achieved by selectively localizing GO in one of the blend components. For example, Botlhoko et al. [59] noted that when thermally exfoliated and partially reduced GO (TERGO) was used to prepare nanocomposites of PLA/PCL blends with a 60:40 weight ratio of PLA to PCL, the filler particles localized in the dispersed PCL phase, resulting in significant improvements in morphology, ductility, and thermal conductivity. Nanocomposites were prepared by melt mixing all components simultaneously in a rheomix at a rotor speed of 60 rpm and a temperature of 195 °C for 8 min. Both GO and TERGO were utilized for comparative studies. Scanning electron microscopy (SEM) studies of cryofractured samples [Fig. 6.7A (a)–(c)] revealed a higher quality of dispersion (smaller droplet size and more uniform size distribution) of PCL droplets in TERGO-filled nanocomposites compared to GO-filled nanocomposites. TEM studies [Fig. 6.7A (d), (e)] showed that the TERGO-filled nanocomposites contained TERGO particles that were localized and homogeneously dispersed within the PCL phase, while the GO-filled nanocomposites displayed most of the GO particles randomly dispersed in the PLA phase. The enhanced quality of dispersion of PCL droplets was attributed to the selective localization of TERGO particles within the PCL phase, leading to an increased viscosity ratio and an improvement in blend morphology.

The mechanical properties (Fig. 6.7B) and thermal conductivity (Fig. 6.7C) of TERGO-filled nanocomposites were significantly superior to those of the GO-filled

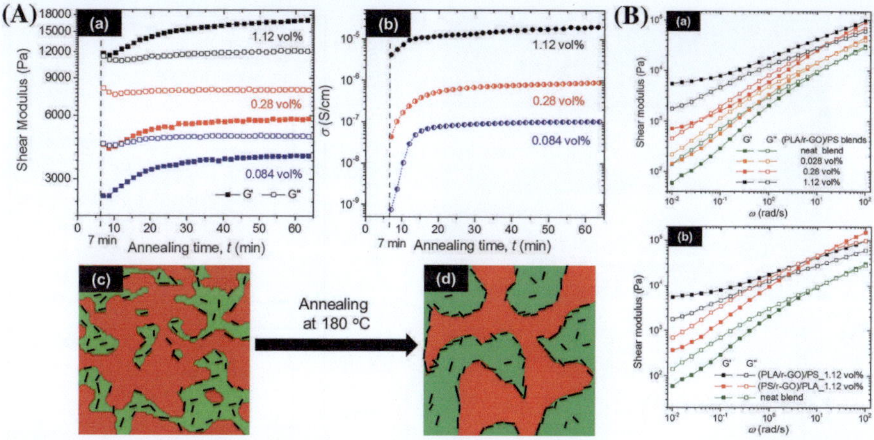

Fig. 6.6 **A** (**a**) Shear modulus as a function of annealing time at 180 °C for (PLA/r-GO)/PS blends with different filler contents. The solid and open squares represent the storage modulus, G′, and loss modulus, G″, respectively, which were measured at 1 rad/s and 1% strain. (**b**) The real part of conductivity at $\omega_{AC} = 20$ Hz as a function of annealing time at 180 °C for (PLA/r-GO)/PS blends with different filler contents. Both the rheology and conductivity time sweeps were taken simultaneously. Cartoons of (PLA/r-GO)/PS blends showing the morphology change and interfacial localization of r-GO during annealing: (**c**) ~7 min annealing; (**d**) after ~30 min annealing where the morphology has been stabilized. The darker (red) region is PS, while the lighter (green) region is PLA. The short black lines represent r-GO sheets. **B** Shear modulus, G′ and G″, as a function of frequency after 1 h annealing at 180 °C of: (**a**) (PLA/r-GO)/PS blends with different filler contents; (**b**) a neat blend, a (PLA/r-GO)/PS blend and a (PS/r-GO)/PLA blend at the same filler content of 1.12 vol%. All data are measured at 1% strain and 180 °C. Reproduced with permission from [57]. Copyright 2017, American Institute of Physics

composites. The enhancement in mechanical properties was attributed to their superior morphology. The improved thermal conductivity was linked to the predominance of TERGO particles in the PCL phase, resulting in increased particle–particle interactions that created more phonon scattering interfaces and facilitated short-distance phonon transport within the blend. In contrast, the GO-filled nanocomposites had greater distances between GO particles, which led to lower particle–particle interactions and insufficient phonon scattering interfaces at the boundaries, producing decreased thermal conductivity. On the other hand, the electrical conductivity of the TERGO-filled nanocomposites was lower than that of the GO-filled counterparts (Fig. 6.7C). This was due to the localization of the TERGO particles within the PCL droplets, which hindered the formation of a 3D network within the PLA matrix necessary for electrical transport. The researchers clearly demonstrated the impact of the localization of graphene particles on the mechanical, thermal, and electrical conductivity of graphene-reinforced immiscible polymer blends. A significant improvement in mechanical properties and thermal conductivity, with little change in electrical conductivity, due to the localization of TERGO in the minor phase of immiscible blends, has been reported by the same research group in a separate

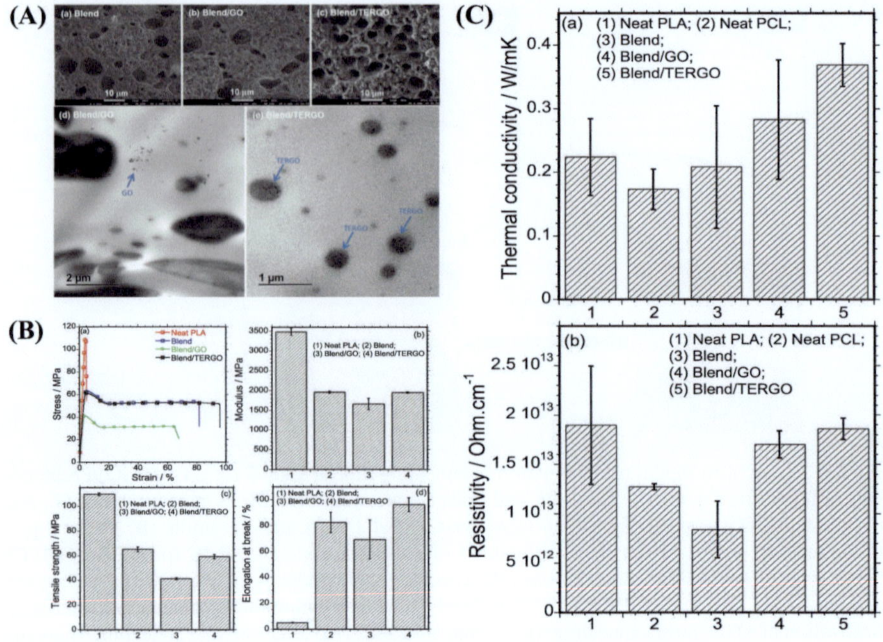

Fig. 6.7 **A** The freeze-fractured PCL-etched surface morphology of the (**a**) neat blend, (**b**) GO-filled blend composite, and (**c**) TERGO-filled blend composite. TEM images of (**d**) GO filled blend composite and (**e**) TERGO-filled blend composite. The arrows show the position of GO and TERGO particles in blend matrix. **B** (**a**) Stress–strain curves, (**b**) tensile modulus, (**c**) tensile strength, and (**d**) elongation-at-break of neat polymers, blend, and GO- and TERGO-filled blend composites. **C** (**a**) The thermal conductivity and (**b**) the electrical resistivity behaviours of neat PLA, PCL, blend, GO- or TERGO-filled blend composites. The filler loading was fixed at 0.05 wt%. Reproduced from [59]. Copyright 2017, the Authors (Open Source Publication)

study [62]. However, when the researchers substituted TERGO with functionalized TERGO, they observed improvements in both thermal and electrical conductivity at relatively high loadings of the filler; the latter being attributed to increased particle–particle interactions and the partially elongated domains of the dispersed phase [56].

Thus, it may be summarised that, as in the case of CNT-reinforced immiscible polymer blends, the key to achieving superior electrical conductivity in graphene/GO-filled immiscible polymer blends is to achieve a percolating network of filler particles. Selective localisation of graphene/GO nanoparticles within the minor phase of the blend may lead to improved blend morphology, resulting in superior mechanical properties. This may also lead to improved thermal conductivity. But, a combination of thermal and electrical conductivity, along with stabilisation again coarsening, is usually achieved by the selective interfacial localisation of nanoparticles in a co-continuous structure.

6.1.4 Carbon Black

The effect of the migration and selective localisation of carbon black on the properties of carbon black (CB)-reinforced polymer blends has been studied by several research groups [63–70]. For example, Yang et al. [64] investigated the impact of blend component proportions and mixing sequence on the conductivity and mechanical properties of 2 vol% CB reinforced PP/ styrene–butadiene–styrene (SBS) blends. Composites with PP:SBS volume ratios of 30:70, 50:50 and 70:30 were prepared by melt mixing. Details of the mixing sequences employed have been discussed in Sect. 2.2.3.1.3 of Chap. 2. Premixing PP and CB, followed by blending the mixture with SBS, resulted in higher electrical conductivities (Fig. 6.8A), with P50C2S50 and P70C2S30 (compositions prepared using PP/CB premixture and having PP:SBS volume ratios 50:50 and 70:30, respectively) exhibiting volume resistivities of 1.46 $\times$ 10^4 Ωm and 2.15 $\times$ 10^4 Ωm, respectively. This was ascribed to the fact that premixing PP with CB initially localised the CB particles in the PP phase, which resulted in the full utilisation of a low percolation threshold.

Furthermore, the high PP content ensured the desired phase continuity, enabling the construction of a complete conductive network, as confirmed from rheological behaviour (Fig. 6.8B). Tensile tests (Fig. 6.8C) revealed that compositions having a higher SBS content possessed higher values of tensile strength and elongation at break. This was ascribed to the fact that the CB-filled PP particles fortified the SBS phase. Thus, P50C2S50 was found to be superior to P70C2S30 in applications requiring medium deformation.

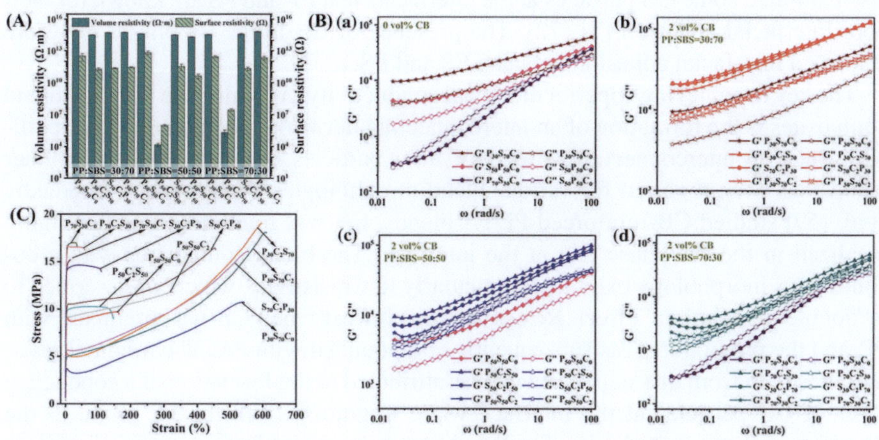

Fig. 6.8 **A** The conductivity performance comparison of composites. **B** The storage modulus (G′) and loss modulus (G″) variation with angular frequency for a the unfilled composites; b the composites with PP: SBS = 30:70; c the composites with PP: SBS = 50:50; and d the composites with PP:SBS = 70:30. **C** The Stress–Strain curves of the studied composites. Reproduced with permission from [64]. Copyright 2023, Springer

In another study, conducted by Li et al. [63] on vulcanised CB-reinforced NR/epoxidized NR (ENR) co-continuous blends, the CB was found to be preferentially localised in the NR phase. The CB-filled blends were found to exhibit a lower electrical percolation threshold, in addition to higher electrical conductivity, a higher microwave absorption capacity and a broader absorption bandwidth than NR/CB and ENR/CB composites having the same CB loading. The superior electromagnetic properties were ascribed to the preferential localisation of the CB particles, resulting in increased conduction loss, stronger dielectric relaxation and greater interface scattering in the ternary composites.

A lower percolation threshold in CB-reinforced blends, compared to the individual CB-filled blend components, was also reported by Al-Saleh and Sundararaj [65]. Their studies on CB-reinforced (60/40) PP/PS blends indicated that CB particles were preferentially localized in the PS phase. At the same CB loading, the blend composite exhibited an EMI SE value between that of the PS composite (lower limit) and the PP composite (upper limit). In the PP/PS volume ratio ranging from 75/25 to 25/75, the electrical conductivity and EMI SE of the blend composites with 10 vol% CB content were independent of the PP/PS volume ratio. However, as the PP/PS volume ratio increased from 75/25 to 95/5, there was a steady rise in electrical conductivity. This increase was attributed to the small amount of PS being inadequate to accommodate the entire CB content. Consequently, CB particles also entered the PP phase, thereby enhancing the conductivity of the blend.

Furthermore, SEM micrographs revealed that the interfaces of the (50/50) and (75/25) PP/PS blend composites containing 10 vol% CB had an irregular shape, indicating good interfacial adhesion between the PP and PS phases. These blends also exhibited some CB particles at the interface. Neat PP and PS are known for their poor interfacial adhesion [71, 72]. The presence of CB in the PS phases therefore enhanced interfacial adhesion between PP and PS.

The key to achieving superior electrical conductivity in conductive polymer blend composites is the formation of an interconnecting network of conductive filler particles. Such an interconnected network of filler particles in CB-reinforced polymer blends has been reported for various blend morphologies. For example, Toudakov et al. [69] studied CB-reinforced PP/PA blends. CB was found to be preferentially localized in the PA phase and at the interface. The blend composites with a co-continuous morphology exhibited particularly low resistivity, which was ascribed to a "double percolation" effect. Resistivity was lowest when CB was premixed with PP, and the premixture was subsequently compounded with PA. The minimal resistivity resulted from this sequence and was attributed to the formation of a conductive layer of CB particles at the interface while migrating from the PP phase to the PA phase. CB-reinforced microfibrillated blends have also been reported to possess conducting pathways of CB particles. Reports exist on in situ microfibrillation of CB/poly(ethylene terephthalate) (PET) compound in a PE matrix, achieved through slit die extrusion and hot stretching [66–68, 70]. In this context, Xu et al. [70] conducted studies where CB was premixed with PET, followed by melting blending with PE and extrusion through a slit die. This process was then followed by hot stretching. SEM studies revealed that the CB/PET/PE composites with lower CB loadings (below the maximum packing fraction) had smooth CB/PET microfibrils, with all the CB particles localized within the microfibrils. At higher CB loadings, CB particles were observed on the surface of the microfibrils. An increase in CB content resulted in a greater number of these surface CB particles, which led to an increase in the surface roughness of the microfibrils. The composites with low CB content exhibited high electrical resistivity, attributed to the surface microstructure of the in-situ CB/PET microfibrils. As CB particles were localized within the microfibrils, a pure PET layer existed on the surface of the microfibrils. Thus, electrons were unable to jump from the outermost CB particles of one microfibril to the adjacent microfibril, even when these microfibrils were in contact. This resulted in low electrical conductivity. As the CB content increased and the CB particles appeared on the surface of the microfibrils, they could encounter CB particles on the surfaces of neighboring microfibrils. This led to the formation of a conducting pathway, causing a sharp drop in electrical resistivity and the occurrence of percolation. In a separate study by the same research group [66], the positive temperature coefficient of resistivity (PTCR) was found to attenuate sharply after an adequate time of isothermal treatment without oxygen above the melting temperature range of PE. The non-uniform surface microstructure and the large dimensions of the microfibrils were identified as the main factors influencing PTCR attenuation.

In yet another report, Dai et al. [67] aimed to produce in situ microfibrillar CB/PET/PE composites with selectively localized CB particles on the surface of the microfibrils, achieving a low percolation threshold. To accomplish this, the researchers pre-mixed CB particles with PE and then compounded the pre-mixture with PET. Slit die extrusion, hot stretching, and quenching of the ternary composition led to the formation of in situ PET microfibrils. The CB particles were largely found to have migrated from the PE phase to the surfaces of the PET microfibrils. This resulted in a very low percolation threshold and enhanced conductivity. The conductive network formed through the contact and overlap of the PET microfibrils coated with conductive CB particles. Furthermore, at high filler loadings, the CB particles remaining in the PE matrix also created conductive pathways.

Thus, it may be summarised that the selective localisation of CB particles in CB-reinforced immiscible polymer blends may have an impact on the electrical properties, EMI SE, and the mechanical properties of the composites. The key to achieving superior electrical conductivity is the generation of a percolating network of CB particles. The electrical properties of in situ generated CB-reinforced microfibrillar PET/PE composites are governed by the surface microstructure of the microfibrils. Selective localisation of CB particles on the surfaces of PET microfibrils leads to superior electrical conductivity and a low percolation threshold.

6.1.5 Hybrid Effect

Researchers [73–85] have fabricated hybrid nanocomposites to achieve a synergistic effect. For example, Pötschke et al. [74] developed a hybrid polymer blend system

PP + MMT/PC + CNT aimed at attaining the highest possible conductivity in PC/PP blends while using the least amount of CNT. To accomplish this, the CNTs needed to be localized in the PC phase, since earlier studies on CNT-reinforced PC/PE blends [86] revealed that the migration of CNTs from PC to PE decreased the composites' electrical conductivity. In their current studies, the blends were tailored to achieve co-continuity. Co-continuity was achieved when the filled PC content ranged from 40 to 80 vol%. MMT acted as a barrier to prevent the CNTs from migrating toward the PP phase by localizing at the interface between PC and PP. The blends prepared using this method exhibited significant conductivity even with CNT loading as low as 0.6 vol%. In a related study, Chen et al. [79] used 0.1 wt% organoclay to confine CNTs at the interface of (40/60) PC/PVDF blends, achieving a percolation threshold as low as 0.06 wt%.

In another study, Chen et al. [78] utilized GO particles to confine CNTs at the interface of a (60/40) PLLA/ethylene-co-vinyl acetate (EVA) blend. The goal was to achieve a substantial enhancement in the electrical conductivity of the blend at a very low percolation threshold. Three distinct mechanisms were proposed for confining CNTs at the interface: (i) Interfacially localized GO particles attracted CNTs from the PLLA phase to the interface through π–π interactions. (ii) A GO/CNT complex formed during the initial processing stages, with both GOs and CNTs in the complex migrating to the interface. (iii) GO particles at the interface served as a barrier to prevent the migration of CNTs from the interface to the EVA phase. The PLLA/EVA/CNT/GO blends demonstrated a significant reduction of about seven orders of magnitude in volume resistivity compared to those blends reinforced solely with CNTs or GO.

CNTs alone cannot provide very high electromagnetic interference (EMI) shielding effectiveness (SE). Therefore, they are often used alongside magnetic nanomaterials to absorb incident electromagnetic (EM) radiation [30]. For instance, Pawar et al. [80] conducted studies on polycarbonate/poly(styrene-acrylonitrile) (SAN) blends reinforced with CNTs grafted onto dopamine-anchored Fe_3O_4 nanoparticles. The hybrid nanofiller not only demonstrated better dispersion quality compared to CNTs but also provided superior EMI SE. The best results were achieved with the 60/40 blend because of the double percolation of the conductive filler particles and the enhanced dispersion quality of the hybrid nanofiller. The same research group examined the effects of various magnetic nanomaterials decorated on graphene sheets (rGO-Ni [81, 82], rGO-Fe_3O_4 [83]), used in conjunction with CNTs in polycarbonate/SAN blends. In all cases, the nanoparticles were preferentially localized in the polycarbonate phase, forming an interconnected network that improved electrical conductivity and EMI SE. Additionally, high dielectric and magnetic permeability losses contributed to the attenuation of incident EM waves through absorption. A loading of 10 wt% rGO-Fe_3O_4 combined with 3 wt% CNTs effectively achieved a total SE of 50.7 dB. The enhancement in SE via absorption was attributed to the high saturation magnetization of Fe_3O_4.

Similarly, Menon et al. [76] demonstrated the superior EMI SE of PVDF/TPU blends reinforced with FeNi alloy particles coated with MWNTs. The coating of FeNi particles with MWNTs was achieved through ball milling. The hybrid filler particles were then dispersed in 50/50 PVDF/TPU blends. MWNT-reinforced blends with various MWNT loadings were also prepared for comparative studies. SEM [Fig. 6.9A (a)] and EDS studies [Figs. 6.9A (b, c)] following selective etching of the TPU phase using pyridine revealed the selective localization of the hybrid filler particles in the TPU phase. This was confirmed in selective dissolution studies [Fig. 6.9A (d)], where the PVDF phase extracted in acetone was colorless. In contrast, the TPU phase extracted in pyridine was colored due to the presence of the filler particles in TPU. The PVDF/TPU/MWNT composites also exhibited selective localization of the MWNTs in the TPU phase.

The SE of the blends reinforced with MWNTs and those reinforced with hybrid fillers is shown in Fig. 6.9B. As expected, increasing the MWNT content in PVDF/TPU/MWNT composites led to a rise in total SE (SE_T) [Fig. 6.9B (a)] due to the formation of a well-connected CNT network. However, when the blends were reinforced with 30 wt% uncoated FeNi particles, SE_T was poor [Fig. 6.9B (b)]. The hybrid composites exhibited significantly superior SE_T values [Fig. 6.9B (b)]. The primary mechanism of EMI shielding in the hybrid composites was found to be absorption [Fig. 6.9B (c–e)]. Furthermore, SE_T depended on the FeNi/MWNT ratio. A decrease in the ratio resulted in an increase in SE_T, attributed to the establishment of a well-connected MWNT network that enables the attenuation of EM waves, as shown in Fig. 6.9C.

Another demonstration of the superior electrical conductivity and EMI shielding effectiveness of polymer blend hybrid nanocomposites is presented in the work of Kar et al. [84]. The researchers conducted studies on co-continuous PVDF/acrylonitrile–butadiene–styrene (ABS) (50/50, wt/wt) blends reinforced with CNTs, along with barium titanate (BT) nanoparticles that were chemically stitched onto GO sheets. To achieve high-quality dispersion and facilitate network formation, the CNTs were modified with amine-terminated ionic liquid (IL). The nanoparticles were selectively localized in the PVDF phase. The blend nanocomposites exhibited an increase in electrical conductivity of approximately six orders of magnitude higher than that of the neat blends. Furthermore, the EMI SE was about 10 dB higher than that of the blends reinforced solely with IL-modified CNTs and 3 dB higher than that achieved when IL-modified CNTs were combined with Fe_3O_4. Thus, the researchers demonstrated the synergistic effect of hybrid nanofillers and established the superiority of BT-GO over Fe_3O_4 in enhancing EMI SE.

Therefore, it can be summarized that the key to achieving high electrical conductivity in polymers is creating a percolating network of conductive nanoparticles. The formation of such a network in immiscible polymer blends can sometimes be facilitated by selectively localizing the nanofiller in one of the phases. This may, in turn, be achieved by using another nanofiller as a barrier. The use of a magnetic nanomaterial in combination with CNTs may help attain superior EMI SE along with enhanced electrical conductivity.

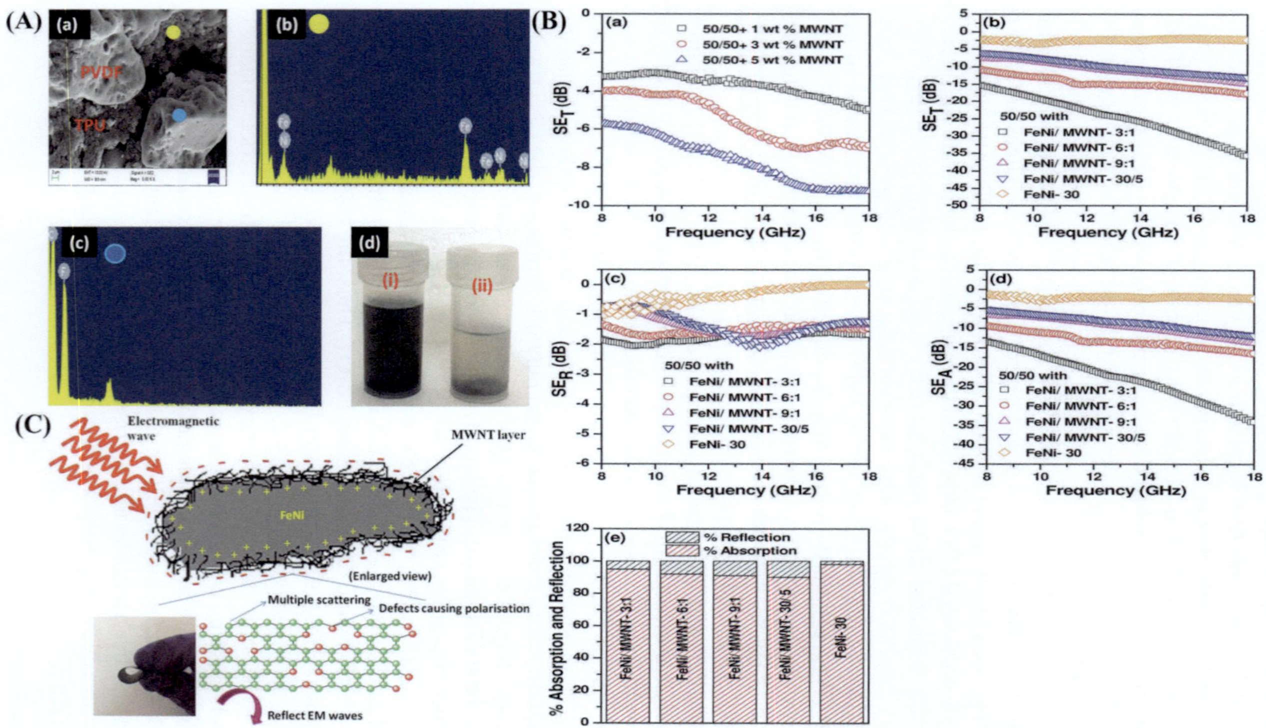

Fig. 6.9 **A** SEM image of (**a**) Blends etched with pyridine to remove TPU phase (**b**) EDS spectra showing presence of FeNi in etched area (**c**) EDS spectra showing presence of PVDF in unetched area (**d**) Selective dissolution in (i) pyridine (ii) acetone. **B** Total shielding effectiveness for (**a**) PVDF/TPU blends with MWNTs (**b**) Blends with FeNi/MWNTs ball milled at different ratios (**c**) Shielding effectiveness due to reflection for blends (**d**) Shielding effectiveness due to absorption for blends (h) Absorption and reflection components of SE_T at 18 GHz. (**C**) Various mechanisms for EM wave attenuation (inset showing flexible composites derived from PVDF/TPU blends with magnetic alloy-MWNT heterostructures). Reproduced with permission from [76]. Copyright 2017, Wiley Periodicals LLC

6.1.6 Other Fillers

Researchers have also conducted studies on the migration and preferential localisation of various other nanofillers, such as boehmite [87–90], silica [91–96], and calcium carbonate [97], in one of the phases or at the interface of immiscible polymer blends.

Li et al. [87] investigated the morphology and mechanical properties of boehmite nanorods reinforced PLA/poly(1,4-butylene succinate) (PBSU) (70/30 wt/wt) blends prepared by melt processing all components in a single step. Both pristine boehmite nanorods (p–BNRs) and epoxide group-modified boehmite nanorods (m-BNRs) were utilized for the studies. TEM analyses (Fig. 6.10) revealed that p–BNRs and m–BNRs were located at the interface, and the morphologies of the composites with the two types of nanofillers were very similar at the same filler loadings. However, their mechanical properties differed. The m–BNR reinforced blends exhibited superior elongation at break, notched impact strength, and yield strength compared to the unfilled blends, while the p–BNR reinforced blends displayed a notable deterioration in mechanical properties, despite the filler particles being localized at the interface. The researchers attributed this deterioration to the lack of interaction between the p–BNR particles and the blend components, leading to poor interfacial adhesion. This contrasted with the m–BNR reinforced blends, where the epoxide groups of the nanorods reacted with the end carboxylic acid groups of both PLA and PBSU, facilitating the chemical bonding of the interfacial polymer chains to the surface of the nanoparticles. The interfacial polymer chains that chemically bonded to the nanorods became entangled with the molecular chains of the two polymers, resulting in enhanced interfacial adhesion and improved mechanical properties. Thus, the researchers demonstrated that the interfacial localization of nanoparticles in immiscible polymer blends, without interfacial adhesion, does not enhance mechanical properties. The necessity of interaction between boehmite particles at the interface and the blend components for improved mechanical properties was also illustrated by Agwuncha et al. [90] in their studies on PLA/PCL/boehmite nanocomposites.

Similarly, Saiprasit and Schlarb [91] conducted studies on the morphology and mechanical properties of nanosilica-reinforced PLA/PBAT blends. The researchers investigated the effect of mixing sequence on the migration and localization of silica particles. Nanocomposites with composition S1 (79.2 vol% PLA, 19.8 vol% PBAT, and 1 vol% nanosilica) were prepared through masterbatch dilution in a TSE, utilizing four types of mixing sequences (P1, P2, P3, and P4), as presented in Fig. 6.11a. The nanocomposites prepared using P1, P2, P3, and P4 were designated as S1-P1, S1-P2, S1-P3, and S1-P4, respectively. For a comparative study on properties, a blend with composition S0 (80 vol% PLA and 20 vol%) was also prepared.

SEM studies and focused ion beam SEM studies on the nanocomposites revealed that the nanosilica was localized in the polymer phase in which it was originally incorporated in the masterbatch and at the interface. However, an increase in residence time, caused by a greater number of steps in compounding, led to the migration of

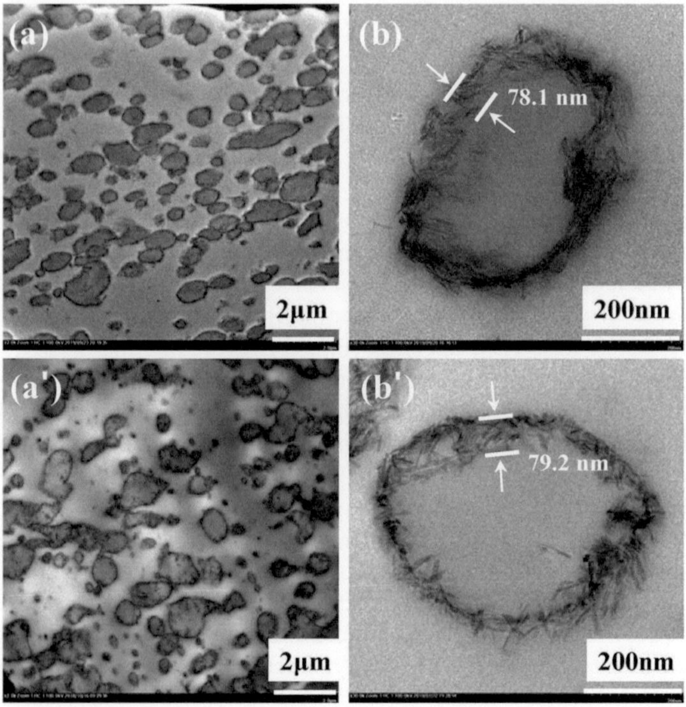

Fig. 6.10 TEM images for the PLLA/PBSU (70/30) blend sample with 3 wt% p-BNRs (**a**) (**b**) and 3 wt% m-BNRs (**a**′) (**b**′) with different magnifications. Reproduced with permission from [87]. Copyright 2020, Elsevier Science Ltd

more nanosilica particles to the interface, as presented schematically in Fig. 2.15 (Chap. 2).

The tensile properties of PLA, the PLA/PBAT blend, and the nanocomposites are presented in Fig. 6.11(b-e). Blending PLA with PBAT resulted in a significant enhancement of elongation at break, but a reduction in Young's modulus and tensile strength. The incorporation of the nanofiller improved Young's modulus and tensile strength, but decreased elongation at break. However, the overall mechanical properties of the blend nanocomposites were superior to those of both neat PLA and the blend. The best balance of tensile properties (Young's modulus of 2970 MPa, tensile strength of 48.7 MPa, and elongation at break of 21.6%) was exhibited by S1-P3, which contained the highest concentration of silica nanoparticles localized within the PBAT droplets. The researchers demonstrated the influence of mixing sequence on the localization and mechanical properties of silica nanoparticle reinforced polymer blends.

A recent study [96] on nanosilica reinforced PLLA/ethylene–acrylic ester-glycidyl methacrylate terpolymer (EGMA) blends demonstrates how controlling the localization of nanoparticles can tailor the thermal, mechanical, and rheological properties of these multiphase materials. The localization of silica nanoparticles was

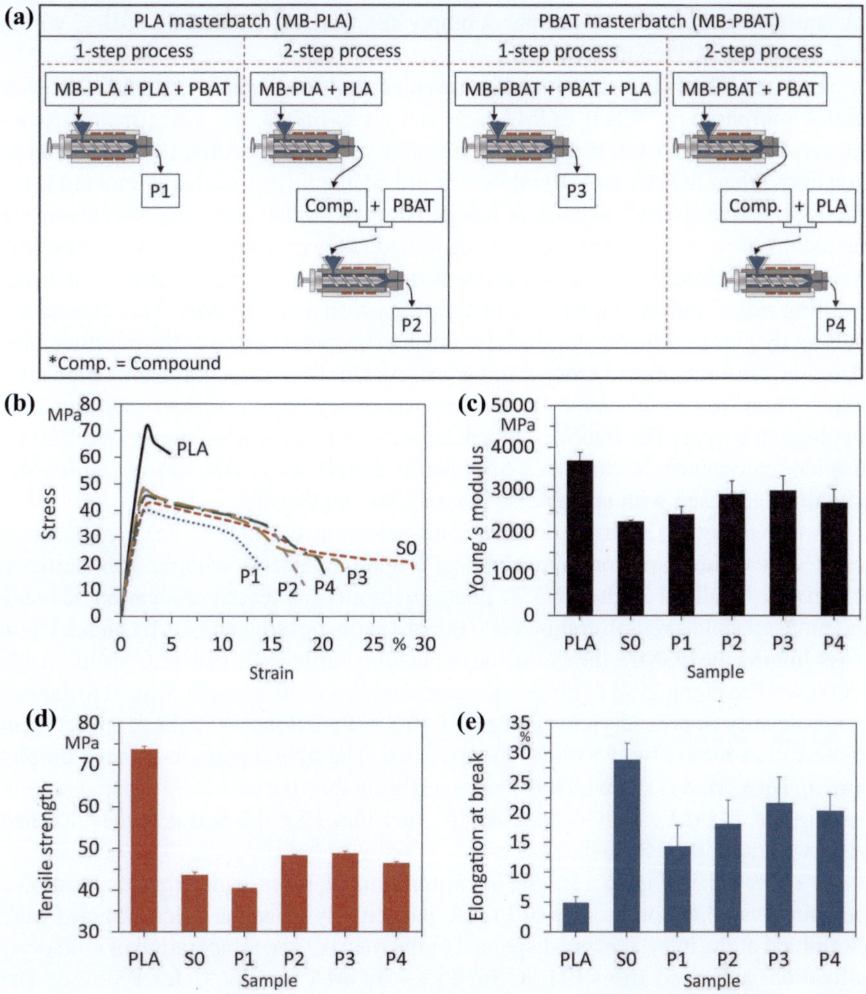

Fig. 6.11 **a** Blending procedures. **b** Average stress–strain curves, **c** Young's modulus, **d** tensile strength, and **e** elongation at break of neat PLA, PLA/PBAT-S0 (S0), and PLA/PBAT/SiO_2 nanocomposites with different blending procedures (P1, P2, P3, and P4). Reproduced with permission from [91]. Copyright 2021, De Gruyter/Polymer Processing Society (Open Source Publication though Polymer Processing Society)

achieved by cross-linking the EGMA phase. Studies were conducted on nanocomposites with a PLLA/EGMA weight ratio of 85/15 and silica content of 4.5 wt%, prepared in a torque rheometer at a temperature of 190 °C and a screw speed of 50 rpm. The EGMA was cross-linked by melt processing with various amounts of dicumyl peroxide (DCP) for 10 min. The silica was compounded with PLLA for 10 min, and then EGMA or cross-linked EGMA was processed with the premix for another 10 min. The blend nanocomposites were designated as PSE-x, with x indicating the DCP content in wt%.

SEM and TEM (Fig. 6.12) studies revealed that the blends exhibited a droplet-matrix morphology. When EGMA was not cross-linked, the silica nanoparticles migrated from the PLLA phase to the thermodynamically favored EGMA domains. However, when EGMA was slightly cross-linked, the filler particles assembled at the interface, forming a hybrid particle-bridged droplet domain network. This change in the localization of silica particles was attributed to the enhanced elasticity of EGMA due to cross-linking, which restricted the formation of the interfacial tension-induced "wetting ridge" during the wetting of the silica surface by EGMA. This slowed the wetting dynamics, causing the silica particles to become trapped at the interface. The increased elasticity due to cross-linking was evident from the significantly higher G' values compared to G" values in the low-frequency region of melt-state dynamic rheological curves. The interfacial localization of silica particles suppressed EGMA droplet coalescence, leading to a finer blend morphology. The size of the EGMA domains decreased with an increase in cross-linking density.

As seen in Fig. 6.13a, b, the difference in the localization of the silica nanoparticles resulted in a wide variation in mechanical properties. PSE-0, with the nanoparticles selectively localized in the EGMA phase, exhibited increased elongation at break and reduced yield strength compared to the PLLA/uncrosslinked EGMA blend. Upon cross-linking the EGMA, the exclusion of the filler particles from the EGMA domains decreased the elongation at break but increased the yield strength. PSE-0.8 showed a significantly improved yield strength of 46.8 MPa, attributed to the reinforcement of the PLLA matrix by the silica nanoparticles. The hybrid particle-bridged droplet domain network was more effective at toughening than the particle-inclusion droplet domains of EGMA, as evidenced by the fact that PSE-0.2 had a higher notched impact strength than PSE-0.

As observed in Fig. 6.13c, the incorporation of silica nanoparticles promoted the cold crystallization process of PLLA, particularly when the silica particles were assembled at the interface and dispersed in the matrix. The temperature of cold crystallization decreased from 121 °C for PLLA/EGMA to 105 °C for PSE-0.8. The reduced half-width of the crystallization peak of PLLA in PSE-x also indicated an increased crystallization rate of PLLA in the blend nanocomposite containing interfacially assembled nanoparticles. As shown in Fig. 6.13d, the glass transition temperature (T_g) of PLLA in the blend decreased with the addition of the nanoparticles. The T_g of PLLA in PSE-0 was 8.1 °C lower than that in PLLA/EGMA. However, the interfacial assembly of nanoparticles, along with their presence in PLLA, mitigated this decrease. Therefore, the researchers demonstrated through their extensive studies that the properties of polymer blend nanocomposites can be tailored by controlling the migration and localization of nanoparticles.

Another demonstration of the effect of selective localization of nanoparticles on the properties of nanosilica-reinforced polymer blends can be found in the work of Dil et al. [93]. The researchers studied the effect of interfacially localized spherical silica nanoparticles on the morphology, rheology, and mechanical properties of PLA/PBAT blends prepared by melt processing in an internal mixer. Interfacial

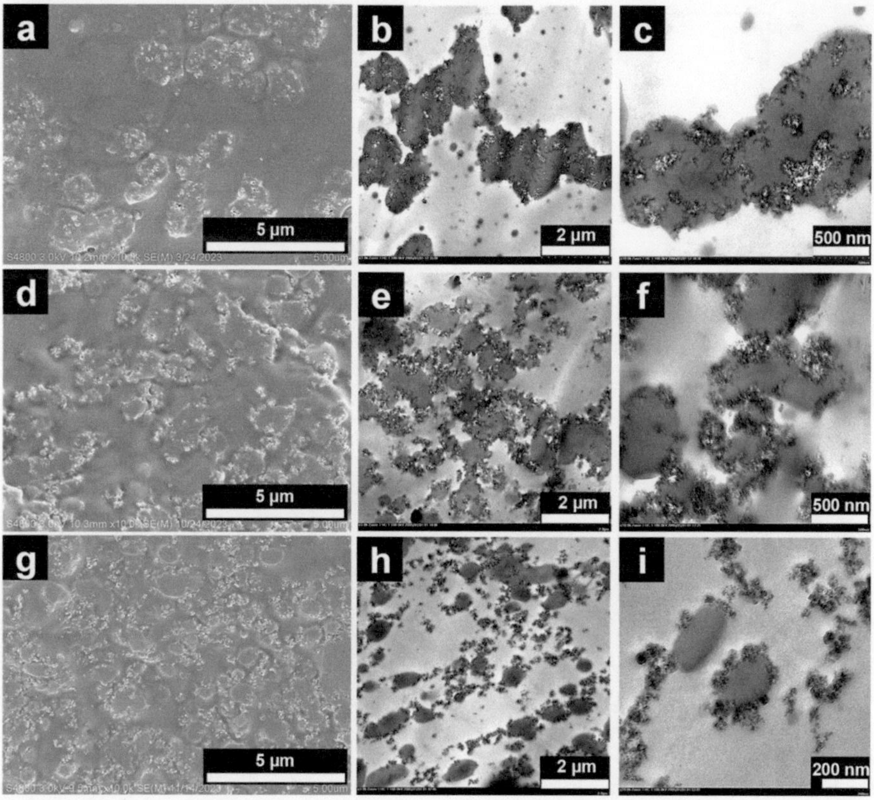

Fig. 6.12 SEM and TEM images of PSE-*x* (x denotes the DCP content in wt%) with 4.5 wt% SiO_2 nanoparticles: **a–c** PSE-0, **d–f** PSE-0.2, and **g–i** PSE-0.8. Note that the panels **a**, **d**, and **g** are from SEM, while the rest is from TEM. Reproduced with permission from [96]. Copyright 2024, American Chemical Society

localization of nanosilica in the blends, as evident in AFM studies, was achieved by premixing the filler with PLA, followed by compounding the premixture with PBAT. Rheological studies revealed that the interfacial assembly of nanosilica significantly reduced the relaxation of the dispersed PBAT phase. At 3 wt% silica loading, the blend with 30 vol% PBAT content exhibited a transition from a matrix-dispersed phase morphology to a highly co-continuous morphology. There was also a rheological transition from liquid-like to gel-like behavior. This change in the morphological and rheological characteristics was attributed to the interfacial localization of silica. The co-continuous blend with interfacially localized silica nanoparticles also demonstrated strong stability against coarsening during annealing without any deterioration in mechanical properties, unlike unfilled PLA/PBAT blends, which coarsened during annealing with a decline in mechanical properties. The researchers thus demonstrated the potential of interfacially localized silica to create a stable co-continuous morphology.

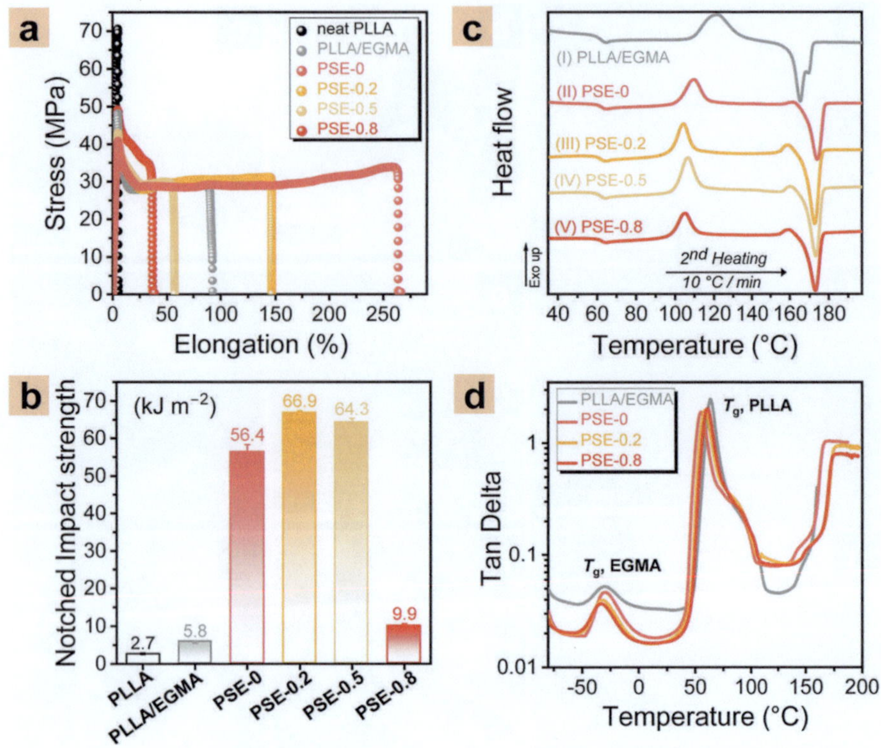

Fig. 6.13 **a** Stress–elongation curves, **b** notched impact strength, **c** second differential scanning calorimetry (DSC) heating curves (10 °C/min), and **d** dependence of tan δ on temperature of PLLA/EGMA and **PSE-***x* with varying cross-linking densities of EGMA. Reproduced with permission from [96]. Copyright 2024, American Chemical Society

A similar demonstration of the effect of selective localization of nanoparticles in polymer blend nanocomposites can be found in the work of Gong et al. [97]. The researchers employed nano-calcium carbonate for their studies. They investigated the effect of migration of nano-calcium carbonate from dispersed ethylene–propylene–diene monomer (EPDM) droplets to the PP matrix on the breakup dynamics of the EPDM droplets during melt extrusion. Premixed nanofiller with EPDM containing 0.5 wt%, 2 wt%, and 4 wt% filler content was melt processed with PP in a weight ratio of 80:20. Mixing occurred in a twin screw extruder. The microstructural evolution of the droplets was studied for the different compositions by quenching and removing the screw immediately after the melt was blended homogeneously. The breakup process of EPDM droplets was found to be significantly expedited by the migration of $CaCO_3$ nanoparticles from the dispersed droplets to the matrix. A reduction in both total breakup time and shape stability of the dispersed droplets was observed with an increase in calcium carbonate loading. Additionally, the maximum value of the equivalent diameter and aspect ratio of droplets decreased with increasing filler content. The researchers attributed the enhanced breakup of EPDM droplets in the filled blends primarily to the reduction of interfacial tension by calcium carbonate nanoparticles when the filler loading was less than 2 wt%. At loadings greater than 2 wt%, the filler formed aggregates in the EPDM phase, which acted as regions of stress concentration, facilitating the breakup of droplets. The researchers thus demonstrated that the migration of nanoparticles from the dispersed phase to the matrix during two-step melt processing of polymer blend nanocomposites also expedites morphology evolution and improves the quality of dispersion of the dispersed droplets.

6.2 Conclusions

Customizing the localization of nanoparticles enables the achievement of desired properties in blend composites. For example, to create conductive polymer composites with a low percolation threshold, it is crucial to establish an interconnected network of conductive filler particles. This can be effectively achieved by placing these particles at the interface. Additionally, filler particles positioned at the interface that exhibit strong interfacial adhesion can result in improved mechanical properties. The regulation of nanoparticle localization can be controlled through various thermodynamic and kinetic factors, as detailed in Chap. 2. A comprehensive understanding of the relationship between structure, property, and processing is vital for developing polymer blend nanocomposites with enhanced characteristics.

References

1. T. Gcwabaza, S. Sinha Ray, W.W. Focke, A. Maity, Morphology and properties of nanostructured materials based on polypropylene/poly(butylene succinate) blend and organoclay. Eur. Polym. J. **45**, 353–367 (2009). https://doi.org/10.1016/J.EURPOLYMJ.2008.11.007
2. T. Malwela, S.S. Ray, Study of morphology and crystal growth behaviour of nanoclay-containing biodegradable polymer blend thin films using atomic force microscopy. Polymer (Guildf) **53**, 2705–2716 (2012). https://doi.org/10.1016/J.POLYMER.2012.04.025
3. T.G. Mofokeng, S.S. Ray, V. Ojijo, Structure–property relationship in PP/LDPE blend composites: the role of nanoclay localization. J. Appl. Polym. Sci. **135**, 46193 (2018). https://doi.org/10.1002/APP.46193
4. S.S. Ray, J. Bandyopadhyay, M. Bousmina, Effect of organoclay on the morphology and properties of poly (propylene)/poly [(butylene succinate)-co-adipate] blends. Macromol. Mater. Eng. **292**, 729–747 (2007)
5. V. Ojijo, S.S. Ray, R. Sadiku, Concurrent enhancement of multiple properties in reactively processed nanocomposites of polylactide/poly[(butylene succinate)-co-adipate] blend and organoclay. Macromol. Mater. Eng. **299**, 596–608 (2014). https://doi.org/10.1002/MAME.201300306
6. S.S. Ray, M. Bousmina, Compatibilization efficiency of organoclay in an immiscible polycarbonate/poly (methyl methacrylate) blend. Macromol. Rapid Commun. **26**, 450–455 (2005)

7. T. Mhlabeni, S. Kesavan Pillai, S.S. Ray, Effect of organically modified layered double hydroxides on the properties of poly(lactic acid)/poly[(butylene succinate)-co-adipate] immiscible blends. J. Appl. Polym. Sci. **137**, 48654 (2020). https://doi.org/10.1002/APP.48654
8. H. Eslami, M.R. Kamal, Elongational rheology of biodegradable poly(lactic acid)/poly[(butylene succinate)-co-adipate] binary blends and poly(lactic acid)/poly[(butylene succinate)-co-adipate]/clay ternary nanocomposites. J. Appl. Polym. Sci. **127**, 2290–2306 (2013). https://doi.org/10.1002/APP.37928
9. N. Bitinis, R. Verdejo, E.M. Maya, E. Espuche, P. Cassagnau, M.A. Lopez-Manchado, Physicochemical properties of organoclay filled polylactic acid/natural rubber blend bionanocomposites. Compos. Sci. Technol. **72**, 305–313 (2012). https://doi.org/10.1016/J.COMPSCITECH.2011.11.018
10. R. Salehiyan, S.S. Ray, J. Bandyopadhyay, V. Ojijo, The distribution of nanoclay particles at the interface and their influence on the microstructure development and rheological properties of reactively processed biodegradable polylactide/poly(butylene succinate) blend nanocomposites. Polymers (Basel) **9**, 350 (2017). https://doi.org/10.3390/polym9080350
11. T. Wu, D. Yuan, F. Qiu, R. Chen, G. Zhang, J. Qu, Polypropylene/polystyrene/clay blends prepared by an innovative eccentric rotor extruder based on continuous elongational flow: Analysis of morphology, rheology property, and crystallization behavior. Polym. Test. **63**, 73–83 (2017). https://doi.org/10.1016/J.POLYMERTESTING.2017.07.012
12. M. Nofar, R. Salehiyan, U. Ciftci, A. Jalali, A. Durmuş, Ductility improvements of PLA-based binary and ternary blends with controlled morphology using PBAT, PBSA, and nanoclay. Compos. Part B Eng. **182**, 107661 (2020). https://doi.org/10.1016/J.COMPOSITESB.2019.107661
13. V. Ojijo, T. Malwela, S.S. Ray, R. Sadiku, Unique isothermal crystallization phenomenon in the ternary blends of biopolymers polactide and poly [(butylene succinate)-co-adipate] and nano-clay. Polymer (Guildf) **53**, 505–518 (2012)
14. X. Zhang, H. Zhang, M. Zuo, J. Zhong, X. Shi, S. Sun et al., Effects of selective distribution and migration of poly(methyl methacrylate)-grafted nanoclays on the phase behavior of poly(methyl methacrylate)/poly (styrene-co-acrylonitrile) blends. Polymer (Guildf) **252**, 124965 (2022). https://doi.org/10.1016/J.POLYMER.2022.124965
15. T.G. Mofokeng, S.S. Ray, V. Ojijo, Influence of selectively localised nanoclay particles on non-isothermal crystallisation and degradation behaviour of PP/LDPE blend composites. Polymers **10**, 245 (2018). https://doi.org/10.3390/POLYM10030245
16. V. Ojijo, H. Cele, R.S. Sinha, Morphology and properties of polymer composites based on biodegradable polylactide/poly[(butylene succinate)-co-adipate] blend and nanoclay. Macromol. Mater. Eng. **296**, 865–877 (2011). https://doi.org/10.1002/MAME.201100042
17. V. Ojijo, S.S. Ray, R. Sadiku, Effect of nanoclay loading on the thermal and mechanical properties of biodegradable polylactide/poly[(butylene succinate)-co-adipate] blend composites. ACS Appl. Mater. Interfaces **4**, 2395–2405 (2012). https://doi.org/10.1021/AM201850M/SUPPL_FILE/AM201850M_SI_001.PDF
18. S. Sinha Ray, S. Pouliot, M. Bousmina, L.A. Utracki, Role of organically modified layered silicate as an active interfacial modifier in immiscible polystyrene/polypropylene blends. Polymer (Guildf) **45**, 8403–8413 (2004). https://doi.org/10.1016/J.POLYMER.2004.10.009
19. R. Salehiyan, S.S. Ray, J. Bandyopadhyay, V. Ojijo, The distribution of nanoclay particles at the interface and their influence on the microstructure development and rheological properties of reactively processed biodegradable polylactide/poly(butylene succinate) blend nanocomposites. Polymers **9**, 350 (2017). https://doi.org/10.3390/POLYM9080350
20. S. Sinha Ray, M. Okamoto, Polymer/layered silicate nanocomposites: a review from preparation to processing. Prog. Polym. Sci. **28**, 1539–1641 (2003). https://doi.org/10.1016/j.progpolymsci.2003.08.002
21. E. Makhado, S.S. Ray, Mechanical, barrier and antimicrobial properties of biodegradable poly (ε-caprolactone) nanocomposites. Adv. Sci. Eng. Med. **7**, 351–360 (2015)
22. Y. Cui, S. Kumar, B.R. Kona, D. van Houcke, Gas barrier properties of polymer/clay nanocomposites. RSC Adv. **5**, 63669–63690 (2015)

23. R. Banerjee, S.S. Ray, An overview of the recent advances in polylactide-based sustainable nanocomposites. Polym. Eng. Sci. **61**, 617–649 (2021). https://doi.org/10.1002/PEN.25623
24. M.J. Mochane, J.S. Sefadi, T.S. Motsoeneng, T.E. Mokoena, T.G. Mofokeng, T.C. Mokhena, The effect of filler localization on the properties of biopolymer blends, recent advances: a review. Polym. Compos. **41**, 2958–2979 (2020). https://doi.org/10.1002/PC.25590
25. R. Salehiyan, S.S. Ray, Influence of nanoclay localization on structure-property relationships of polylactide-based biodegradable blend nanocomposites. Macromol. Mater. Eng. **303**, 1800134 (2018). https://doi.org/10.1002/MAME.201800134
26. O.C. Wokadala, S.S. Ray, J. Bandyopadhyay, J. Wesley-Smith, N.M. Emmambux, Morphology, thermal properties and crystallization kinetics of ternary blends of the polylactide and starch biopolymers and nanoclay: the role of nanoclay hydrophobicity. Polymer (Guildf) **71**, 82–92 (2015)
27. R. Salehiyan, M. Nofar, D. Makwakwa, S.S. Ray, Shear-induced carbon nanotube migration and morphological development in polylactide/poly(vinylidene fluoride) blend nanocomposites and their impact on dielectric constants and rheological properties. J. Phys. Chem. C **124**, 9536–9547 (2020). https://doi.org/10.1021/ACS.JPCC.0C00891/ASSET/IMAGES/MEDIUM/JP0C00891_0013.GIF
28. S.M.N. Sultana, S.P. Pawar, U. Sundararaj, Effect of processing techniques on EMI SE of immiscible PS/PMMA blends containing MWCNT: enhanced intertube and interphase scattering. Ind. Eng. Chem. Res. **58**, 11576–11584 (2019). https://doi.org/10.1021/ACS.IECR.8B05957/SUPPL_FILE/IE8B05957_SI_001.PDF
29. Y. Li, Y. Zheng, P. Zhan, G. Zheng, K. Dai, C. Liu et al., Vapor sensing performance as a diagnosis probe to estimate the distribution of multi-walled carbon nanotubes in poly(lactic acid)/polypropylene conductive composites. Sens. Actuators B Chem. **255**, 2809–2819 (2018). https://doi.org/10.1016/J.SNB.2017.09.098
30. S. Biswas, S.S. Panja, S. Bose, Tailored distribution of nanoparticles in bi-phasic polymeric blends as emerging materials for suppressing electromagnetic radiation: challenges and prospects. J. Mater. Chem. C **6**, 3120–3142 (2018). https://doi.org/10.1039/C8TC00002F
31. R. Banerjee, A. Gebrekrstos, J.T. Orasugh, S.S. Ray, Nanocarbon-containing polymer composite foams: a review of systems for applications in electromagnetic interference shielding, energy storage, and piezoresistive sensors. Ind. Eng. Chem. Res. (2023). https://doi.org/10.1021/ACS.IECR.3C00089
32. F. Morshedi, S. Tarashi, H. Nazockdast, Effect of multi-walled carbon nanotube localization on toughening mechanism and electrical properties of compatibilized PP/EOC immiscible blend. Polym. Compos. **44**, 650–662 (2023). https://doi.org/10.1002/PC.27125
33. R. Salehiyan, M. Nofar, K. Malkappa, S.S. Ray, Effect of nanofillers characteristics and their selective localization on morphology development and rheological properties of melt-processed polylactide/poly(butylene adipate-co-terephthalate) blend composites. Polym. Eng. Sci. **60**, 2749–2760 (2020). https://doi.org/10.1002/PEN.25505
34. X. Qi, J. Yang, N. Zhang, T. Huang, Z. Zhou, I. Kühnert et al., Selective localization of carbon nanotubes and its effect on the structure and properties of polymer blends. Prog. Polym. Sci. **123**, 101471 (2021). https://doi.org/10.1016/J.PROGPOLYMSCI.2021.101471
35. R. Rohini, S. Bose, Electromagnetic interference shielding materials derived from gelation of multiwall carbon nanotubes in polystyrene/poly(methyl methacrylate) blends. ACS Appl. Mater. Interfaces **6**, 11302–11310 (2014). https://doi.org/10.1021/AM502641H/ASSET/IMAGES/MEDIUM/AM-2014-02641H_0005.GIF
36. G.P. Kar, S. Biswas, S. Bose, Simultaneous enhancement in mechanical strength, electrical conductivity, and electromagnetic shielding properties in PVDF–ABS blends containing PMMA wrapped multiwall carbon nanotubes. Phys. Chem. Chem. Phys. **17**, 14856–14865 (2015)
37. P. Xavier, S. Bose, Electromagnetic shielding materials and coatings derived from gelation of multiwall carbon nanotubes in an LCST mixture. RSC Adv. **4**, 55341–55348 (2014)

38. K. Zhang, H.O. Yu, Y.D. Shi, Y.F. Chen, J.B. Zeng, J. Guo et al., Morphological regulation improved electrical conductivity and electromagnetic interference shielding in poly(l-lactide)/poly(ε-caprolactone)/carbon nanotube nanocomposites via constructing stereocomplex crystallites. J. Mater. Chem. C **5**, 2807–2817 (2017). https://doi.org/10.1039/c7tc00389g
39. R. Salehiyan, M. Nofar, S.S. Ray, V. Ojijo, Kinetically controlled localization of carbon nanotubes in polylactide/poly(vinylidene fluoride) blend nanocomposites and their influence on electromagnetic interference shielding, electrical conductivity, and rheological properties. J. Phys. Chem. C **123**, 19195–19207 (2019). https://doi.org/10.1021/acs.jpcc.9b04494
40. A. Bharati, R. Cardinaels, J.W. Seo, M. Wübbenhorst, P. Moldenaers, Enhancing the conductivity of carbon nanotube filled blends by tuning their phase separated morphology with a copolymer. Polymer (Guildf) **79**, 271–282 (2015). https://doi.org/10.1016/J.POLYMER.2015.09.080
41. L. Zhou, Y. Tian, P. Xu, H. Wei, Y. Li, H.X. Peng et al., Effect of the selective localization of carbon nanotubes and phase domain in immiscible blends on tunable microwave dielectric properties. Compos. Sci. Technol. **213**, 108919 (2021). https://doi.org/10.1016/J.COMPSCITECH.2021.108919
42. X. Zhao, H. Wang, Z. Fu, Y. Li, Enhanced interfacial adhesion by reactive carbon nanotubes: new route to high-performance immiscible polymer blend nanocomposites with simultaneously enhanced toughness, tensile strength, and electrical conductivity. ACS Appl. Mater. Interfaces **10**, 8411–8416 (2018). https://doi.org/10.1021/ACSAMI.8B01704/SUPPL_FILE/AM8B01704_SI_001.PDF
43. E.J. Dil, M. Arjmand, I.O. Navas, U. Sundararaj, B.D. Favis, Interface bridging of multiwalled carbon nanotubes in polylactic acid/poly(butylene adipate-co-terephthalate): morphology, rheology, and electrical conductivity. Macromolecules **53**, 10267–10277 (2020). https://doi.org/10.1021/ACS.MACROMOL.0C01525/ASSET/IMAGES/MEDIUM/MA0C01525_0011.GIF
44. F. Tao, D. Auhl, A.C. Baudouin, F.J. Stadler, C. Bailly, Influence of multiwall carbon nanotubes trapped at the interface of an immiscible polymer blend on interfacial tension. Macromol. Chem. Phys. **214**, 350–360 (2013). https://doi.org/10.1002/MACP.201200518
45. J. Chen, X. Cui, Y. Zhu, W. Jiang, K. Sui, Design of superior conductive polymer composite with precisely controlling carbon nanotubes at the interface of a co-continuous polymer blend via a balance of π-π interactions and dipole-dipole interactions. Carbon N Y **114**, 441–448 (2017). https://doi.org/10.1016/J.CARBON.2016.12.048
46. T. Gong, M.Q. Liu, H. Liu, S.P. Peng, T. Li, R.Y. Bao et al., Selective distribution and migration of carbon nanotubes enhanced electrical and mechanical performances in polyolefin elastomers. Polymer (Guildf) **110**, 1–11 (2017). https://doi.org/10.1016/J.POLYMER.2016.12.056
47. Y.Y. Shi, W.B. Zhang, J.H. Yang, T. Huang, N. Zhang, Y. Wang et al., Super toughening of the poly (L-lactide)/thermoplastic polyurethane blends by carbon nanotubes. RSC Adv. **3**, 26271–26282 (2013)
48. A.C. Baudouin, J. Devaux, C. Bailly, Localization of carbon nanotubes at the interface in blends of polyamide and ethylene–acrylate copolymer. Polymer (Guildf) **51**, 1341–1354 (2010). https://doi.org/10.1016/J.POLYMER.2010.01.050
49. L. Zonder, A. Ophir, S. Kenig, S. McCarthy, The effect of carbon nanotubes on the rheology and electrical resistivity of polyamide 12/high density polyethylene blends. Polymer (Guildf) **52**, 5085–5091 (2011). https://doi.org/10.1016/J.POLYMER.2011.08.048
50. J. Huang, C. Mao, Y. Zhu, W. Jiang, X. Yang, Control of carbon nanotubes at the interface of a co-continuous immiscible polymer blend to fabricate conductive composites with ultralow percolation thresholds. Carbon N Y **73**, 267–274 (2014). https://doi.org/10.1016/j.carbon.2014.02.063
51. E. Cohen, L. Zonder, A. Ophir, S. Kenig, S. McCarthy, C. Barry et al., Hierarchical structures composed of confined carbon nanotubes in cocontinuous ternary polymer blends.. Macromolecules **46**(5), 1851–1859 (2013)
52. M. Gültner, A. Göldel, P. Pötschke, Tuning the localization of functionalized MWCNTs in SAN/PC blends by a reactive component. Compos. Sci. Technol. **72**, 41–48 (2011)

53. X. Zhao, J. Zhao, J.P. Cao, D. Wang, G.H. Hu, F. Chen et al., Effect of the selective localization of carbon nanotubes in polystyrene/poly (vinylidene fluoride) blends on their dielectric, thermal, and mechanical properties. Mater. Des. **56**, 807–815 (2014)
54. Y. Li, H. Shimizu, Conductive PVDF/PA6/CNTs nanocomposites fabricated by dual formation of cocontinuous and nanodispersion structures. Macromolecules **41**, 5339–5344 (2008). https://doi.org/10.1021/MA8006834/SUPPL_FILE/MA8006834-FILE001.PDF
55. J.F. Gao, D.X. Yan, B. Yuan, H.D. Huang, Z.M. Li, Large-scale fabrication and electrical properties of an anisotropic conductive polymer composite utilizing preferable location of carbon nanotubes in a polymer blend. Compos. Sci. Technol. **70**, 1973–1979 (2010)
56. O.J. Botlhoko, S.S. Ray, J. Ramontja, Influence of functionalized exfoliated reduced graphene oxide nanoparticle localization on mechanical, thermal and electronic properties of nanobiocomposites. Eur. Polym. J. **102**, 130–140 (2018). https://doi.org/10.1016/J.EURPOLYMJ.2018.03.012
57. L. Bai, S. He, J.W. Fruehwirth, A. Stein, C.W. Macosko, X. Cheng, Localizing graphene at the interface of cocontinuous polymer blends: Morphology, rheology, and conductivity of cocontinuous conductive polymer composites. J. Rheol. (N Y N Y) **61**, 575–587 (2017). https://doi.org/10.1122/1.4982702
58. G. Vleminckx, S. Bose, J. Leys, J. Vermant, M. Wübbenhorst, A.A. Abdala et al., Effect of thermally reduced graphene sheets on the phase behavior, morphology, and electrical conductivity in poly[(αmethyl styrene)-co-(acrylonitrile)/poly(methyl-methacrylate) blends. ACS Appl. Mater. Interfaces **3**, 3172–3180 (2011). https://doi.org/10.1021/AM200669W/ASSET/IMAGES/MEDIUM/AM-2011-00669W_0001.GIF
59. O.J. Botlhoko, J. Ramontja, S.S. Ray, Thermally shocked graphene oxide-containing biocomposite for thermal management applications. RSC Adv. **7**, 33751–33756 (2017). https://doi.org/10.1039/C7RA05421A
60. K. Kultravut, K. Kuboyama, V. Sedlarik, M. Mrlík, J. Osička, P. Dröhsler et al., Localization of poly(glycidyl methacrylate) grafted on reduced graphene oxide in poly(lactic acid)/poly(trimethylene terephthalate) blends for composites with enhanced electrical and thermal conductivities. ACS Appl. Nano Mater. **4**, 8511–8519 (2021). https://doi.org/10.1021/ACSANM.1C01843/SUPPL_FILE/AN1C01843_SI_001.PDF
61. O.J. Botlhoko, J. Ramontja, S.S. Ray, Thermal, mechanical, and rheological properties of graphite- and graphene oxide-filled biodegradable polylactide/poly(ε-caprolactone) blend composites. J. Appl. Polym. Sci. **134**, 45373 (2017). https://doi.org/10.1002/APP.45373
62. O.J. Botlhoko, J. Ramontja, S.S. Ray, Morphological development and enhancement of thermal, mechanical, and electronic properties of thermally exfoliated graphene oxide-filled biodegradable polylactide/poly (ε-caprolactone) blend composites. Polymer (Guildf) **139**, 188–200 (2018)
63. L. Li, Z. Wang, P. Zhao, Y. Luo, L. Liao, K. Xu et al., Thermodynamics favoured preferential location of nanoparticles in co-continuous rubber blend toward improved electromagnetic properties. Eur. Polym. J. **92**, 275–286 (2017). https://doi.org/10.1016/J.EURPOLYMJ.2017.05.005
64. B. Yang, J. Zhang, J. Bao, Effect of polymeric ratio in matrix and blending sequence on electrical resistivity of carbon black-filled multi-phase composites. J. Mater. Sci. Mater. Electron. **34**, 625 (2023). https://doi.org/10.1007/S10854-023-10027-3/METRICS
65. M.H. Al-Saleh, U. Sundararaj, Electromagnetic interference (EMI) shielding effectiveness of PP/PS polymer blends containing high structure carbon black. Macromol. Mater. Eng. **2008**(293), 621–630 (2008)
66. X.B. Xu, Z.M. Li, K. Dai, M.B. Yang, Anomalous attenuation of the positive temperature coefficient of resistivity in a carbon-black-filled polymer composite with electrically conductive in situ microfibrils. Appl. Phys. Lett. **89**, 917249 (2006). https://doi.org/10.1063/1.2222339/917249
67. K. Dai, X.B. Xu, Z.M. Li, Electrically conductive carbon black (CB) filled in situ microfibrillar poly (ethylene terephthalate)(PET)/polyethylene (PE) composite with a selective CB distribution. Polymer (Guildf) **48**, 849–859 (2007)

68. Z.M. Li, X.B. Xu, A. Lu, K.Z. Shen, R. Huang, M.B. Yang, Carbon black/poly (ethylene terephthalate)/polyethylene composite with electrically conductive in situ microfiber network. Carbon N Y **42**, 428–432 (2004)
69. R. Tchoudakov, O. Breuer, M. Narkis, A. Siegmann, Conductive polymer blends with low carbon black loading: polypropylene/polyamide. Polym. Eng. Sci. **36**, 1336–1346 (1996)
70. X.B. Xu, Z.M. Li, M.B. Yang, S. Jiang, R. Huang, The role of the surface microstructure of the microfibrils in an electrically conductive microfibrillar carbon black/poly(ethylene terephthalate)/polyethylene composite. Carbon N Y **43**, 1479–87 (2005). https://doi.org/10.1016/J.CARBON.2005.01.039
71. G. Radonjič, V. Musil, I. Šmit, Compatibilization of polypropylene/polystyrene blends with poly (styrene-b-butadiene-b-styrene) block copolymer. J. Appl. Polym. Sci. **69**, 2625–2639 (1998)
72. Y. Wang, Y. Xiao, Q. Zhang, X.L. Gao, Q. Fu, The morphology and mechanical properties of dynamic packing injection molded PP/PS blends. Polymer (Guildf) **44**, 1469–1480 (2003)
73. I. Sagalianov, L. Vovchenko, L. Matzui, O. Lazarenko, Synergistic enhancement of the percolation threshold in hybrid polymeric nanocomposites based on carbon nanotubes and graphite nanoplatelets. Nanoscale Res. Lett. **12**, 1–7 (2017). https://doi.org/10.1186/S11671-017-1909-Z/FIGURES/4
74. P. Pötschke, B. Kretzschmar, A. Janke, Use of carbon nanotube filled polycarbonate in blends with montmorillonite filled polypropylene. Compos. Sci. Technol. **67**, 855–860 (2007)
75. V.A. Menon, G. Madras, S. Bose, Phase specific dispersion of functional nanoparticles in soft nanocomposites resulting in enhanced electromagnetic screening ability dominated by absorption. Phys. Chem. Chem. Phys. **19**, 467–479 (2016). https://doi.org/10.1039/C6CP07355G
76. A.V. Menon, G. Madras, S. Bose, Magnetic alloy-MWNT heterostructure as efficient electromagnetic wave suppressors in soft nanocomposites. ChemistrySelect **2**, 7831–7844 (2017). https://doi.org/10.1002/SLCT.201700986
77. S. Biswas, I. Arief, S.S. Panja, S. Bose, Electromagnetic screening in soft conducting composite-containing ferrites: the key role of size and shape anisotropy. Mater. Chem. Front. **1**, 2574–2589 (2017). https://doi.org/10.1039/C7QM00305F
78. J. Chen, Y. Shen, J.H. Yang, N. Zhang, T. Huang, Y. Wang et al., Trapping carbon nanotubes at the interface of a polymer blend through adding graphene oxide: a facile strategy to reduce electrical resistivity. J. Mater. Chem. C **1**, 7808–7811 (2013). https://doi.org/10.1039/C3TC31702A
79. J. Chen, H. Lu, J. Yang, Y. Wang, X. Zheng, C. Zhang et al., Effect of organoclay on morphology and electrical conductivity of PC/PVDF/CNT blend composites. Compos. Sci. Technol. **94**, 30–38 (2014). https://doi.org/10.1016/J.COMPSCITECH.2014.01.010
80. S.P. Pawar, D.A. Marathe, K. Pattabhi, S. Bose, Electromagnetic interference shielding through MWNT grafted Fe_3O_4 nanoparticles in PC/SAN blends. J. Mater. Chem. A **3**, 656–669 (2015)
81. S.P. Pawar, S. Stephen, S. Bose, V. Mittal, Tailored electrical conductivity, electromagnetic shielding and thermal transport in polymeric blends with graphene sheets decorated with nickel nanoparticles. Phys. Chem. Chem. Phys. **17**, 14922–14930 (2015). https://doi.org/10.1039/C5CP00899A
82. S.P. Pawar, M. Gandi, C. Saraf, S. Bose, Exceptional microwave absorption in soft polymeric nanocomposites facilitated by engineered nanostructures. J. Mater. Chem. C **4**, 4954–4966 (2016)
83. S.P. Pawar, M. Gandi, S. Bose, High performance electromagnetic wave absorbers derived from PC/SAN blends containing multiwall carbon nanotubes and Fe_3O_4 decorated onto graphene oxide sheets. RSC Adv. **6**, 37633–37645 (2016)
84. G.P. Kar, S. Biswas, R. Rohini, S. Bose, Tailoring the dispersion of multiwall carbon nanotubes in co-continuous PVDF/ABS blends to design materials with enhanced electromagnetic interference shielding. J. Mater. Chem. A **3**, 7974–7985 (2015)

85. P.K.S. Mural, S.P. Pawar, S. Jayanthi, G. Madras, A.K. Sood, S. Bose, Engineering nanostructures by decorating magnetic nanoparticles onto graphene oxide sheets to shield electromagnetic radiations. ACS Appl. Mater. Interfaces **7**, 16266–16278 (2015). https://doi.org/10.1021/ACSAMI.5B02703/ASSET/IMAGES/MEDIUM/AM-2015-02703D_0015.GIF
86. P. Pötschke, A.R. Bhattacharyya, A. Janke, Morphology and electrical resistivity of melt mixed blends of polyethylene and carbon nanotube filled polycarbonate. Polymer (Guildf) **44**, 8061–8069 (2003)
87. X. Li, Z. Fu, X. Gu, H. Liu, H. Wang, Y. Li, Interfacially located nanoparticles: barren nanorods versus polymer grafted nanorods. Compos. Part B Eng. **198**, 108153 (2020). https://doi.org/10.1016/J.COMPOSITESB.2020.108153
88. E.S. Ogunniran, R. Sadiku, S. Sinha Ray, N. Luruli, Morphology and thermal properties of compatibilized PA12/PP blends with boehmite alumina nanofiller inclusions. Macromol. Mater. Eng. **297**, 627–638 (2012). https://doi.org/10.1002/MAME.201100254
89. E.S. Ogunniran, R. Sadiku, S. Sinha Ray, N. Luruli, Effect of boehmite alumina nanofiller incorporation on the morphology and thermal properties of functionalized poly(propylene)/polyamide 12 blends. Macromol. Mater. Eng. **297**, 237–248 (2012). https://doi.org/10.1002/MAME.201100148
90. S.C. Agwuncha, S.S. Ray, J. Jayaramudu, C. Khoathane, R. Sadiku, Influence of boehmite nanoparticle loading on the mechanical, thermal, and rheological properties of biodegradable polylactide/poly(ϵ-caprolactone) blends. Macromol. Mater. Eng. **300**, 31–47 (2015). https://doi.org/10.1002/MAME.201400212
91. P. Saiprasit, A.K. Schlarb, The effect of the compounding procedure on the morphology and mechanical properties of PLA/PBAT-based nanocomposites. Int. Polym. Process. **36**, 219–227 (2021). https://doi.org/10.1515/IPP-2020-4032/MACHINEREADABLECITATION/RIS
92. R.S. Shah, S. Bryant, M. Trifkovic, Decoupling the interplay of polymer properties and particle size in stability of co-continuous blend composites. Phys. Fluids **35**, 2890022 (2023). https://doi.org/10.1063/5.0151178/2890022
93. E. Jalali Dil, N. Virgilio, B.D. Favis, The effect of the interfacial assembly of nano-silica in poly(lactic acid)/poly(butylene adipate-co-terephthalate) blends on morphology, rheology and mechanical properties. Eur. Polym. J. **85**, 635–646 (2016). https://doi.org/10.1016/J.EURPOLYMJ.2016.07.022
94. M. Hoseini, A. Haghtalab, M.H.N. Famili, Rheology and morphology study of immiscible linear low-density polyethylene/poly(lactic acid) blends filled with nanosilica particles. J. Appl. Polym. Sci. **134**, 45526 (2017). https://doi.org/10.1002/APP.45526
95. R. Salehiyan, M. Nofar, K. Malkappa, S.S. Ray, Effect of nanofillers characteristics and their selective localization on morphology development and rheological properties of melt-processed polylactide/poly (butylene adipate-co-terephthalate) blend composites. Polym. Eng. Sci. **60**, 2749–2760 (2020)
96. J. Zhang, G. Wu, S. Huang, M. Sang, H. Wang, L. Ye et al., Wetting kinetics-controlled interplay between nanoparticles and polymer domains in multiphase polymer blends. ACS Appl. Polym. Mater. (2024). https://doi.org/10.1021/ACSAPM.4C01905
97. L. Gong, S. Chen, Y. Yu, B. Yin, M. Yang, Breakup promotion of deformed EPDM droplets by the migration of nanoparticles during extrusion. Polym. Test. **86**, 106445 (2020). https://doi.org/10.1016/J.POLYMERTESTING.2020.106445

Chapter 7
Degradation Studies on Biodegradable Polymer Blend Nanocomposites

Abstract In the case of biodegradable polymer blend nanocomposite, it is important to understand how each component of the blend, the attributes of the filler particles (such as composition, shape, and size), and their placement affect the composite degradability. By understanding these elements, we can gain insights into how to adjust the degradation rate to achieve a balance between durability and degradability in line with product specifications. This chapter reports the overview of biodegradability of various polymer blend nanocomposites and the factors that influence their biodegradability.

7.1 Introduction

Research on the degradation characteristics is vital for assessing the quality of biodegradable materials. When considering a biodegradable polymer blend nanocomposite, it is important to analyse how each component of the blend, the attributes of the filler particles (such as composition, shape, and size), and their placement affect the material's degradability. By understanding these elements, we can gain insights into how to adjust the degradation rate to achieve a balance between durability and degradability in line with product specifications [1, 2]. For example, hydrolytic degradation of PLA-based packaging is beneficial during composting but not during storage [3]. Likewise, the degradation rate of scaffolds should align with the rate of tissue formation [4]. This chapter is divided into two sections. The first section provides a foundational overview of polymer degradation, addressing hydrolytic degradation, weathering, photodegradation, thermal degradation, and biodegradation-types of degradation that often occur during storage, usage, and composting. The second section explores the degradation behaviour of various polymer blend nanocomposites, emphasising the diverse factors that influence their degradability.

S. S. Ray and R. Banerjee, *Polymer Blend Nanocomposites*, Springer Series in Materials Science 354, https://doi.org/10.1007/978-3-032-00309-6_7

7.2 Fundamentals of Polymer Degradation

Degradation is an irreversible process in which the useful properties of a polymer are lost due to photo-oxidation, chemical reactions, thermal, and biological agents [5, 6]. It typically leads to a decrease in molecular weight and results in mechanical failure. The material becomes brittle, and its half-life is reduced [7]. The degradation process occurs primarily through chain scission. In the environment, the degradation of polymers is caused by the combined action of various biotic and abiotic factors. Biotic factors include microbes, enzymes, or a combination of both, which conduct biological activity using polymers as a carbon source. Abiotic factors, such as pH, temperature, humidity, salinity, and oxygen availability, influence the growth of microbes as well as their metabolic activity. The biodegradability of different polymers under aerobic conditions is significantly affected by the structure of the polymer molecules [8]. In addition to biological processes, physical and chemical processes, such as thermal activation, oxidation, hydrolysis, photolysis, or radiolysis, contribute to the degradation of polymeric materials [6]. Table 7.1 presents an overview of the different types of polymer degradation.

7.2.1 *Weathering*

The ageing behaviour of polyesters is often influenced by abiotic factors, such as temperature, pH, soil texture, oxygen, salinity, radiation, and rainfall. The changes in molecular structure during the ageing of polyesters are attributed to a series of molecular scission and crosslinking reactions. Both reactions result in the loss of properties.

Table 7.1 Overview of different types of polymer degradation

Degradation: Degradation is an irreversible phenomenon in which the properties of the polymer deteriorate	
Type of degradation	Causative agents
Photo degradation caused by electromagnetic radiation	*High energy*: X-ray, α, β, γ-rays *Photo-oxidation*: O_2 + UV and visible light
Thermal degradation is initiated via heat and temperature	*Thermal oxidation*: O_2 + temperature *Thermal hydrolysis*: H_2O + temperature
Chemical degradation	*Hydrolytic*: H_2O *Etching*: Acid, alkali, salt or reactive agent
Weathering	*Physical wear*: Cracking and erosion *Abrasive force*: stress and fatigue
Biodegradation mediated by biological agents	*Enzymatic degradation*: Lipase, proteinase *Microbial degradation*: bacteria or fungi

Reproduced with permission from [6]. Copyright 2016, Royal Society of Chemistry

Thus, ageing characteristics influence the useful life of polymeric products [9]. Photo-ageing is regarded as an oxidative process that involves chain scission and recombination. The two types of reactions compete with one another, with one gaining predominance, characterising the overall behaviour of the polymer undergoing ageing [10]. For example, in studies on the ageing behaviour of carbon black reinforced bio-based anti-ageing polyester elastomer nanocomposites [9], crosslinking phenomena were found to be predominant in the initial stage, but chain scission gained predominance later. For co-polyesters with a higher aromatic content, photoxidation is governed by crosslinking [11]. However, the weathering of PLA is governed by chain scission [11]. The influence of climatic conditions and soil factors on the degradation of polymers has also been studied by weight loss measurement after placing poly(3-hydroxybutyrate-co-valerate) (PHBV), poly(ε-caprolactone) (PCL), poly(butylene succinate) (PBS), poly(butylene succinate and adipate) (PBSA), and polylactide (PLA) in soil in different locations, and the following order of degradation was observed [12]:

$$\text{PBSA} = \text{PHBV} = \text{PCL} > \text{PBS} > \text{PLA}.$$

7.2.2 Photo-Degradation

Photo-degradation of polymers is a physical phenomenon caused by exposure to high-frequency electromagnetic radiation, particularly ultraviolet and gamma rays. The absorption of photons results in the chain scission of polymer molecules, leading to a reduction in molar mass. The condition for photochemical reactions to occur is the presence of chromophoric groups within the molecules [6]. The photo-oxidation of polymers has been explained using the widely accepted singlet oxygen mechanism [13]. The incorporation of certain organic and inorganic molecules into polymers is believed to facilitate the production of free radicals in the presence of light, resulting in the formation of hydroperoxide [14, 15].

7.2.3 Thermal Degradation

High temperatures increase the thermal energy of a system and, thereby, accelerate the process of degradation at elevated temperatures. Several studies have been conducted in this area. Hyon et al. [16] reported that PLA filaments remained intact for 6 months at 37 °C in phosphate buffer, while they degraded by 50% in 30 h at 100 °C. Fukushima et al. [17], in their studies on the effect of temperature on PLA degradation in phosphate buffer, observed that faster degradation occurred at 58 °C compared to 37 °C. The biodegradation of poly(L-lactide) (PLLA) has been reported to be higher at 52 °C than at 37 °C under both anaerobic and aerobic conditions [18]. The degree of mineralisation was found to be approximately 60% in 40 days at 52 °C, while a similar extent of degradation occurred in 100 days at 37 °C in the anaerobic aquatic test. The degradation rate of PLLA has been found to increase significantly at temperatures close to the glass transition temperature (Tg) (55 °C) [18]. Poly(lactic-co-glycolic acid) (PLGA) has been reported to exhibit different activation energies for degradation at temperatures below and above its T_g [19], despite the fact that the degradation of PLLA exhibits the same mechanism [20]. The thermal degradation of oligo-L-lactide at 200 °C under a nitrogen atmosphere involves various reactions such as intra- and intermolecular ester exchange, cis elimination, radical and concerted non-radical reactions, and selective Sn-catalysed depolymerisation via the generation of various carbon- and oxygen-centred macro-radicals [6, 21].

7.2.4 Hydrolytic Degradation

Hydrolytic degradation of a polymer matrix causes a change in the refractive index of the sample due to water absorption [22]. Any factor that increases the hydrolytic tendency of a polymer would accelerate hydrolytic degradation. For instance, the presence of hydroxyl groups in the silicate layers of nanoclay facilitates the hydrolysis of the ester linkages of PLA [23, 24]. The hydrolysis of PLA in aqueous media occurs more rapidly in amorphous regions compared to crystalline regions, resulting in the production of lactic acid oligomers that enhance the concentration of carboxylic acid end groups in the medium, which in turn catalyses further hydrolytic degradation. Thus, the hydrolytic degradation of PLA primarily occurs in the bulk of the material rather than on its surface. The formation of degradation products is influenced by the pH and the dielectric constant of the medium. In acidic media, hydrolysis occurs via chain-end scission, whereas in alkaline media, it is characterised by the splitting of lactoyl lactate [25]. Depolymerisation of PLA results from a random alkaline attack on the ester group, followed by the hydrolysis of the polymer [26]. Basic hydrolysis of polyesters occurs more rapidly than acidic hydrolysis. An increase in hydrolysis time has been reported to lead to a higher concentration of lactic acid, hydrolysed oligomers, and alkali decomposition products [27].

7.2.5 Biodegradation

Biodegradation is a biochemical phenomenon in which organic waste is converted into its original chemical compound or other compounds. This process occurs in the presence of microbes or enzymes. Biodegradable polymers, especially polyesters, have received significant attention due to their hydrolysable bonds, which are converted into water and carbon dioxide under aerobic conditions. Methane is produced under anaerobic conditions, in addition to water and carbon dioxide. Figure 7.1 presents a scheme of biodegradation in the presence and absence of oxygen. There are three different ways in which the biodegradation of a polymer

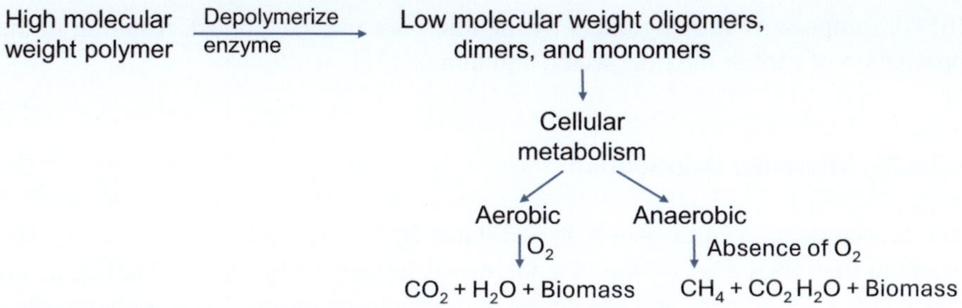

Fig. 7.1 Scheme for biodegradation of polymers in the presence or absence of oxygen. Reproduced with permission from [6]. Copyright 2016, Royal Society of Chemistry

Table 7.2 Factors affecting the biodegradation of polymers

Abiotic	pH, temperature, moisture, electromagnetic radiation, salinity
Biotic	Saprophytic diversity (bacteria and fungi) type of enzymes and its concentration
Polymer characteristics	Molecular weight, monomeric composition, pattern of cross linking, morphology (structure, shape and size), route of processing, percentage of crystal and amorphous content, presence of active functional groups. Type of copolymer and composite hydrophobicity

Reproduced with permission from [6]. Copyright 2016, Royal Society of Chemistry

may occur: (a) compost, (b) microbial, and (c) enzymatic. The various factors that influence the biodegradability of polymers have been presented in Table 7.2.

7.2.5.1 Composting

Composting is a biological process of decomposition and transformation that is used for the treatment and recycling of biodegradable organic waste, converting it into a humus-like substance. Compost provides the conditions required for waste decomposition, namely microbes and a mineral salts solution under anaerobic or aerobic conditions. The actual composting conditions are governed by the type of compost and environmental parameters such as temperature, moisture, and pH of the medium [6]. Elevated temperatures, high moisture content, and a high concentration of microbes in compost are found to be favourable for the biodegradation of PLA [28, 29]. The disintegration of plastic by composting is regarded as satisfactory when 90% of the original dry weight of the material passes through a 2.0 mm sieve [30]. The rate of biodegradation of PHBV by composting shows remarkable improvement in the presence of nanoclay or nanohydroxyapatite [31]. Biodegradation by composting comprises two steps: (i) hydrolysis of high molecular weight polymer molecules to lower molecular weight oligomers due to the action of acids, bases, or moisture and

(ii) consumption of the oligomers by the microbes in the compost, resulting in the production of carbon dioxide, water, and humus [31].

7.2.5.2 Microbial Degradation

The biodegradation of polymers is facilitated by a wide variety of microbes. The extent of microbial degradation of a polymer is influenced by the distribution, abundance, diversity, activity, and adaptation of the microorganisms, along with the structural characteristics of the polymer. The population of aliphatic polyester-degrading microbes decreases in the order of poly(β-hydroxybutyrate) (PHB) = PCL > PBS > PLA [32, 33]. However, different microbes require either intracellular or extracellular enzymes to facilitate polymer degradation. Therefore, assessing the microbial degradation process requires an understanding of how enzymes function [6].

7.2.5.3 Enzymatic Degradation

Enzymes are biological catalysts. Enzymatic degradation is a sustainable option for the waste management of plastics. The enzymatic degradation of aliphatic polyesters consists of two steps. The first step involves the adsorption of the enzyme onto the substrate, leading to the formation of an enzyme–substrate complex. The second step entails the hydrolysis of the ester bonds, resulting in soluble degradation products. PHB depolymerase, PCL depolymerase, proteinase K, cutinase, serine proteases, trypsin, elastase, esterase, and lipases are enzymes that have been extensively studied and employed for biopolymer degradation. The stereochemistry of the polymer profoundly influences its enzymatic degradation; for instance, proteinase K degrades the stereo-isomers of PLA in the following order: L-L > L-D > D-L > D-D at 37 °C and pH 8.6. Similarly, the rate of degradation of PLA50-mes is higher than that of PLA50-rac due to a higher water uptake ratio, which is favourable for enzymatic attack [34]. The recombinant enzyme DL-PLA depolymerase catalyses various biodegradable polyesters, such as PLA, PBS, PBSA, and PCL [35]. Furthermore, nanoparticles, owing to their high surface area, offer the advantage of increased enzyme adsorption [36]. Thus, bulk modification of biopolymers and surface functionalisation of nanofillers facilitate the enzymatic degradation of polymers containing hydrolysable ester bonds [6].

7.3 Degradation Behavior of Polymer Blend Nanocomposites

The degradability of various biodegradable polymer blend nanocomposites has been extensively investigated in research. In this context, Campo et al. [37] conducted studies on the influence of nano- and micro-sized ZnO particles on the degradability of Ecovio® F blend C2224 (55/45 wt ratio PBAT/PLA produced by BASF). Micro- and nano-sized ZnO particles, of two types each, designated as MsZnO, microZnO, NPZnO, and nanoZnO, were used for the studies. MsZnO comprised star-shaped particles with a particle size of around 1.5 μm, whereas microZnO consisted of hexagonal rods measuring 1 μm in length. NPZnO and nanoZnO consisted of nanoparticles with diameters of 56 nm and 20 nm, respectively. Composites containing 2 wt% ZnO and 15 wt% Technipol® compatibiliser (an aliphatic–aromatic polyester produced by Sipol) were prepared by melt processing in a twin-screw micro-compounder at 180 °C and 100 rpm for 3 min. The extruded composites were compressed into films of approximately 200 μm thickness and subjected to the disintegrability test using simulated composting conditions (58 °C, 50% humidity, and aerobic conditions) at a laboratory scale, in accordance with ISO 20,200 standard [38].

Raman studies (Fig. 7.2) on the unfilled blend and composite films before degradation revealed that the presence of ZnO particles led to a finer blend morphology, with smaller sized PLA domains dispersed in a PBAT matrix. The effect was more pronounced in the nanocomposites as compared to the microcomposites. SEM-EDX elemental analysis revealed a uniform dispersion of the ZnO particles. The researchers proposed that the ZnO particles served as nucleating agents for the PLA crystals, leading to the formation of smaller PLA domains, which were better dispersed in the PLA matrix. This was supported by the fact that in differential scanning calorimetry (DSC) studies, the crystallisation temperature obtained by melt cooling was found to be higher in the composites in comparison with the unfilled blend. There was no comment on the localisation of the filler particles. However, aggregates of ZnO were visible in PBAT phase of the Raman micrographs.

The visual appearance of the film samples recovered at various stages of degradation in compost is presented in Fig. 7.3. During the first week of composting, all samples exhibited slight deformation and a yellowish tint, accompanied by a minor increase in opacity. The researchers attributed these changes to water absorption and the formation of low molecular weight compounds due to hydrolytic degradation. The colour of the compost medium changed to dark brown within a week and released an ammonia odour, as noted in the ISO 20,200 standard. As disintegration progressed, the colour of the film samples shifted from yellowish and brown to dark brown. Ultimately, the compost medium transformed into a humus-rich soil, confirming the success of degradation facilitated by thermophilic microbes. The composite samples disintegrated significantly faster than the samples of the unfilled blend. The composite of star-shaped micro-particles was the most rapid in disintegration, with noticeable changes in the film after 21 days. This suggested that the surface area in contact with

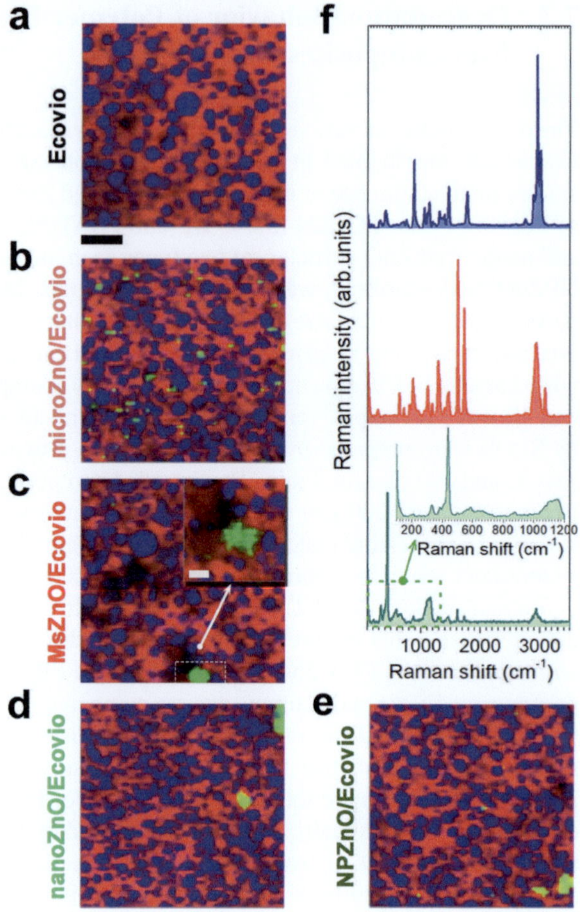

Fig. 7.2 Evaluation of the phase morphology of the Ecovio blend and its components. Raman **micrographs** (XY maps) of (**a**) Ecovio, (**b**) microZnO/Ecovio composite, (**c**) MsZnO/Ecovio composite, (**d**) nanoZnO/Ecovio composite and (**e**) NPZnO/Ecovio composite films. Average **Raman spectra** of the corresponding PLA (typical bands at 2947, 1770, 1454, 393 cm^{-1}), PBAT (bands at 1720 and 1615 cm^{-1}) and a representative spectrum in which is clearly appreciated the band at 440 cm^{-1} associated to the E_2 mode of ZnO materials. Blue: PLA, red: PBAT and green: ZnO particles. *Scale bar in black, 10 μm, and scale bar in white, 3 μm.* Reproduced with permission from [37]. Copyright 2021, Elsevier Science Ltd.

the ZnO particles was not a critical factor in the disintegration rate, as the largest particles resulted in the highest rate of disintegration.

The film's degradability was also assessed in terms of mass loss as a function of degradation time (Fig. 7.4). The ZnO filler particles were found to accelerate the degradation process during the first 21 days. Afterwards, the star-shaped ZnO microparticles produced significantly faster degradation, although higher levels of degradation were observed for all composite samples in comparison with the unfilled blend sample. This difference in degradability was attributed to the catalytic effect of ZnO on the hydrolytic degradation of the blend. Furthermore, based on the well-established fact that PLA degrades much faster than PBAT [39, 40], the researchers concluded that the film disintegration in the initial 21 days, which was significantly catalysed by the ZnO particles, was largely due to the hydrolytic degradation of PLA. This was indeed confirmed by the results of Confocal Raman spectroscopy, conducted on the film samples at different incubation times in the compost, wherein

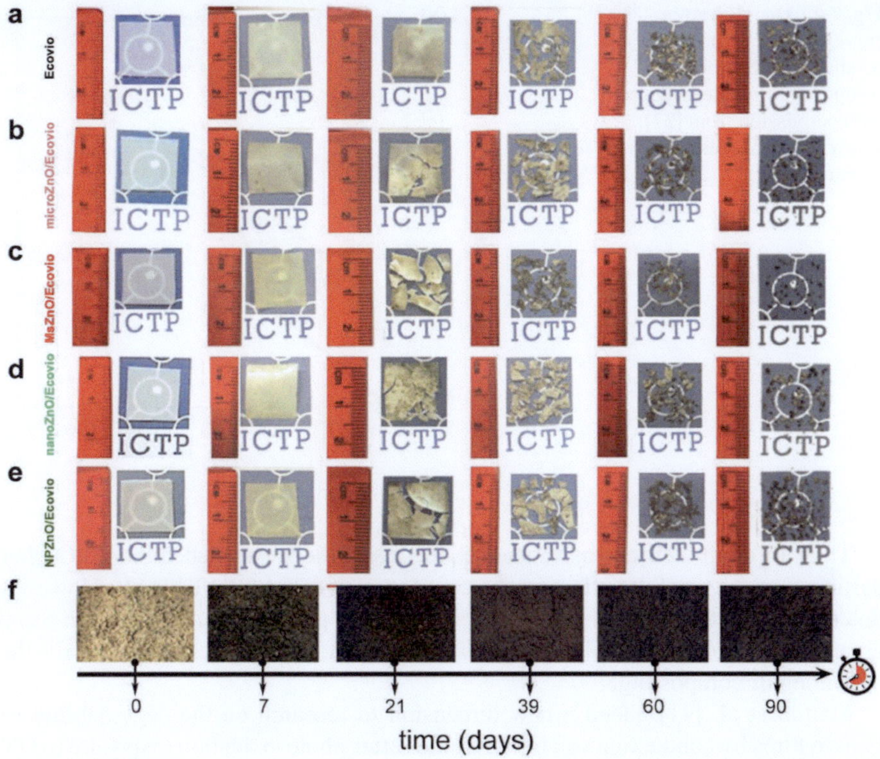

Fig. 7.3 Photographs of neat Ecovio and composite films recovered at different incubation times in composting conditions (to highlight the loss of transparency and the fragmentation of the polymer films, they have been photographed on the top of blue ICTP-CSIC logo). Reproduced with permission from [37]. Copyright 2021, Elsevier Science Ltd.

PLA was found to disappear faster than PBAT. Interestingly, no trace of ZnO was found in the samples after 7 days of composting, though physical disintegration was minimal during that period. This was also evident in X-ray diffraction studies on the samples, where the peak due to ZnO particles was absent in samples that were composted for more than 7 days. EDX studies revealed that after 7 days of composting, not only was there no ZnO, there was no trace of Zn at all. The researchers concluded that the ZnO particles are chemically dissolved, leading to the release of Zn^{+2} ions into the compost medium, where they underwent changes due to interactions with the natural macromolecules and humic substances in the compost. Furthermore, degradation commenced through the hydrolysis of PLA. This was followed by microbial degradation, which resulted in the erosion of the film surface. Microbial degradation was also confirmed by the smell of ammonia, which was detected in the later stages of degradation, as microbes use nitrogen for cell construction.

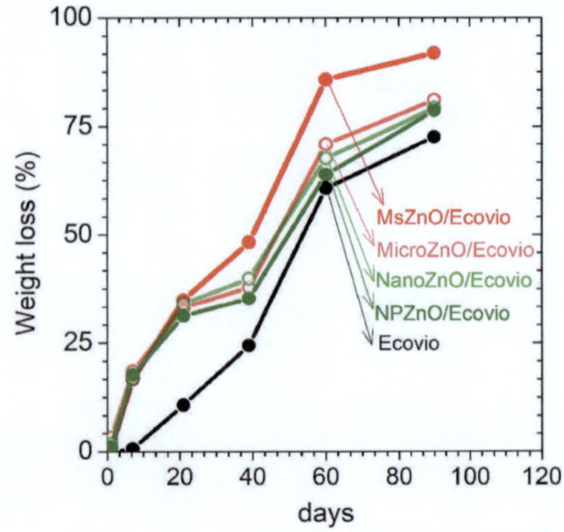

Fig. 7.4 Disintegration curves under composting conditions of Ecovio and composite films. Reproduced with permission from [37]. Copyright 2021, Elsevier Science Ltd.

Thus, through their extensive studies, the researchers established that ZnO filler particles exert a catalytic effect on the composting of PLA/PBAT blends. However, this effect is not dependent on the size of the filler particles. Furthermore, it has a greater impact on the hydrolytic degradation of PLA, which takes place towards the beginning of composting.

Mantia et al. [41] added a new dimension to research on the degradability of Ecovio films by conducting soil burial studies after photo-oxidation (exposure to UV radiation). Biodegradation was accelerated following photo-oxidation. Furthermore, the incorporation of carbon black (CB) particles (1 wt%) into the films resulted in a decrease in the rates of both photo-oxidation and biodegradation, the latter being ascribed to the former. The films ultimately completely disintegrated in the soil. This finding can be applied to outdoor applications of PLA/PBAT blends, such as agricultural mulch films, where the inclusion of a small amount of CB particles may enhance the durability of the films. Although the rate of biodegradation will be reduced, eventual composting will still lead to complete disintegration.

Scaffaro et al. [42] explored a new aspect of the biodegradability of blend nanocomposites by investigating the combined influence of a nanofiller [Dellite 72T, an organomodified montmorillonite (OMMT)] and an antimicrobial agent [Carvacrol (2-methyl-5-(1-methylethyl)-phenol)] on the biodegradability of blown film samples of a PLA/PBAT blend (Bio-Flex® F2110). The formulations, operating parameters for sample preparation, and film characteristics are presented in Table 7.3.

Degradation tests were conducted on disk-shaped samples (diameter 25 mm, thickness ~20 μm, weight approximately 8–10 mg) by immersing them in a buffer solution at pH 10 and 37 °C for up to 864 h. At regular time intervals, the samples were washed with distilled water, dried, and weighed. Degradability was evaluated by measuring weight loss as a function of time. The results are presented in Fig. 7.5. Both Carvacrol and nanoclay enhanced the degradability of the PLA/PBAT blend.

Table 7.3 Formulation, operating parameters used, and characteristics of the films prepared

Sample	Carvacrol wt%[a]	D72T wt%[a]	Screw speed (rpm)	Profile temperature (°C)	BUR[b]	Thickness (mm)
BF	–	–	60	120–130–140–150–160	4	16 ± 3
BF-C	5	–	60	120–130–140–150–160	3.9	20 ± 2
BF-D	–	5	60	120–130–140–150–160	4.1	16 ± 2
BF-D-C	5	5	60	120–130–140–150–160	4	21 ± 2

[a] Based on the total weight of the compounded materials. [b] Calculated as the ratio between the diameters of the bubble and of the die

Reproduced from [42]. Copyright 2020, the Authors (Open Access Publication)

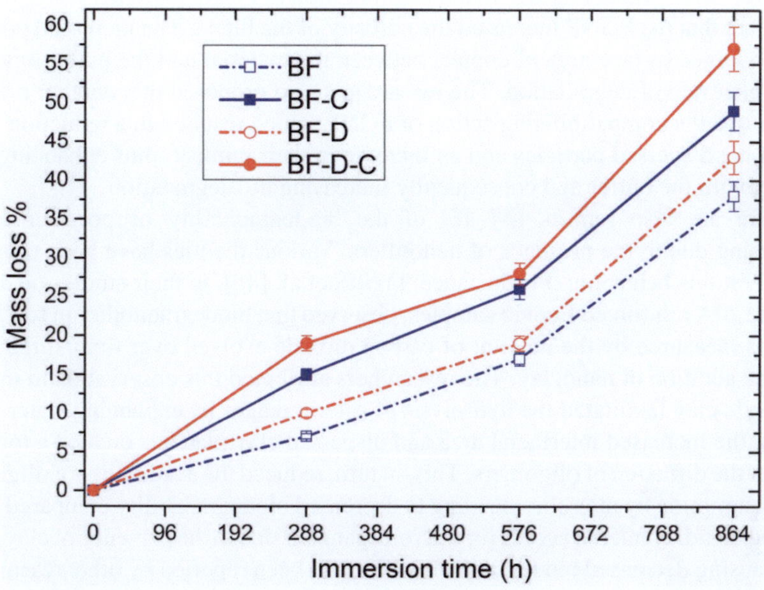

Fig. 7.5 Mass loss percentage of the samples as a function of immersion time. Reproduced from [42]. Copyright 2020, the Authors (Open Access Publication)

This was attributed to the increased hydrophilicity of the blend due to the presence of the additive(s), which facilitated the hydrolysis of the ester linkages. The increase in hydrophilicity due to the presence of the additive(s) was confirmed by water contact angle (WCA) measurements. The enhancement in degradability was greatest when Carvacrol and nanoclay were used in combination. The researchers thus demonstrated that the presence of a hydrophilic nanofiller (OMMT) enhances the hydrolytic degradability of biodegradable polyester blends. Furthermore, OMMT in combination with another hydrophilic additive may exhibit a synergistic effect in increasing hydrophilicity.

An increase in the biodegradability of PLA-based blends due to the incorporation of nanofillers has been reported by other researchers. For example, Jalalvandi et al. [43] conducted soil simulation studies on PLA/starch/MMT nanocomposites, in which maleic anhydride and glycerol were used as a compatibiliser and a plasticiser, respectively. Biodegradability was determined by measuring weight loss as a function of time. The increased biodegradation rate due to the incorporation of MMT was attributed to greater contact between the polymer matrix and the edges and surfaces of MMT. This led to the easy fragmentation of PLA, thereby increasing the rate of biodegradation. Similarly, the incorporation of functionalised graphene nanoplatelets (FGNP) was found to enhance the biodegradability of a (75/25) PLA/chitosan blend. In this case, biodegradability was determined by immersing samples in a phosphate buffer solution and measuring weight loss as a function of time. The improved biodegradability due to the incorporation of FGNP particles was attributed to the fact that the FGNP increased the porosity of the blend. The increased porosity led to a larger surface area of contact between the material and the buffer, resulting in a higher rate of degradation. The researchers also proposed that another possible reason was the compatibilising action of FGNP, which resulted in a reduction in the size of the dispersed particles and an increase in their number, thus enhancing their contact with the buffer and consequently increasing the degradation.

There are also reports [44–46] of the biodegradability of polymer blends decreasing due to the presence of nanofillers. Various theories have been proposed to explain this behaviour. For instance, Freitas et al. [46], in their studies on 5 wt% Cloisite 20A reinforced Ecovio samples, observed that biodegradability in soil burial tests, as measured by the amount of carbon dioxide evolved over time, diminished with the addition of nanoclay. The researchers attributed this observation to the fact that while clay facilitated the hydrolysis of ester linkages by enhancing water penetration, the increased interfacial area and dispersed clay platelets created a tortuous path for the diffusion of oligomers. This, in turn, reduced the availability of oligomers for consumption by microbes, leading to decreased biodegradability compared to the unfilled blend. Hindered access for micro-organisms due to the presence of clay particles, causing decreased biodegradability, has also been reported by other researchers [44, 47]. Spiridon et al. [45], in their studies on poly(vinyl alcohol) (PVA)/starch/clay nanocomposites, reported that the water uptake and mass loss rate in enzymatic degradation studies depended on the type of clay. Nanoparticles hindered degradation, and the enzymatic degradability varied in the order PVA/starch/nanocore > PVA/starch/bentonite > PVA/starch/Peruvian clay. Another explanation for the reduction in biodegradability of polymer blends due to the incorporation of filler particles can be found in the work of Rocha et al. [48], where the addition of $CaCO_3$ particles was found to decrease water uptake and the degradability of Ecovio films in simulated soil degradation tests. These findings were attributed to the coupling effect of the filler particles, which reduced the empty interfacial spaces. The $CaCO_3$ particles used were, however, micrometre sized (30 μm).

Contrary to most studies which reported a change in the biodegradability of blends due to the addition of nanofillers, Tang et al. [49], in their studies on nano-SiO_2 reinforced PVA/starch blend films, observed that the presence of the nano-SiO_2

particles had no effect on the biodegradability of the blend in soil burial test. The researchers attributed this observation to two opposing effects. The filler particles enhanced the miscibility and compatibility of the blend components, forming a dense structure between them. This lowered the infiltration velocity of the microbes towards the beginning of the soil burial test. But with time, the compactness of the films was destroyed. The net effect was the same biodegradability as that of the unfilled blend.

As'habi et al. [50] added new dimensions to knowledge on the biodegradability of polymer blend nanocomposites by investigating the effects of both clay type and mixing sequence on the degradation behaviour of 3, 4.5 and 6 wt% nanoclay reinforced (75/25) PLA/linear low-density polyethylene (LLDPE) blend, compatibilised with 5 wt% glycidylmethacrylate (GMA). Two types of clay (Cloisite® 15A (C15A) and Cloisite® 30B (C30B)) were used for the studies. The effect of mixing sequence was investigated by preparing the nanocomposites in two ways: the one-step mixing process and the two-step mixing process. In the one-step mixing process, all components were added to the extruder in one go. In the two-step mixing process, the nanoclay was premixed with the polymer to which it had a higher affinity. This was followed by compounding the pre-mix with the second polymer and the compatibiliser in the second extrusion step. Thus, the C30B-based nanocomposites were prepared by producing a mix of C30B in PLA and compounding this pre-mix with LLDPE and the compatibiliser. In the case of C15A-based nanocomposites, the premix used was that of C15A and LLDPE. Processing was done using a temperature profile of 160–190 °C, a screw speed of 150 rpm and a feed rate of 10 kg/h. The clay content and its type were indicated by a number and a letter in the sample designation, e.g., 3% 15A indicated 3 wt% C15A.

Transmission electron microscopy (TEM) images of the nanocomposites of Cloisite 30 B and Cloisite 15A, with 4.5 wt% clay content prepared by different methods, are displayed in Figs. 7.6 and 7.7, respectively. The morphological characteristics were found to be independent of clay content for the clay loadings used in the studies.

The sequence of mixing was found to have a profound influence on morphology in the case of the C30B nanocomposites. In single-step mixing, the nanoparticles were observed to be localised at the interface (Fig. 7.6a–c). This was attributed to the fact that the compatibiliser had a significantly lower (~75 °C) melting point compared to the blend components, hence the nanoparticles were encompassed by the molten compatibiliser. However, some nanoparticles were also found to be localised in the thermodynamically preferred PLA phase, thus indicating that some interface crossing had occurred during the short mixing time. In the case of 2-step processing of Cloisite 30B nanocomposites (Fig. 7.6d–f), the nanoparticles were found to be selectively localised in the PLA phase. This was attributed to their premixing in PLA, which was the thermodynamically preferred phase. The particles continued to remain in the thermodynamically preferred phase.

In the case of C15A nanocomposites, the mixing sequence was found to have little influence on the morphological characteristics. Nanoparticles were predominantly localised at the interface in both single-stage (Fig. 7.7a–c) and two-stage mixing (Fig. 7.7d–f). Some clay was also detected in the PLA phase, but there was no

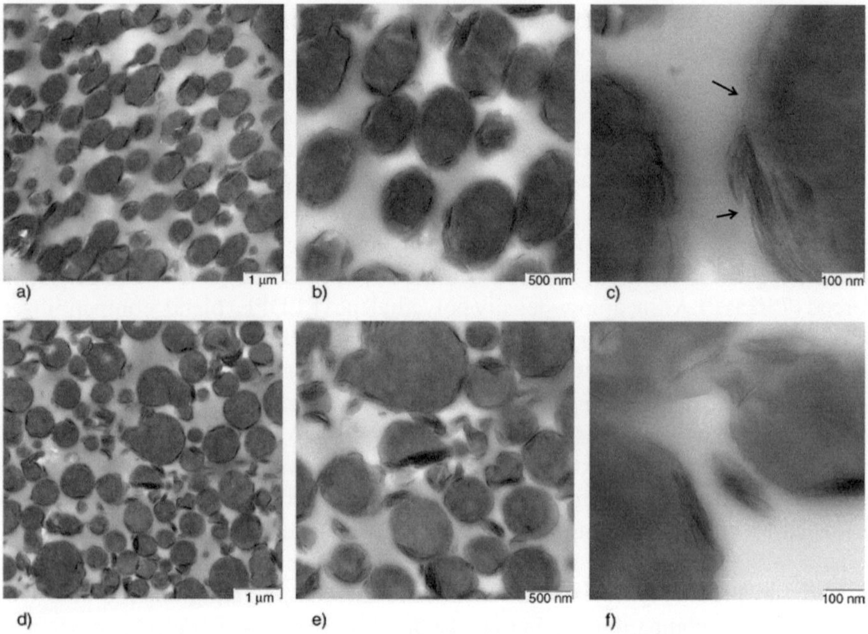

Fig. 7.6 TEM micrographs of compatibilized PLA/LLDPE/30B nanocomposites **a–c** prepared through one step mixing at different magnifications, **d–f** prepared through two steps mixing at different magnifications (all samples contain 4.5 wt% 30B). Reproduced with permission from [50]. Copyright 2013, the Authors (Open Access Publication)

localisation in the LLDPE phase. The presence of C15A at the interface, despite being premixed in the thermodynamically more favourable LLDPE phase during two-stage mixing, was possibly attributed to the higher viscosity of LLDPE compared to PLA, which allowed PLA chains to migrate more readily into the clay tactoids. Furthermore, during the mixing flow, the clay particles tended to move towards the interface because the carboxylic group of the organic modifier of the clay exhibited good affinity for both the compatibiliser and the PLA phase. The net result was a morphology in which LLDPE droplets were encapsulated by clay particles, which were largely arranged along the surfaces of the LLDPE droplets. These nanoparticles served as emulsifiers, preventing the coalescence of LLDPE droplets and enhancing interfacial adhesion. The increased interfacial adhesion was evident in the SEM studies.

Biodegradability was assessed by composting samples of $3 \times 3 \times 0.1$ cm^3 at approximately 58 °C, measuring the residual mass and the amount of CO_2 evolved every 7 days for about 5 months. The results of the biodegradability test are presented in Fig. 7.8.

For all samples, including neat PLA, the initial weight loss was minimal. For PLA and the PLA/LLDPE blend, this period lasted two weeks. Following this initial stage (induction period), the weight loss rate increased rapidly, and neat PLA degraded

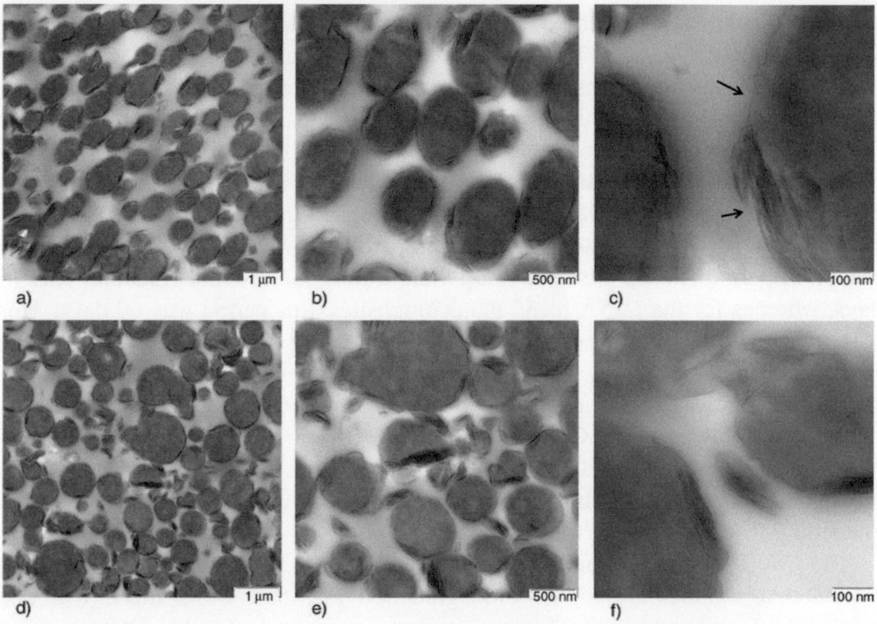

Fig. 7.7 TEM Micrographs of PLA/LLDPE/15A based nanocomposites **a**–**c** prepared through one step mixing at different magnifications, **d**–**f** prepared through two steps mixing at different magnifications (all samples contain 4.5 wt% 15A). Reproduced with permission from [50]. Copyright 2013, the Authors (Open Access Publication)

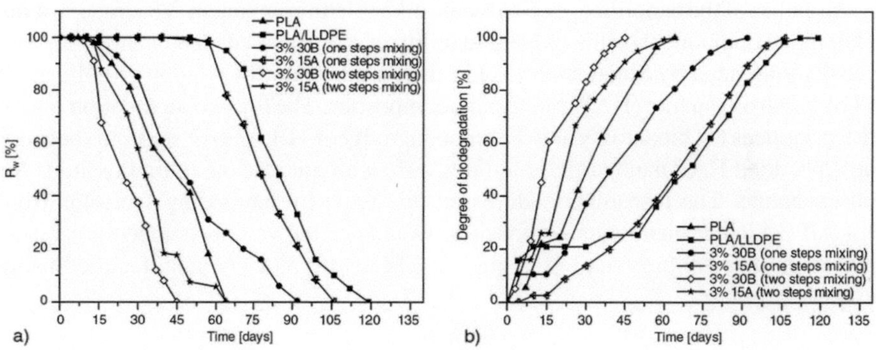

Fig. 7.8 Time-dependence of: **a** weight residual percentage (*R*w); **b** degree of biodegradation (CO_2 evolution) of pure PLA, PLA/LLDPE and its nanocomposites. Reproduced with permission from [50]. Copyright 2013, the Authors (Open Source Publication)

completely within two months. The researchers attributed these observations to the fact that the degradation of PLA comprised two stages: the first stage involved the hydrolysis of ester bonds, resulting in a structural change and a slight loss of weight. The second stage involved microbial attack, which led to the release of carbon dioxide (Fig. 7.8b) and a rapid loss of weight. The biodegradability of the blend was significantly enhanced with the addition of the clay, which was attributed to an increase in hydrophilicity due to the hydroxylated edge groups of the silicate layers.

For the C30B blend nanocomposites prepared through single-step mixing, the induction period for the onset of sharp weight loss due to biodegradation was 15 days, whereas for the C15A blend nanocomposites, this period was 60 days. This was attributed to the difference in hydrophilicity of the two types of clay resulting from the different modifiers. Another explanation for this behaviour, as proposed by the researchers, was the variation in the extent of crystallinity of PLA achieved by the two types of clay particles. C15A was more effective in enhancing the crystallinity of PLA compared to C30B. Since amorphous regions were more susceptible to biodegradation, the C30B blend nanocomposites degraded faster than the C15A nanocomposites.

The C30B blend nanocomposites prepared using the two-stage mixing process exhibited a significantly higher biodegradation rate than those made with the single-stage mixing process. This was attributed to the selective localisation of the hydrophilic clay particles in the PLA phase, which is the phase undergoing biodegradation. Consequently, both the type of clay and its localisation were found to influence the biodegradability of the blend composites.

The biodegradability of a PLA-based blend nanocomposite is influenced not only by the nature of the nanofiller but also by the other blend component. The impact of the other blend component on the biodegradability of a PLA-based blend nanocomposite was demonstrated by Shuai et al. [51] in their studies on PLLA/poly(glycolic acid) (PGA)/hydroxyapatite (HAP) blend nanocomposites. The low biodegradation rate of PLLA reduces the bioactivity and osteoconductivity of PLLA/HAP scaffolds because the embedded HAP nanoparticles are not sufficiently accessible to bodily fluids for ion exchange. The researchers addressed this issue by fabricating scaffolds from PLLA/PGA/HAP blend nanocomposites. Details of the various compositions used in their studies are presented in Table 7.4. The scaffolds were manufactured using laser 3D printing.

Table 7.4 Composition of the composite scaffolds

Label	Scaffold	PLLA (wt%)	PGA (wt%)	HAP (wt%)
1#	PLLA/HAP	90	0	10
2#	3PLLA/1PGA/HAP	67.5	22.5	10
3#	1PLLA/1PGA/HAP	45	45	10
4#	1PLLA/3PGA/HAP	22.5	67.5	10
5#	PGA/HAP	0	90	10

Reproduced with permission from [51]. Copyright 2021, the Authors (Open Access Publication)

Biodegradability was evaluated by immersing specimens (13 mm in diameter and 10 mm in thickness) cut from the scaffolds in phosphate buffer (PBS) solution (pH 7.4 at 37 °C) for a period of 28 days, examining the weight loss, pH variation, and changes in morphology at regular intervals. All the scaffolds exhibited weight loss during the degradation test. However, the PLLA/HAP scaffold showed the lowest weight loss (3.3 ± 2.6 wt% over 28 days). The presence of PGA significantly enhanced the degradation rate, with the 1 PLLA/1PGA/HAP composite displaying a weight loss of 25 ± 4.2% over 28 days. This was attributed to the strong hydrophilicity of PGA, which facilitated self-degradation and increased the area of contact of PLLA with water, thus catalysing the hydrolysis of PLLA and accelerating the overall degradation rate. The pH for the PLLA/HAP scaffold changed very little over 28 days. In contrast, the incorporation of PGA resulted in a rapid decrease in pH during degradation. This was attributed to the release of lactic acid and glycolic acid during the hydrolysis of the polymer.

The SEM micrographs of the specimens before and after the degradation test are presented in Fig. 7.9. Immersion of the PLA/HAP specimen in the PBS solution for 28 days resulted in the appearance of some tiny pores. The incorporation of PGA into the scaffold increased both the number and size of the pores, enhanced surface roughness, and led to the development of cracks. These results clearly demonstrate the role of PGA in enhancing the biodegradability of the scaffolds. The researchers, therefore, established through their extensive studies that the biodegradability of a PLA-based blend nanocomposite is profoundly influenced by the nature of the other blend component. In the case of PGA, it not only undergoes degradation itself but also facilitates the degradation of PLA by increasing its area of contact with water during hydrolysis.

7.4 Conclusions

The biodegradability of a polymer blend nanocomposite depends on the components, their composition, and the placement of filler particles. For polyesters, biodegradation occurs through the hydrolysis of ester linkages, followed by microbial attack on smaller oligomers. Fillers that boost hydrophilicity or promote microbial activity enhance polyester degradability. For instance, hydrophilic functional groups in fillers like nanoclays often improve the degradation of PLA and its blends. However, increasing the path tortuosity for microbial attack can sometimes slow down biodegradation. In blends containing a biodegradable component, strategically placing hydrophilic filler particles within that component fosters biodegradation. Additionally, the characteristics of the second blend component are crucial. For example, in a PLA-based blend nanocomposite featuring a hydrophilic polymer like PGA, both components will undergo hydrolysis, thereby accelerating the degradation rate of PLA. By judiciously choosing the blend components, filler types, and their placement, one can customise the degradation rate to achieve an ideal balance between durability and biodegradability.

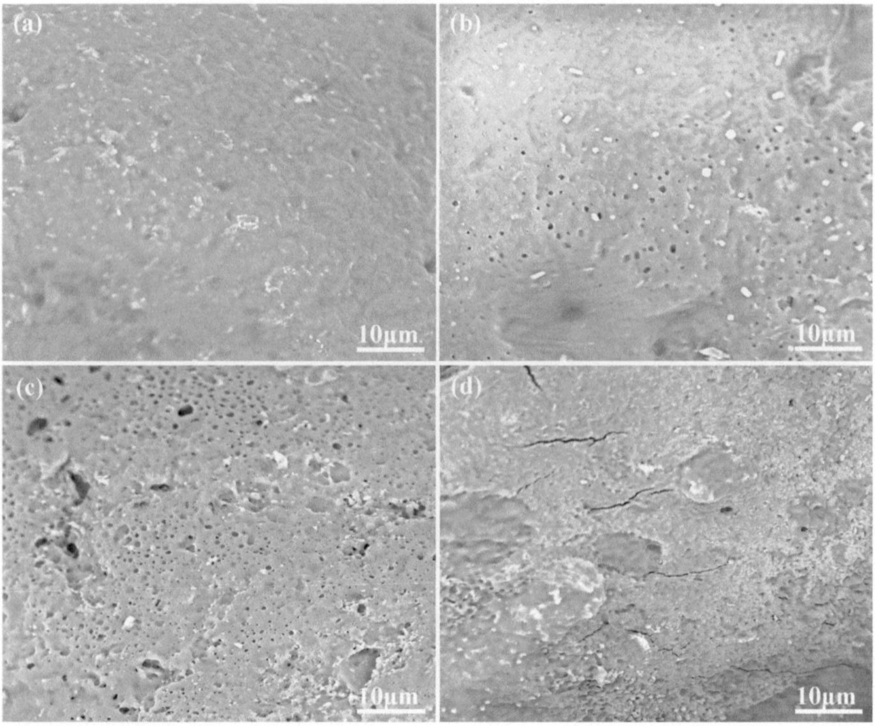

Fig. 7.9 SEM micrographs of surface morphologies of the 1PLLA/1PGA/HAP scaffold **a** before PBS immersion and **b–d** after PBS immersion for 28 d, **b** the PLLA/HAP scaffold, **c** the 3PLLA/1PGA/HAP scaffold, **d** the 1PLLA/1PGA/HAP scaffold. Reproduced with permission from [51]. Copyright 2021, the Authors (Open Access Publication)

References

1. T. Glaskova-Kuzmina, O. Starkova, S. Gaidukovs, O. Platnieks, G. Gaidukova, Durability of biodegradable polymer nanocomposites. Polym **13**, 3375 (2021). https://doi.org/10.3390/POLYM13193375
2. N. Tripathi, M. Misra, A.K. Mohanty, Durable polylactic acid (PLA)-based sustainable engineered blends and biocomposites: recent developments, challenges, and opportunities. ACS Eng. Au. **1**, 7–38 (2021). https://doi.org/10.1021/ACSENGINEERINGAU.1C00011
3. R. Banerjee, S.S. Ray, An overview of the recent advances in polylactide-based sustainable nanocomposites. Polym. Eng. Sci. **61**, 617–649 (2021). https://doi.org/10.1002/PEN.25623
4. S.S. Ray, R. Banerjee, *Foamability of Thermoplastic Polymeric Materials* (Elsevier, Amsterdam, 2021)
5. P. Ghosh, *Polymer Science and Technology* (Tata McGraw-Hill Education, Noida, India, 1990)
6. S. Kumar, P. Maiti, Controlled biodegradation of polymers using nanoparticles and its application. RSC Adv. **6**, 67449–67480 (2016). https://doi.org/10.1039/C6RA08641A
7. J.K. Pandey, K. Raghunatha Reddy, A. Pratheep Kumar, R.P. Singh, An overview on the degradability of polymer nanocomposites. Polym. Degrad. Stab. **88**, 234–250 (2005). https://doi.org/10.1016/J.POLYMDEGRADSTAB.2004.09.013

8. D. Adamcová, M. Vaverková, Biodegradation of degradable/biodegradableplastic material in controlled composting environment. Polish J. Environ. Stud. **23**, 1465–1474 (2014).
9. B. Fang, H. Kang, R. Wang, Z. Wang, W. Wang, L. Zhang, Aging behavior and mechanism of bio-based engineering polyester elastomer nanocomposites. J. Appl. Polym. Sci. **131**, 40862 (2014). https://doi.org/10.1002/APP.40862
10. S. Commereuc, H. Askanian, V. Verney, A. Celli, P. Marchese, C. Berti, About the end life of novel aliphatic and aliphatic-aromatic (co)polyesters after UV-weathering: Structure/ degradability relationships. Polym. Degrad. Stab. **98**, 1321–1328 (2013). https://doi.org/10.1016/J.POLYMDEGRADSTAB.2013.03.030
11. P. Stloukal, V. Verney, S. Commereuc, J. Rychly, L. Matisova-Rychlá, V. Pis et al., Assessment of the interrelation between photooxidation and biodegradation of selected polyesters after artificial weathering. Chemosphere **88**, 1214–1219 (2012). https://doi.org/10.1016/J.CHEMOSPHERE.2012.03.072
12. A. Hoshino, H. Sawada, M. Yokota, M. Tsuji, K. Fukuda, M. Kimura, Influence of weather conditions and soil properties on degradation of biodegradable plastics in soil. Soil. Sci. Plant. Nutr. **47**, 35–43 (2001). https://doi.org/10.1080/00380768.2001.10408366
13. E. Yousif, R. Haddad, Photodegradation and photostabilization of polymers, especially polystyrene: review. Springerplus **2**, 1–32 (2013). https://doi.org/10.1186/2193-1801-2-398/FIGURES/49
14. A.P. Kumar, D. Depan, N. Singh Tomer, R.P. Singh, Nanoscale particles for polymer degradation and stabilization—trends and future perspectives. Prog. Polym. Sci. **34**, 479–515 (2009). https://doi.org/10.1016/J.PROGPOLYMSCI.2009.01.002
15. J.F. Rabek, B. R°anby, Role of singlet oxygen in photo-oxidative degradation and photostabilization of polymers. Polym. Eng. Sci. **15**, 40–43 (1975). https://doi.org/10.1002/PEN.760150107
16. S.-H. Hyon, K. Jamshidi, Y. Ikada, Melt spinning of poly-L-lactide and hydrolysis of the fiber in vitro, in *Polymers as Biomaterials*, ed. by W. Shalaby, A. Hoffman, B. Ratner, T. Horbett (Springer, Seattle, Washington, 1984), pp. 51–65
17. K. Fukushima, D. Tabuani, M. Dottori, I. Armentano, J.M. Kenny, G. Camino, Effect of temperature and nanoparticle type on hydrolytic degradation of poly(lactic acid) nanocomposites. Polym. Degrad. Stab. **96**, 2120–2129 (2011). https://doi.org/10.1016/J.POLYMDEGRADSTAB.2011.09.018
18. M. Itävaara, S. Karjomaa, J.F. Selin, Biodegradation of polylactide in aerobic and anaerobic thermophilic conditions. Chemosphere **46**, 879–885 (2002). https://doi.org/10.1016/S0045-6535(01)00163-1
19. C.M. Agrawal, D. Huang, J.P. Schmitz, K.A. Athanasiou, Elevated temperature degradation of a 50:50 copolymer of PLA-PGA. **3**, 345–352 (2007). https://doi.org/10.1089/TEN.1997.3.345
20. N.A. Weir, F.J. Buchanan, J.F. Orr, D.F. Farrar, G.R. Dickson, Degradation of poly-L-lactide. Part 2: increased temperature accelerated degradation. **218**, 321–330 (2004). https://doi.org/10.1243/0954411041932809
21. F.D. Kopinke, M. Remmler, K. Mackenzie, M. Möder, O. Wachsen, Thermal decomposition of biodegradable polyesters—II. Poly(lactic acid). Polym. Degrad. Stab. **53**, 329–342 (1996). https://doi.org/10.1016/0141-3910(96)00102-4
22. S. Li, S. McCarthy, Further investigations on the hydrolytic degradation of poly (DL-lactide). Biomaterials **20**, 35–44 (1999). https://doi.org/10.1016/S0142-9612(97)00226-3
23. K. Fukushima, C. Abbate, D. Tabuani, M. Gennari, G. Camino, Biodegradation of poly(lactic acid) and its nanocomposites. Polym. Degrad. Stab. **94**, 1646–1655 (2009). https://doi.org/10.1016/J.POLYMDEGRADSTAB.2009.07.001
24. N.K. Singh, B.P.D. Purkayastha, M. Panigrahi, R.K. Gautam, R.M. Banik, P. Maiti, Enzymatic degradation of polylactide/layered silicate nanocomposites: effect of organic modifiers. J. Appl. Polym. Sci. **127**, 2465–2474 (2013). https://doi.org/10.1002/APP.37954
25. S.J. De Jong, E.R. Arias, D.T.S. Rijkers, C.F. Van Nostrum, J.J. Kettenes-Van Den Bosch, W.E. Hennink, New insights into the hydrolytic degradation of poly(lactic acid): participation of the alcohol terminus. Polymer (Guildf) **42**, 2795–2802 (2001). https://doi.org/10.1016/S0032-3861(00)00646-7

26. N. Lucas, C. Bienaime, C. Belloy, M. Queneudec, F. Silvestre, J.E. Nava-Saucedo, Polymer biodegradation: mechanisms and estimation techniques – a review. Chemosphere **73**, 429–442 (2008). https://doi.org/10.1016/J.CHEMOSPHERE.2008.06.064
27. S. Dopico-García, A. Ares-Pernas, J. Otero-Canabal, M. Castro-López, J.M. López-Vilariño, V. González-Rodríguez et al., Insight into industrial PLA aging process by complementary use of rheology, HPLC, and MALDI. Polym. Adv. Technol. **24**, 723–731 (2013). https://doi.org/10.1002/PAT.3136
28. G. Kale, R. Auras, S.P. Singh, Degradation of commercial biodegradable packages under real composting and ambient exposure conditions. J. Polym. Environ. **14**, 317–334 (2006). https://doi.org/10.1007/S10924-006-0015-6/TABLES/5
29. G. Kale, R. Auras, S.P. Singh, Comparison of the degradability of poly(lactide) packages in composting and ambient exposure conditions. Packag. Technol. Sci. **20**, 49–70 (2007). https://doi.org/10.1002/PTS.742
30. A.A. Shah, F. Hasan, A. Hameed, S. Ahmed, Biological degradation of plastics: a comprehensive review. Biotechnol. Adv. **26**, 246–265 (2008). https://doi.org/10.1016/J.BIOTECHADV.2007.12.005
31. P. Maiti, P.J.P. Yadav, Biodegradable nanocomposites of poly(hydroxybutyrate-co-hydroxyvalerate): the effect of nanoparticles. J. Nanosci. Nanotechnol. **8**, 1858–1866 (2008). https://doi.org/10.1166/JNN.2008.18251
32. H. Pranamuda, Y. Tokiwa, Degradation of poly(L-lactide) by strains belonging to genus Amycolatopsis. Biotechnol. Lett. **21**, 901–905 (1999). https://doi.org/10.1023/A:1005547326434/METRICS
33. Y. Tokiwa, B.P. Calabia, Biodegradability and biodegradation of poly(lactide). Appl. Microbiol. Biotechnol. **72**, 244–251 (2006). https://doi.org/10.1007/S00253-006-0488-1/TABLES/2
34. S. Li, M. Tenon, H. Garreau, C. Braud, M. Vert, Enzymatic degradation of stereocopolymers derived from l-, dl- and meso-lactides. Polym. Degrad. Stab. **67**, 85–90 (2000). https://doi.org/10.1016/S0141-3910(99)00091-9
35. Y. Akutsu-Shigeno, T. Teeraphatpornchai, K. Teamtisong, U. Nomura, H. Uchiyama, T. Nakahara et al., Cloning and sequencing of a poly(DL-lactic acid) depolymerase gene from Paenibacillus amylolyticus strain TB-13 and its functional expression in Escherichia coli. Appl. Environ. Microbiol. **69**, 2498–2504 (2003). https://doi.org/10.1128/AEM.69.5.2498-2504.2003/ASSET/4739293A-8F8B-4BAA-9A64-1A3557903CD5/ASSETS/GRAPHIC/AM0531996005.JPEG
36. B. Eker, P. Asuri, S. Murugesan, R.J. Linhardt, J.S. Dordick, Enzyme-carbon nanotube conjugates in room-temperature ionic liquids. Appl. Biochem. Biotechnol. **143**, 153–163 (2007). https://doi.org/10.1007/S12010-007-0035-2/FIGURES/5
37. A. del Campo, E. de Lucas-Gil, F. Rubio-Marcos, M.P. Arrieta, M. Fernández-García, J.F. Fernández et al., Accelerated disintegration of compostable Ecovio polymer by using ZnO particles as filler. Polym. Degrad. Stab. **185**, 109501 (2021). https://doi.org/10.1016/J.POLYMDEGRADSTAB.2021.109501
38. ISO 20200:2015, Plastics—determination of the degree of disintegration of plastic materials under simulated composting conditions in a laboratory-scale test (n.d.), https://www.iso.org/standard/63367.html. Accessed 2 October 2024
39. I. Taiatele, T.C. Dal Bosco, P.C.S. Faria-Tischer, A.P. Bilck, F. Yamashita, J. Bertozzi et al., Abiotic hydrolysis and compostability of blends based on cassava starch and biodegradable polymers. J. Polym. Environ. **27**, 2577–2587 (2019). https://doi.org/10.1007/S10924-019-01541-9/TABLES/3
40. A. Carbonell-Verdu, J.M. Ferri, F. Dominici, T. Boronat, L. Sanchez-Nacher, R. Balart et al., Manufacturing and compatibilization of PLA/PBAT binary blends by cottonseed oil-based derivatives. **12**, 808–823 (2018). https://doi.org/10.3144/expresspolymlett.2018.69
41. F.P.L. Mantia, L. Ascione, M.C. Mistretta, M. Rapisarda, P. Rizzarelli, Comparative investigation on the soil burial degradation behaviour of polymer films for agriculture before and after photo-oxidation. Polym **12**, 753 (2020). https://doi.org/10.3390/POLYM12040753

42. R. Scaffaro, A. Maio, E.F. Gulino, M. Morreale, F.P. La Mantia, The effects of nanoclay on the mechanical properties, carvacrol release and degradation of a PLA/PBAT blend. Materials **13**, 983 (2020). https://doi.org/10.3390/MA13040983
43. E. Jalalvandi, R.A. Majid, T. Ghanbari, H. Ilbeygi, Effects of montmorillonite (MMT) on morphological, tensile, physical barrier properties and biodegradability of polylactic acid/starch/MMT nanocomposites. **28**, 496–509 (2013). https://doi.org/10.1177/0892705713486129
44. M.P. Guarás, V.A. Alvarez, L.N. Ludueña, Biodegradable nanocomposites based on starch/polycaprolactone/compatibilizer ternary blends reinforced with natural and organo-modified montmorillonite. J. Appl. Polym. Sci. **133**, 44163 (2016). https://doi.org/10.1002/APP.44163
45. I. Spiridon, M.C. Popescu, R. Bodârlău, C. Vasile, Enzymatic degradation of some nanocomposites of poly(vinyl alcohol) with starch. Polym. Degrad. Stab. **93**, 1884–1890 (2008). https://doi.org/10.1016/J.POLYMDEGRADSTAB.2008.07.017
46. A.L.P. de L. Freitas, L.R. Tonini Filho, P.S. Calvão, A.M.C. de Souza, Effect of montmorillonite and chain extender on rheological, morphological and biodegradation behavior of PLA/PBAT blends. Polym. Test **62**, 189–195 (2017). https://doi.org/10.1016/J.POLYMERTESTING.2017.06.030
47. K. Fukushima, C. Abbate, D. Tabuani, M. Gennari, P. Rizzarelli, G. Camino, Biodegradation trend of poly(ε-caprolactone) and nanocomposites. Mater. Sci. Eng. C **30**, 566–574 (2010). https://doi.org/10.1016/J.MSEC.2010.02.012
48. D.B. Rocha, J. Souza de Carvalho, S.A. de Oliveira, D. dos Santos Rosa, A new approach for flexible PBAT/PLA/$CaCO_3$ films into agriculture. J. Appl. Polym. Sci. **135**, 46660 (2018). https://doi.org/10.1002/APP.46660
49. S. Tang, P. Zou, H. Xiong, H. Tang, Effect of nano-SiO_2 on the performance of starch/polyvinyl alcohol blend films. Carbohydr. Polym. **72**, 521–526 (2008). https://doi.org/10.1016/J.CARBPOL.2007.09.019
50. L. As'habi, S.H. Jafari, H.A. Khonakdar, R. Boldt, U. Wagenknecht, G. Heinrich, Tuning the processability, morphology and biodegradability of clay incorporated PLA/LLDPE blends via selective localization of nanoclay induced by melt mixing sequence. Express Polym. Lett. **7**, 21–39 (2013). https://doi.org/10.3144/expresspolymlett.2013.3
51. C. Shuai, W. Yang, P. Feng, S. Peng, H. Pan, Accelerated degradation of HAP/PLLA bone scaffold by PGA blending facilitates bioactivity and osteoconductivity. Bioact. Mater. **6**, 490–502 (2021). https://doi.org/10.1016/J.BIOACTMAT.2020.09.001

Chapter 8
Applications of Polymer Blend Nanocomposites

Abstract This chapter explores the creation of polymer blend nanocomposites designed for scaffolds, food packaging, EMI shielding, and energy storage applications. It highlights the importance of understanding the relationship between structure, properties, and processing to enhance material development for these advanced applications.

8.1 Introduction

The key to successful product development using multiphase polymeric materials lies in the proper selection of components and achieving the desired morphology that meets product requirements. For example, obtaining an open cell structure is essential for the fabrication of scaffolds. Additionally, biocompatibility, biodegradability, and cytocompatibility are crucial [1, 2]. A porous structure comprising well-connected open pores, created using a polylactide (PLA)/poly(glycolic acid) (PGA)/hydroxyapatite (HAP or HA) nanocomposite, may fulfil these requirements [3]. Barrier properties and adequate mechanical characteristics, such as puncture resistance, tear strength, and toughness, are vital in flexible food packaging applications. A nanoclay reinforced immiscible polymer blend with well-dispersed clay particles selectively localised in the continuous phase may meet these criteria [4]. Biodegradability and anti-microbial properties provide additional value. Likewise, an interconnected network of conductive nanoparticles interfacially localised in immiscible polymer blends aids in fulfilling the requirements for electromagnetic interference (EMI) shielding applications [5]. The selection of an appropriate processing technique and processing conditions is critical for achieving the desired morphology of multiphase polymeric materials. This chapter discusses the development of polymer blend nanocomposites for scaffolds, food packaging, EMI shielding, and energy storage applications. Emphasis is placed on how an understanding of the structure-property-processing relationship can be utilised in material development for advanced applications.

S. S. Ray and R. Banerjee, *Polymer Blend Nanocomposites*, Springer Series in Materials Science 354, https://doi.org/10.1007/978-3-032-00309-6_8

8.2 Scaffolds

The objective of tissue engineering is to replace damaged tissue with an artificial substitute. Porous synthetic biodegradable polymers are employed in tissue engineering because of their microstructure, biocompatibility, biodegradability, and economic viability. Scaffolds are a crucial component of tissue engineering, serving as an artificial extracellular matrix that provides support for cell migration, differentiation, and reproduction. A scaffold replicates the normal process of tissue development, offering essential support for cell attachment, migration, proliferation, and the maintenance of their differentiated function in generating new tissue. Once tissue formation is complete, the scaffold is degraded by the cells in the media or in the in vivo environment. Thus, the fabrication of scaffolds is a vital procedure in tissue engineering [6].

To fulfil the requirements of tissue engineering, an ideal scaffold should possess the following features [6]: (1) a 3-D, highly porous structure with an interconnected network to facilitate cell movement, reproduction, and the diffusion of nutrients, metabolic waste, and paracrine factors [7]; (2) biocompatibility and biodegradability with controllable degradation rates that match cell growth in vitro and in vivo [8, 9]; (3) appropriate surface chemistry that enables cell attachment, migration, proliferation, and differentiation [10, 11]; (4) suitable mechanical properties that align with the surrounding tissue at the implantation site and provide cues for controlling the structure and function of the newly formed tissue [7, 12]; and (5) the ability to be fabricated into various shapes and sizes that correspond to the dimensions of the implantation site while maintaining the desired shape for the necessary time [7]. These characteristics vary according to the requirements of different tissues [2].

A challenge in scaffold fabrication is achieving a porous structure having interconnected microchannels. Such a porous structure is known as an open porous structure.

Various processes have been used for the fabrication of scaffolds. These include textile technologies, solvent casting, phase separation, blending and leaching, gas foaming, freeze drying, electrospinning, UV and laser radiation, salt leaching, and 3D pore architecture designs (CAD/CAM and rapid prototyping). The fabrication method can significantly influence scaffold properties as well as its in vivo performance [13].

In this direction, Shirzaei Sani et al. [14] fabricated poly(ε-caprolactone) (PCL)/chitosan-g-polycaprolactone (CS-g-PCL)/HA nanocomposite electrospun scaffolds, which fulfilled the requirements of bone tissue engineering. PCL is widely employed in biomedical applications, including tissue engineering, due to its biodegradability, superior mechanical properties, biocompatibility, and solubility in a substantial number of solvents [15, 16]. However, PCL also presents certain limitations that impede its use in tissue engineering, including hydrophobicity and a lack of functional groups, the latter of which decreases cell adhesion [17]. This issue is addressed by blending PCL with hydrophilic polymers, such as chitosan (CS). CS, besides being intrinsically hydrophilic, is also biodegradable, biocompatible, and bioactive [17].

The primary challenge in electrospinning the CS/PCL blend stems from the unavailability of co-solvents due to the non-polar nature of PCL and the polar nature of CS. Although acetic acid, trifluoroacetic acid, and hexafluoroisopropanol are commonly used to prepare CS solutions for electrospinning, these solvents exhibit poor compatibility with bioactive molecules. The researchers addressed this challenge by synthesising CS-g-PCL to facilitate the dissolution and electrospinning of its blend with PCL. CS-g-PCL was synthesised by grafting ε-CL monomers onto the hydroxyl groups of CS in a Cs:ε-CL ratio of 1:12. This was followed by the fabrication of PCL/CS-g-PCL blend scaffolds through electrospinning. Blends with varying PCL contents were electrospun using an injection rate of 1.4 ml/h and a voltage of 17 kv. 10 wt% blend solutions in chloroform/DMF (4:1 v/v) were utilised for electrospinning. The electrospun fibres were collected on aluminium foil at a distance of 13 cm from the needle. The fibres thus collected were dried overnight under vacuum at 37 °C.

Scanning electron microscopy (SEM) studies revealed that the CS-g-PCL content of the blend significantly influenced the fibre diameter and its distribution. A notable decrease in fibre diameter from 862 ± 412 nm to 325 ± 107 nm occurred when the composition was altered from neat PCL to a PCL/CS-g-PCL blend containing 5 wt% CS-g-PCL. Furthermore, contact angle measurements indicated that as the CS-g-PCL content increased, the contact angle decreased. The scaffold produced from the blend with 10 wt% CS-g-PCL content consisted of smooth fibres without beads. Therefore, this composition, designated as PCL90, was selected for the preparation of the nanocomposite.

The nanocomposites were prepared by initially dispersing nano HA (nHA) nanoparticles in a chloroform/DMF solvent mixture through ultrasonication for 90 min. The blend components, in the correct weight ratio, were added to this dispersion, and the mixture was homogenised to create nanocomposites containing 1, 3, 5, and 10% (wt/wt) HA nanoparticles, which were electrospun into scaffolds. The nanocomposites were designated as PCL90:nx, where x indicated the nHA content.

The SEM micrographs of the nanocomposite electrospun scaffolds, along with the fibre size distribution and the effect of the HA content on viscosity, fibre diameter, and fibre porosity, are presented in Fig. 8.1. Unlike the nanofibres of PCL90, the nanofibres of the blend nanocomposite exhibit a rough appearance (Fig. 8.1a–d). Figure 8.1e compares the fibre diameter distribution in different nHA contents.

In the scaffold of PCL90:n10, the nanoparticles formed micrometre-sized agglomerates on the scaffold surface. Since these agglomerates were much larger than a single fibre, they were not incorporated within the fibre and thus were observed on the fibre surface. Furthermore, as seen in Fig. 8.1f, an increase in nHA content from 0 to 10 wt% resulted in an increase in solution viscosity from 162 to 215 cp, which in turn led to a rise in the average fibre diameter from 314 to 385 nm. An increase in nHA content led to a greater number of nanoparticles being incorporated within the fibres, which consequently increased fibre diameter and roughness. As illustrated in Fig. 8.1e, the nHA content not only influenced the average fibre diameter but also the width of the fibre diameter distribution. The narrow fibre diameter distribution of

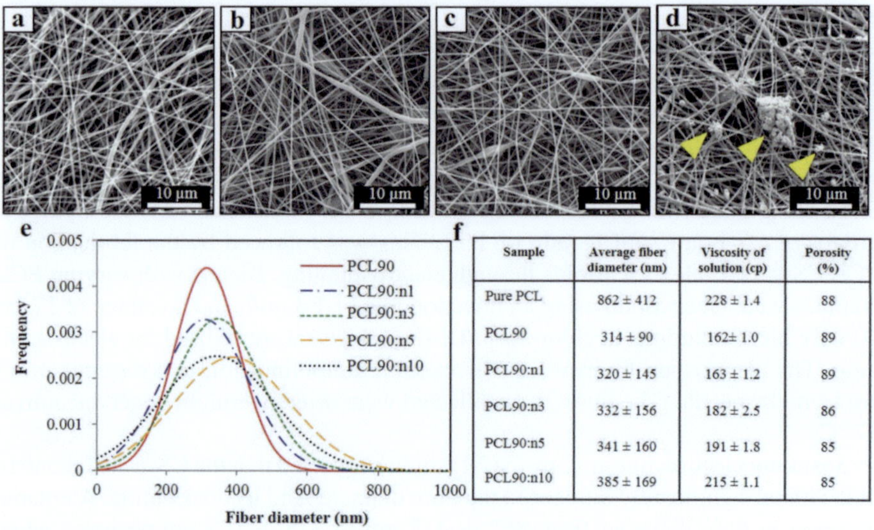

Sample	Average fiber diameter (nm)	Viscosity of solution (cp)	Porosity (%)
Pure PCL	862 ± 412	228 ± 1.4	88
PCL90	314 ± 90	162± 1.0	89
PCL90:n1	320 ± 145	163 ± 1.2	89
PCL90:n3	332 ± 156	182 ± 2.5	86
PCL90:n5	341 ± 160	191 ± 1.8	85
PCL90:n10	385 ± 169	215 ± 1.1	85

Fig. 8.1 SEM images of nanocomposite scaffolds with different nHA contents, **a** 1%, **b** 3%, **c** 5%, **d** 10% nHA; **e** Comparison of fiber diameter distribution in different nHA contents; **f** Solution viscosity, average fiber diameter and porosity in different nanocomposite scaffolds. Reproduced with permission from [14]. Copyright 2021, Elsevier Science Ltd.

the unfilled blend scaffold was rendered wider in the scaffold of the blend nanocomposite containing 10% wt% nHA content. The porosity and mechanical strength of the electrospun scaffolds were within the range reported earlier for trabecular bone (porosity: 50–90%; ultimate tensile strength ~5 MPa) [18]. An increase in nHA content led to a slight decrease in porosity due to pore blocking by agglomerates.

Figure 8.2a–c present results of the degradation studies. Since water absorption is an important factor influencing cell growth, studies on water absorption were conducted, and the results are presented in Fig. 8.2d. The neat PCL scaffold exhibited a water absorption of approximately 40%, while the water absorption of the blend scaffolds reached as high as 500%. The scaffolds of the blend nanocomposites also demonstrated a significantly higher water absorption compared to the neat PCL scaffold. These observations were attributed to the increased polarity due to the presence of the polar CS-g-PCL and HA particles. However, higher contents of nHA resulted in reduced water absorption because of lower porosity, which led to a decrease in the space available for water penetration. Furthermore, the agglomeration of HA particles caused decreased interconnectivity, impeding water diffusion [19].

The scaffolds underwent in vitro degradation studies in PBS (phosphate-buffered saline), and their weight loss was monitored over time. SEM studies were conducted after 8 weeks of degradation. The results of the degradation studies are presented in Fig. 8.2a–c. Blending PCL with hydrophilic CS-g-PCL and nHA resulted in an increased degradation rate, as evidenced by SEM micrographs that showed parts of the blend nanocomposite scaffolds broken off after 8 weeks of immersion in PBS. In

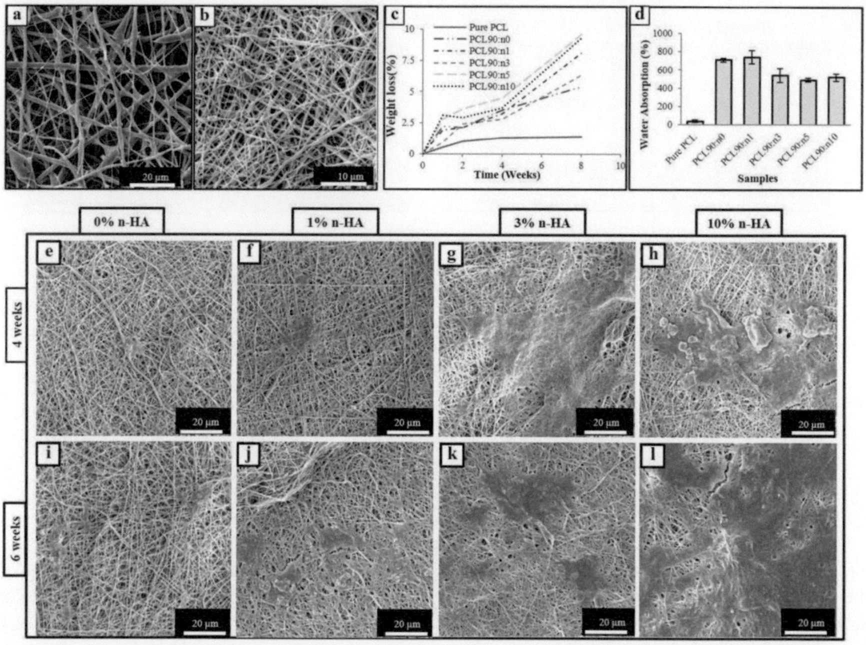

Fig. 8.2 SEM micrographs of nanocomposite scaffolds after degradation for 8 weeks in PBS (pH = 7.4) at 37 °C for **a** pure PCL and **b** PCL90:n5. **c** In vitro degradation rate of fibrous pure PCL and PCL/CS-g-PCL (90:10) nanocomposites with 0, 1, 3, 5 and 10% nHA in PBS (pH = 7.4) at 37 °C. **d** Water absorption of the pure PCL and nanocomposite scaffolds after 24 h immersion in deionised water. SEM micrographs of nanocomposite scaffolds with **e** 0%, **f** 1%, **g** 3%, **h** 10% nHA after 4 weeks soaking and **i** 0%, **j** 1%, **k** 3%, **l** 10% nHA after 6 weeks soaking in SBF (pH = 7.4) at 37 °C. Reproduced with permission from [14]. Copyright 2021, Elsevier Science Ltd.

comparison to the neat PCL scaffold, the blend nanocomposite scaffolds exhibited greater weight loss over time. This was attributed to enhanced hydrophilicity, which facilitated in vitro degradation.

The in vitro bioactivity of the neat blend and the blend nanocomposite scaffolds was investigated by immersing the scaffolds in simulated body fluid (SBF) and incubating them for 4–6 weeks to assess mineralisation. At the conclusion of each incubation period, the scaffolds were removed from the SBF, washed with deionised (DI) water, and lyophilised. Scanning electron microscopy (SEM) studies were conducted to examine the formation of a bone-like apatite layer on the scaffolds. The SEM micrographs from the in vitro bioactivity studies are presented in Fig. 8.2e–l. All samples demonstrated the capacity to form an apatite layer, although this capability was least pronounced in the neat blend scaffold, due to the limited mineralisation associated with the chitosan functional groups. Scaffolds of blend composites with higher contents of nHA exhibited greater formation of apatite layers.

To assess the cytocompatibility of the scaffolds, the in vitro viability, spreading, and metabolic activity of NIH-3T3 mouse embryonic fibroblast cells were studied

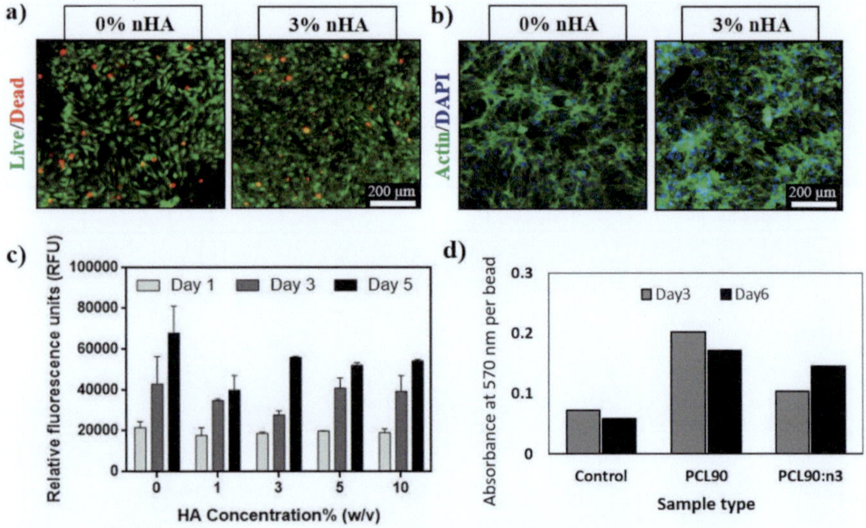

Fig. 8.3 In vitro cytocompatibility of nanocomposite scaffolds. **a** Live (green)/dead (red) assay on the cells seeded, **b** cell spreading and adhesion, **c** cell proliferation on the nanocomposite scaffolds at 1, 3, and 5, days post seeding, **d** MTT assay on MG-63 cells seeded on the PCL90 scaffolds with 0 and 3 wt% nHA after 3 and 6 days. Reproduced with permission from [14]. Copyright 2021, Elsevier Science Ltd.

using surface seeding (2D). Cell viability, spreading, and metabolic activity were measured with the aid of a commercial live/dead kit, Actin/DAPI staining, and Presto Blue assays, respectively. The results are presented in Fig. 8.3. Cell viability, calculated from fluorescent microscope images, exceeded 90% for all scaffolds. Greater osteoblastic cell proliferation was observed in the scaffolds containing nHA.

The bone regenerative capability of the scaffolds was assessed using three genes involved in osteogenesis. For this purpose, mRNA levels of alkaline phosphatase (ALP), collagen type 1 (COL-1), and osteocalcin (OCN) were investigated after three days of cell seeding. ALP is the marker of osteogenic differentiation [20]. An increase in ALP expression was observed after three days of seeding for the scaffolds containing 0 and 3 wt% of nHA compared to the control sample ($P < 0.0001$). Normalised data of the ALP expression indicated that PCL90 and PCL90:n3 scaffolds enhanced the expression level, indicating maturation of MG-63 cells. The cells cultured on the PCL90:n3 scaffold demonstrated better bone-related functions compared to the PCL90 scaffold. The OCN gene is the marker of bone mineralisation [20]. The PCL90:n3 scaffold exhibited a higher OCN gene expression level compared to the PCL90 scaffold, indicating that nHA facilitated mineralisation.

The mRNA expression of COL-1 was similar to the results obtained for the ALP and OCN genes. The mRNA analysis indicated significant expressions of bone-related genes for scaffolds containing nHA, as also observed in other studies [21], demonstrating the osteoconductive properties of nHA.

Thus, the researchers demonstrated through their extensive studies that they had successfully fabricated electrospun scaffolds using PCL/CS-g-PCL/nHA blend nanocomposites, which fulfilled the criteria for bone tissue engineering.

Other researchers have also successfully used biodegradable polymer blend nanocomposites to fabricate scaffolds meeting the criteria of tissue engineering. In all studies, the nanofiller used was nHA; however, the blend components varied. For instance, Shuai et al. [3] blended PGA with poly(L-lactide) (PLLA) to enhance the degradation rate of PLLA. Scaffolds were fabricated using PLLA/PGA/HAP nanocomposites with 10 wt% HAP content through 3D printing. For this purpose, a mixture of 4.5 g of PLLA powder and 4.5 g of PGA powder in distilled water was stirred and sonicated at 37 °C for 30 min. In a separate container, a dispersion of 1.0 g HAP powder in water was stirred and sonicated for 30 min. The HAP suspension was added to the PLLA/PGA suspension, and the mixture was sonicated for an additional 30 min, followed by centrifugation. The composite powder obtained was dried. The dried powder was spread on the build platform of the laser 3D printing machine. The scaffold was fabricated by layer-by-layer printing using a laser power of 2.2 W, a scanning speed of 120 mm/min, a spot diameter of 300 μm, a layer thickness of 0.15 mm, and a scanning distance of 1.0 mm. For comparison, PLLA/HAP and PGA/HAP scaffolds with 10 wt% HAP content were also fabricated in a similar manner, and their optical images are presented in Fig. 8.4.

Studies revealed that the incorporation of PGA increased the degradation rate of the scaffold, resulting in a weight loss that rose from 3.3% (for PLLA/HAP nanocomposite scaffold) to 25.0% (for 1PLLA/1PGA/HAP blend nanocomposite) after immersion in SBF for 28 days. This phenomenon was attributed to the degradation of highly hydrophilic PGA, followed by the accelerated hydrolysis of the PLLA chains. Furthermore, the large number of pores generated by the degradation of the scaffold facilitated the exposure of HAP particles from the matrix. This, in turn, aided the deposition of bone-like apatite on the scaffold and expedited apatite growth. Cytocompatibility tests indicated good osteoblast adhesion, spreading, and proliferation, suggesting that the scaffold provided a suitable environment for cell growth and proliferation.

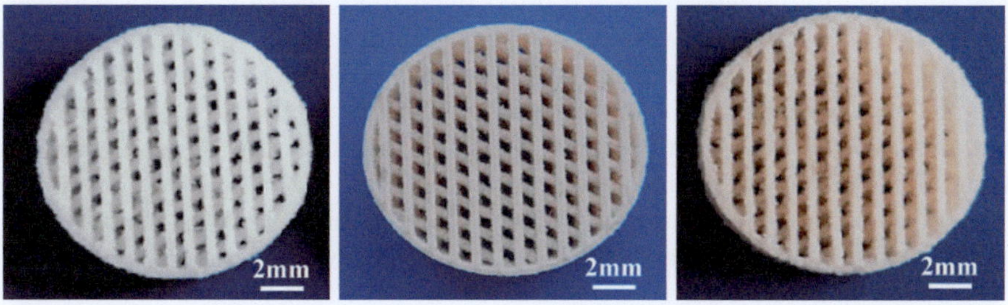

Fig. 8.4 Typical optical images of the PLLA/HAP, 1PLLA/1PGA/HAP and PGA/HAP scaffolds seen from the top. Modified and reproduced from [3]. Copyright 2021, the Authors (This is an open access article)

In vivo bone defect repair studies conducted on rabbits indicated that the PLLA/PGA/HAP scaffold could effectively support the growth of new bone and blood vessel tissue, unlike the control group, which had little impact on healing the bone defect area.

To evaluate bone regeneration capability, 1PLLA/1PGA/HAP and PLLA/HAP scaffolds were utilised as implants in the bone defect site, while the blank group served as a control, with the results presented in Fig. 8.5. All the rabbits survived and showed no signs of inflammation or fracture at the bone defect site. Large blank regions observed in the X-ray radiographs of the blank group demonstrated the formation of a small amount of new bone. In the PLLA/HAP scaffold group, minimal radiopacity was seen surrounding the defects 4 weeks after the surgery, and both ends of the bone defect were partially bridged 8 weeks after the surgery. In contrast, radio-opaque images appeared in the 1PLLA/1PGA /HAP scaffold group, indicating the formation of a significant amount of new bone at the defect site. The new bone tissue grew from the margin of the defect towards the centre, and both ends of the defect site were partially bridged at 8 weeks. The 1PLLA/1PGA/HAP scaffold exhibited significantly greater bone formation at the defect site compared to the PLLA/HAP scaffold and the blank group, as evident in the micro-CT images. Remarkably, no interface could be identified between the scaffold and the host bone 8 weeks after implantation. This clearly indicated that the 1PLLA/1PGA/HAP scaffold expedited the mineralisation of new bone tissue and facilitated tissue integration.

Bone volume fraction and bone mineral density were calculated through quantitative analysis of the micro-CT scans ($^{**}P < 0.01$). The values for the 1PLLA/1PGA/HAP scaffold after 8 weeks were significantly higher than those after 4 weeks. Histological studies on new bone formation in the 1PLLA/1PGA/HAP scaffold were performed using HE and MT staining. The results indicated the formation of a small amount of new bone tissue and fibrous tissue at the defect site with the scaffold, with no signs of inflammation after 4 weeks. After 8 weeks, a considerable amount of new bone tissue was observed alongside the degradation of the scaffold material. Furthermore, a notable number of osteoblasts and new blood vessels appeared. These observations suggested that the 1PLLA/1PGA/HAP scaffold promoted the formation of bone and the repair of bone defects. The enhanced bone regeneration ability was attributed to the bioactivity and osteoconductivity of the HAP nanoparticles. As the HAP nanoparticles were exposed from the polymer matrix during scaffold degradation, they provided an ideal environment for the growth and proliferation of bone cells, thus facilitating the formation of new bone. Consequently, the researchers successfully fabricated 3D printed scaffolds that met the criteria of tissue engineering using PLLA/PGA/HA blend nanocomposites.

Another example of the successful use of biodegradable blend HA nanocomposites in scaffold fabrication can be found in the work of Nahanmoghadam et al. [22]. Poly(hydroxybutyrate-co-hydroxyvalerate) (PHBV)/PCL/HA nanocomposites with a 3:7 weight ratio of PHBV to PCL and 5 wt% HA content were used for studies, and the scaffolds were fabricated using the Dual-Leaching technique. For this purpose, PCL and PHBV, in a 7:3 weight ratio, were dissolved in a (70/30 v/v) mixture of chloroform and dimethylformamide (DMF) at 50 °C. Five wt% HA nanoparticles

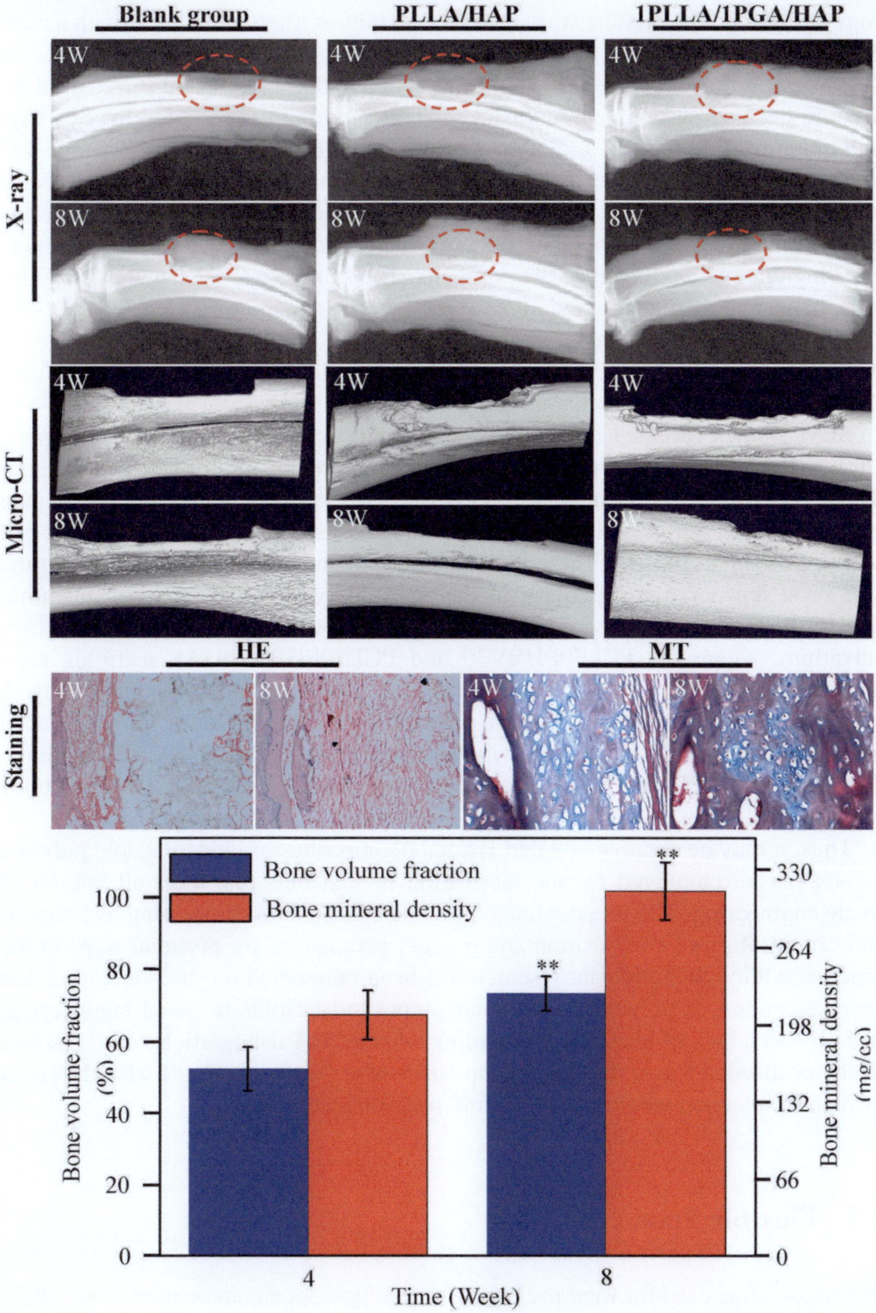

Fig. 8.5 X-ray radiographs and micro-CT images of the 1PLLA/1PGA/HAP scaffold, PLLA/HAP scaffold and blank group after 4 and 8 weeks of implantation. HE and MT images, bone volume fraction and bone mineral density of the 1PLLA/1PGA/HAP scaffold after 4 and 8 weeks of implantation (**$P < 0.01$). Data were presented as mean ± standard deviation. Reproduced with permission from [3]. Copyright 2021, the Authors (This is an open access article)

were added, and the mixture was stirred for 30 min. A mixture of NaCl (with a mass 30 times greater than that of the polymers) and poly(ethylene glycol) (PEG) (with the same mass as that of the polymers) was added to the solution and stirred until the mixture became pasty, thick, and packed. This mixture was air-dried for two days in a lyophilisation flask to allow for solvent evaporation. The resulting sample was immersed in distilled water at 21 °C for five days to leach out the porogens (NaCl and PEG). The water was replaced with fresh water every six to eight hours, and the leaching process continued until the weight of the wet sample became constant, ensuring complete removal of the porogens. The sample was frozen at −70 °C for three hours and then transferred into a cryostat bath containing ethylene glycol (−15 °C) for lyophilisation of the remaining water and chloroform. The scaffold obtained was cut and stored in a dry container. Scaffolds were similarly fabricated with unfilled (70/30) PCL/PHBV for comparative studies.

SEM studies revealed a porous structure with interconnected cells. The presence of HA nanoparticles was found to increase porosity and the rate of biodegradation in PBS. The scaffolds were employed for cell differentiation and animal studies. Cell seeding experiments were conducted using the human osteoblast-like Saos-2 cell line, and ALP activity was assayed on days 7 and 14 post-culture. The bone regenerative capability of the scaffolds was evaluated using quantitative RT-PCR. Critical-sized defects were created on the dorsal region of a rat's calvarium, where the PCL70PHBV30 and PCL70PHBV30HA5 scaffolds were implanted and subsequently examined histologically and histomorphometrically. A greater extent of new bone formation, bone cells and osteons, along with a reduced amount of fibrous connective tissue, was observed in the PCL70PHBV30HA5 group. Thus, the PCL70PHBV30HA5 blend nanocomposite scaffold could accelerate bone regeneration.

Thus, it may be summarised that HA nanocomposites of biodegradable polymer blends can be employed for the fabrication of scaffolds that meet all criteria of tissue engineering. The proper choice of blend components, processing techniques, and optimisation of composition and process parameters are essential steps in the process. Although significant research has been conducted on the fabrication and characterisation of polymer blend nanocomposite scaffolds in tissue engineering, there is still a lack of knowledge regarding whether HA nanoparticles can be selectively localised in one of the blend components or at the interface, and what the effect of such localisation would be on scaffold performance.

8.3 Flexible Food Packaging

Polymeric films used for food packaging should possess a combination of excellent mechanical properties (tensile strength, toughness, flexibility, tear strength, puncture resistance), oxygen and moisture barrier properties, along with additional attributes such as printability and heat sealability, tailored to specific requirements. Biodegradability and anti-microbial properties provide further value. Typically, it is not feasible to achieve all desired properties using a single material, which necessitates the use of multi-layer packaging. The challenge lies in developing materials that fulfil most of the property requirements to minimise the number of layers in multi-layer packaging and, ideally, replace multiple layers with a single layer, resulting in significant cost savings. Polymer blend nanocomposites are promising materials that may serve this purpose, as the appropriate choice of components and processing conditions allows for the tailoring of these multi-phase materials' morphology to meet requirements and achieve a combination of many of these necessary properties.

Various researchers have worked on the development of polymer blend nanocomposites for food packaging applications. In this direction, Pal et al. [23] conducted studies on poly(3-hydroxybutyrate-hydroxyvalerate) (PHBV)/poly(butylene-adipate-*co*-terephthalate) (PBAT)/organically modified montmorillonite (OMMT) films fabricated by two processing techniques, namely compression moulding and cast film extrusion. PHBV was used because of its biodegradability and superior mechanical properties. It was blended with PBAT, another biodegradable polymer, to overcome its limitation of brittleness. Nanoclay was incorporated to achieve superior barrier properties and a further enhancement of mechanical properties.

Nanocomposites blending 40/60 wt/wt PHBV/PBAT with varying clay loadings (0.6, 1.2, and 1.8 wt%) were utilised in the studies. Initially, a pelletised PBAT/nanoclay masterbatch containing 20 wt% clay was produced through melt processing. This masterbatch was then further melt processed with a PHBV/PBAT blend to create nanocomposites of the final composition. The resulting PHBV/PBAT/nanoclay nanocomposite pellets were employed in the fabrication of films via compression moulding and cast film extrusion.

Transmission electron microscopy (TEM) studies (Fig. 8.6) revealed that both types of films exhibited clay intercalated in the PHBV phase and at the interphase. However, the extent of localisation at the interphase was greater in the cast extruded film. This was possibly attributed to the unidirectional orientation achieved during the cast film extrusion process. The significant extent of intercalation was anticipated to result in enhanced barrier properties.

As observed in Fig. 8.7, the cast extruded PHBV/PBAT nanocomposite films demonstrated superior barrier properties (oxygen and water vapour) compared to the compression moulded films. This outcome was attributed to the chain orientation achieved during high-speed stretching. The incorporation of 1.2 wt% nanoclay resulted in approximately 79% and 70% improvements in oxygen and water vapour barrier properties, respectively. This enhancement was attributed to the superior dispersion quality of the clay and the effective interfacial adhesion between the clay and the matrix.

The mechanical properties of the cast extruded films was significantly superior to those of the compression moulded films (Fig. 8.8), which was attributed to the unidirectional orientation of the polymer chains and the clay particles achieved in cast film extrusion.

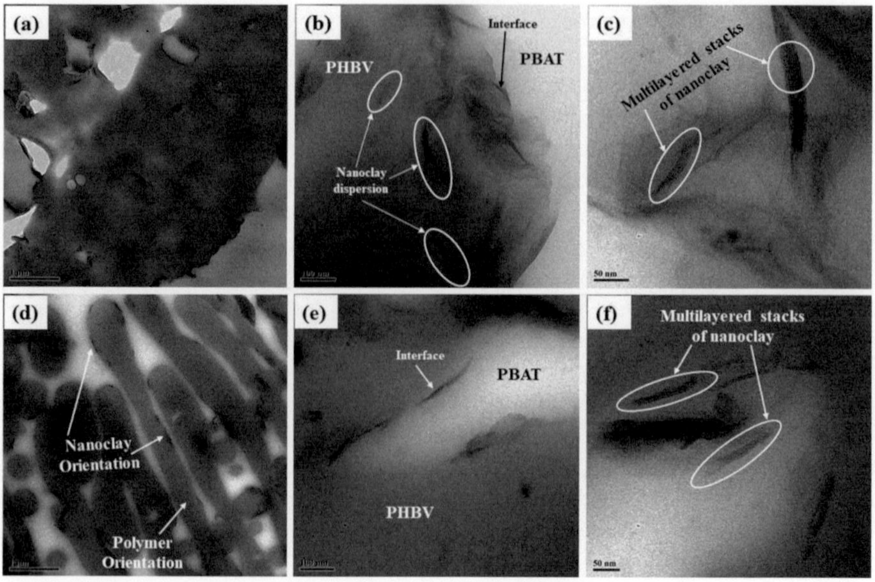

Fig. 8.6 Surface morphology of compression moulded PHBV/PBAT/Nanoclay (1.2%) film at a magnification of 1 μm (**a**), 100 nm (**b**), 50 nm (**c**); and cast extruded PHBV/PBAT/Nanoclay(1.2%) film at a magnification of 1 μm (**d**), 100 nm (**e**), 50 nm (**f**). Reproduced with permission from [23]. Copyright 2020, Elsevier Science Ltd.

Of the samples studied, the cast-extruded film containing 1.2 wt% clay demonstrated the best combination of mechanical and barrier properties, suggesting its potential as a biodegradable flexible food packaging material.

Further value addition to nanoclay reinforced degradable polymer blends suitable for flexible food packaging can be found in the work of Moustafa et al. [24], wherein the researchers imparted antimicrobial properties to organoclay (OC) by modifying it with gum rosin and used the modified clay to produce PLA/PBAT/clay nanocomposites. PBAT was blended with PLA to enhance toughness without compromising biodegradability. OC and gum rosin were first combined in a weight ratio of 1:1 in a ceramic mortar. The mixture was then heated in an oven at 120 °C for 1 h, followed by an additional 20 min of mixing. This step was repeated thrice to ensure the intercalation of rosin into the clay galleries, as confirmed by the increased d-spacing in X-ray diffraction (XRD) studies. The blend nanocomposites were produced via melt mixing. PLA was melted by heating to 170 °C for 3 min in a Brabender mixer. The rosin-modified clay, designated as ROC, was added to the molten PLA and mixed for 5 min. This was followed by compounding the premix with PBAT for another 5 min. The mixing speed was maintained at 80 rpm. Compositions containing 2.5 wt% ROC and PLA/PBAT weight ratios of 25/75, 50/50, and 75/25 were designated as 2575 ROC, 5050 ROC, and 7525 ROC, respectively; these were compression moulded into 1 mm thick sheets, which were used for studies.

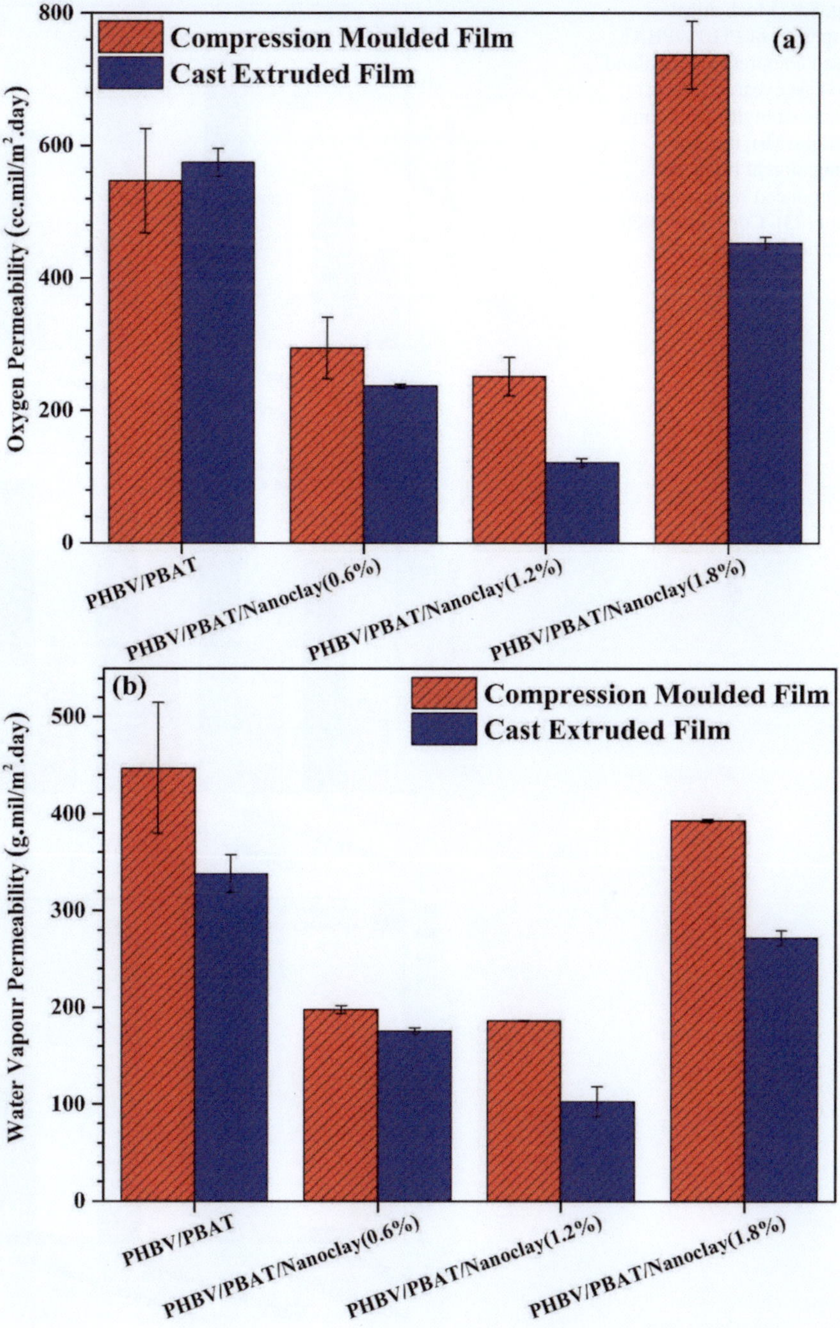

Fig. 8.7 Oxygen permeability (**a**) and water vapour permeability (**b**) of PHBV/PBAT based compression moulded and cast extruded films. Reproduced with permission from [23]. Copyright 2020, Elsevier Science Ltd.

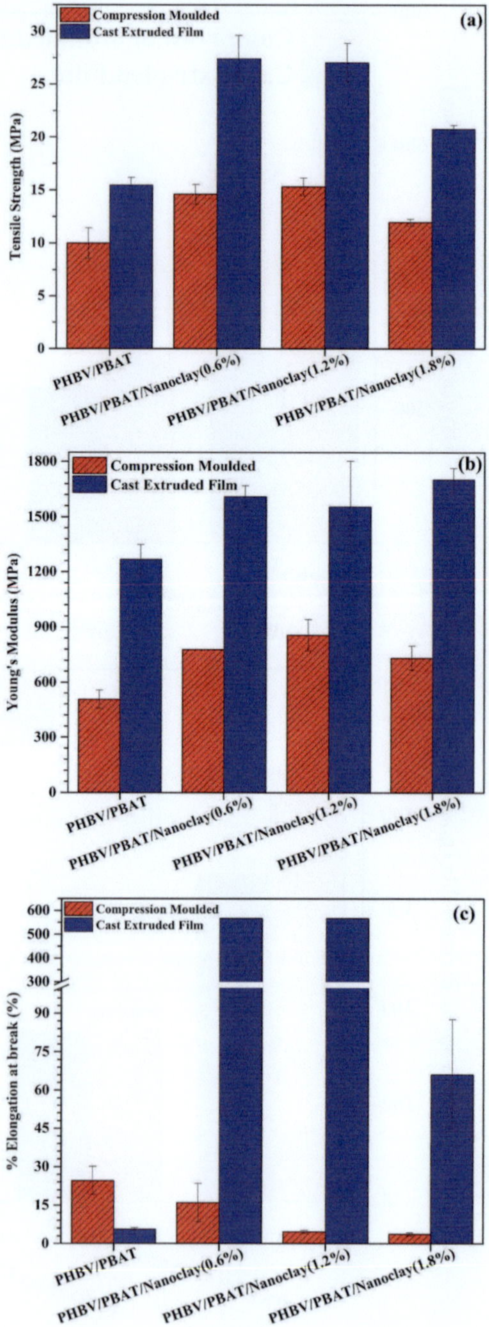

Fig. 8.8 Mechanical properties of PHBV/PBAT based compression moulded and cast extruded films; tensile strength (**a**), Young's modulus (**b**), percent elongation at break (**c**). Reproduced with permission from [23]. Copyright 2020, Elsevier Science Ltd.

TEM studies (Fig. 8.9) revealed the presence of intercalated structures in all the blend nanocomposites. However, an increase in PLA content also resulted in exfoliation, indicating an enhancement in the quality of dispersion, as evident in the TEM micrograph of 7525 ROC.

Neat PLA and the blended nanocomposites were subjected to tensile testing. Neat PLA was found to possess a tensile strength of 116.3 MPa and a strain at break of 15.5%. Blending with 25 wt% ductile PBAT and 2.5 wt% of ROC led to a reduction in tensile strength to ~66.3 MPa for 7525 ROC, while the strain at break increased to 90.3%. When the PBAT content was 50 wt%, both the tensile strength and strain at break decreased, which was possibly attributed to the poor compatibility between PLA and PBAT, which could not be compensated by the clay at the 50/50 ratio of the blend components. 2575 ROC was found to possess a strain at break as high as 1096%. Although the tensile strength was lowered to 19.2 MPa, it remained higher than that of neat PBAT (14.3 MPa [25]). This was attributed to the presence of ROC, which served as both a reinforcement and a compatibiliser. The excess rosin in the clay galleries could possibly react with the hydroxyl terminal groups of PLA and

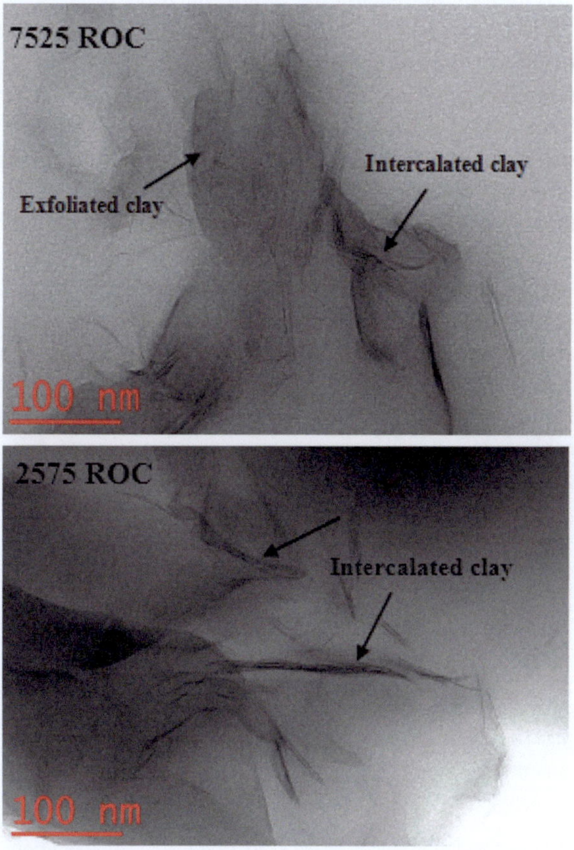

Fig. 8.9 TEM images for 7525 ROC, and 2575 ROC composites. Modified and reproduced with permission from [24]. Copyright 2017, American Chemical Society

PBAT, forming PLA–PBAT mixed chains during melt processing. An increase in the PBAT content resulted in a significant reduction in rigidity, as expected.

The antimicrobial activity of neat PLA and the blend nanocomposites against *Pseudomonas aeruginosa* (Gram-negative), *Staphylococcus aureus* (Gram-positive), and *Candida albicans* (fungi) was evaluated using the disk agar plate technique, following the procedure described in the literature [26, 27]. The test results are presented in Table 8.1 and Fig. 8.10. Neat PLA exhibited no inhibitory effect against any of the microbes tested. In contrast, halos of inhibition were observed for the blend nanocomposites, with diameters varying from 10 to 22 mm, depending on the microbe and the composition of the blend nanocomposite. The enhanced antimicrobial activity of the blend nanocomposites was attributed to the presence of the phenolic structures of the rosin intercalated into the clay galleries. These phenolic structures effectively inhibited the growth of the microbes by interfering with microbial development [28]. The 5050 ROC had a larger inhibitory zone compared to 7525 ROC and 2575 ROC. This was attributed to the poor compatibility of PLA and PBAT, which was not compensated by ROC at this composition. As a result, the ROC nanoparticles migrated outside the polymeric matrix, destroying the microbes and thereby increasing the inhibition zone.

The researchers, through their extensive studies, thus demonstrated that PLA/PBAT/rosin-modified clay nanocomposites possess superior mechanical and antimicrobial properties, making these multi-phase materials suitable for antimicrobial biodegradable food packaging applications. Although the barrier properties of these materials were not evaluated, based on existing literature wherein the barrier properties of clay nanocomposites are well-established [29–32], it can be safely inferred that these materials would also provide superior barrier properties. Furthermore, the combination of mechanical and antimicrobial properties makes these materials suitable for biomembranes and biomedical applications.

Although the enhancement of barrier properties is due to the presence of well-dispersed clay particles, little knowledge exists regarding the effect of selective localisation of clay particles on the barrier properties of polymer blend nanocomposites. In this context, Bitinis et al. [4] investigated the gas permeability of PLA/natural rubber (NR)/C30B nanocomposites with 10 wt% NR and 3 wt% clay contents. The clay particles were located in the continuous PLA phase and at the interface when

Table 8.1 Antimicrobial activity for pristine PLA and its blends with different ratios of PBAT containing 2.5 wt% of ROC or stearic acid against the three microbes

Diameter of inhibitory zone (MM)			
Sample	*Pseudomonas aeruginosa*	*Staphylococcus aureus*	*Canadida albicans*
Pristine PLA	0	0	0
7525ROC	10 ± 0.3	12 ± 0.1	13 ± 0.4
5050ROC	15 ± 0.6	18 ± 0.4	22 ± 0.2
2575ROC	14 ± 0.2	17 ± 0.2	10 ± 0.3

Modified and reproduced with permission from [24]. Copyright 2017, American Chemical Society

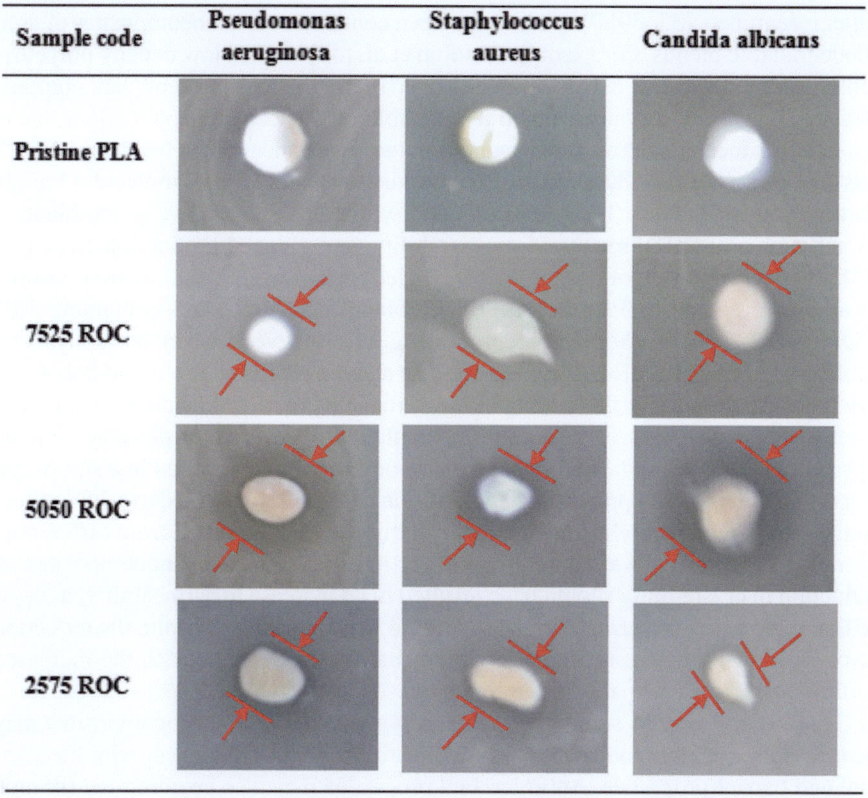

Fig. 8.10 Antimicrobial activity for pristine PLA and its blends with different ratios of PBAT containing 2.5 wt% of ROC against *Pseudomonas aeruginosa*, *Staphylococcus aureus*, and *Candidia albicans*. Modified and reproduced with permission from [24]. Copyright 2017, American Chemical Society

all components were mixed simultaneously, as observed in TEM studies. When C30B was premixed with PLA, and this mixture was compounded with NR, the clay was found to be selectively localised in the continuous PLA phase, resulting in the nanocomposite exhibiting oxygen and carbon dioxide permeability values of 0.68 barrer and 1.86 barrer, respectively. Thus, the researchers demonstrated that selective localisation of nanoclay in the continuous phase can enhance the gas barrier properties of nanoclay-reinforced immiscible polymer blends. However, further research is necessary in this field, particularly related to blend nanocomposites of food-grade materials [33, 34].

The superior properties of biodegradable blend nanocomposites, highlighting their suitability as materials for food packaging applications, have been reported by various other researchers [35–39]. Work has also been conducted on nanocomposites of non-biodegradable blends. For example, Dadbin et al. [40] studied low density polyethylene (LDPE)/linear low density polyethylene (LLDPE)/OMMT blend nanocomposites. LLDPE is a widely used material in flexible food packaging applications due to its superior mechanical properties and puncture resistance [41]. However, LLDPE has low processability due to its high melt viscosity and narrow molecular weight distribution [42]. Thus, LLDPE is blended with LDPE to achieve a combination of superior properties, improved processability, and reduced production costs [41, 43]. The incorporation of clay enhances barrier properties. In this direction, Dadbin et al. [40] prepared LDPE/LLDPE/OMMT blend nanocomposites containing 70% LDPE and 3–7 phr OMMT by melt processing. Five to seven wt% maleic anhydride-grafted LLDPE (LLDPE-g-MA) was used as a compatibiliser. The blend nanocomposites were utilised for fabricating blown films. XRD and TEM studies revealed an intercalated structure, with maximum intercalation observed at 3 phr clay content. Permeability studies indicated that the incorporation of clay resulted in a significant improvement in barrier properties. The addition of only 3 phr clay led to a 50% reduction in oxygen permeability. However, crystallinity, as determined from differential scanning calorimetry, was found to be lower in the case of the nanocomposites. A reduction in crystallinity should have resulted in an increase in permeability, as crystalline regions are impermeable. The decrease in permeability despite the reduction in crystallinity suggests an enhancement of barrier properties due to the increased tortuosity of the travel path caused by the presence of clay particles.

Therefore, it can be summarised that polymer blend clay nanocomposites may be utilised for flexible food packaging applications, which possess superior mechanical and barrier properties. Antimicrobial properties may also be conferred through suitable modification of the clay. The appropriate selection of blend components is essential for meeting product requirements while maintaining cost viability.

8.4 EMI Shielding

8.4.1 Fundamentals of EMI Shielding and EMI Shielding Materials

In the present era of the electronic revolution, the gradual rise in the use of electronic devices has led to the undesirable effect of electronic pollution, specifically electromagnetic interference (EMI). EMI occurs when electromagnetic (EM) waves from one electronic device interfere with another, adversely affecting its performance [44]. This unwanted EM pollution, in addition to impairing the proper functioning of electronic devices, can also lead to radiation-related hazards for living beings [45–49]. Consequently, there is significant interest among researchers in developing materials to shield against these waves and mitigate the harmful effects of EMI [5, 44, 50].

EMI shielding is a phenomenon in which a suitable material acts as a shield and prevents the penetration of high-frequency EM waves through either reflection or absorption. Shielding mechanisms involve absorption, reflection, and multiple reflections. An ideal shielding material should exhibit 100% shielding, resulting in zero transmission. However, the ideal scenario is yet to be achieved.

The critical property of EMI shielding materials is shielding effectiveness (SE). However, mechanical, optical, morphological, and electrical properties (such as permeability and conductivity) are also important [51]. In shielding by reflection, exhibited by electrically conductive shielding materials, the material interacts with mobile charge carriers, leading to the reflection of most EM waves. In shielding by absorption, the radiation is absorbed by the material due to its interaction with electric and/or magnetic dipoles. Shielding by multiple reflections occurs due to the presence of scattering centres, defects, and interfaces within the shielding material. Total SE results from the shielding of all three mechanisms [52]. Reflection occurs due to impedance mismatch, which arises when the incident EM waves strike the material surface. This is due to the presence of mobile charge carriers (electrons and holes) on the surface. The absorption of EM waves results from the loss or dissipation of their energy, subsequent to ohmic and polarization losses. Multiple reflections are a consequence of scattering effects and allow for further attenuation due to absorption [5, 53].

EMI is classified based on the mode of travel and the frequency range of EM waves. Each frequency range has its uses, and to prevent EMI, materials must be designed according to frequency [54]. EM waves of any frequency can cause EMI. However, EMI shielding materials predominantly operate within the microwave and radio wave ranges, as EM waves of these frequencies are commonly used in appliances.

The unit of shielding effectiveness (SE) is the decibel (dB). The greater the SE, the higher the shielding effectiveness. SE values of 20 dB, 30 dB, and 40 dB correspond to shielding efficiency values of 99%, 99.9%, and 99.99% respectively [44].

SE, defined as the ratio of the incident EM radiation power to the transmitted radiated power, is expressed as (8.1) [55–59]:

$$SE = 10\log\left(\frac{P_{\text{in}}}{P_t}\right) = 20\log\left(\frac{E_{\text{in}}}{E_t}\right) = 20\log\left(\frac{H_{\text{in}}}{H_t}\right) \tag{8.1}$$

where P_{in}, E_{in}, and H_{in} refer to the intensities of the incident power, and the electric and magnetic fields, respectively, while P_t, E_t, and H_t represent their corresponding counterparts transmitted through the EMI shielding material. A high EMI SE value indicates reduced energy transmission through the shielding material. According to the Schelkunoff theory [60], the total EMI SE value, SE_T, is the sum of the SE values related to reflection (SE_R), absorption (SE_A), and multiple reflections (SE_M). SE_T is expressed as (8.2) [61–64]:

$$SE_T = SE_R + SE_A + SE_M \tag{8.2}$$

SE_R is given by (8.3) [65, 66]:

$$SE_R = 20\log\frac{(Z+Z_0)}{4ZZ_0} = 39.5 + 10\log\frac{\sigma}{2f\pi\mu} \tag{8.3}$$

where Z and Z_0 represent the impedances of the shield and air, respectively; σ and μ denote the electrical conductivity and magnetic permeability of the shield, respectively; and f signifies the frequency of the incident EM waves.

Metals exhibit EMI shielding through reflection. However, the ability to tailor the EMI SE of metals is limited. Furthermore, electromagnetic pollution is not entirely eliminated or reduced by shielding via reflection. This is because only the area beyond the shield is safe, and all electromagnetic waves are reflected off the shield's surface [67].

SE_A is given by (8.4) [68, 69]:

$$SE_A = 20\left(\frac{d}{\delta}\right)\log_{10} e = 8.68\left(\frac{d}{\delta}\right) = 8.68d\frac{\sqrt{f\mu\sigma}}{2} \tag{8.4}$$

where d represents the thickness of the shielding material, and δ denotes the skin depth of the material. Skin depth refers to the distance at which the intensity of the EM wave is reduced to $1/e$; that is, the energy of EM radiation is lowered to approximately 33% of its original strength [70, 71].

SE_M is given by (8.5) [72, 73]:

$$SE_M = 20\log_{10}\left(1 - e^{-\frac{2d}{\delta}}\right) = 20\log_{10}\left(1 - 10^{\frac{SE_A}{10}}\right) \tag{8.5}$$

The SE_M value is significantly influenced by the thickness of the material. It becomes negligible at thickness values comparable to or greater than the skin depth or when SE_T exceeds 15 db. Internal multiple reflections are crucial in foams, as they reflect the incident EM microwaves multiple times, leading to the dissipation of EM energy [74].

Materials used for EMI shielding include metals [75, 76], carbonaceous materials [75, 77], ceramics [78, 79], cement (or concrete) [80, 81], conducting polymers [82, 83], and conducting polymer composites along with their foams [84, 85]. Various types of conducting polymer composites exist, such as metal particle-reinforced polymers [86, 87], carbon black-reinforced polymers [88, 89], carbon fibre-reinforced polymers [90, 91], carbon nanoparticle-reinforced polymers [84, 85, 92, 93], and their blends [94, 95]. Among conducting polymer composites, there is considerable interest in carbon nanofiller-reinforced polymer blends. This is due to the potential to achieve high electrical conductivity and SE at low filler loading by selectively localising the filler particles within one of the blend components in a co-continuous morphology or at the interface, resulting in the formation of a percolating network of conductive particles despite a low filler loading.

8.4.2 *Polymer Blend Nanocomposites as EMI Shielding Materials*

Researchers have worked towards achieving a very low percolation threshold through the selective localisation of carbon nanoparticles in one of the phases of co-continuous blends. In this regard, Ravindren et al. [96] prepared multiwall carbon nanotube (MWCNT) reinforced co-continuous poly(ethylene-co-methyl acrylate) (EMA)/ethylene-octene copolymer (EOC) blends with the MWCNTs selectively localised in the EMA phase. The blend nanocomposites were produced via solvent casting. The dried solvent cast films were homogenised in a two-roll mill and hot pressed into sheets. Based on the findings of their earlier studies [97], the researchers employed a (50/50) (wt/wt) EMA/EOC composition to achieve a co-continuous morphology. Co-continuity was indeed observed in SEM micrographs of hexane-etched samples of the (50/50) (wt/wt) EMA/EOC blend, with hexane serving as a solvent for EOC, but not for EMA. Various MWCNT loadings were utilised for comparative studies. The process of blend composite preparation has been schematically illustrated in Fig. 8.11.

High resolution TEM (HR TEM) studies (Fig. 8.12) revealed that the MWCNTs were selectively dispersed in one of the phases, which was proven to be the thermodynamically preferred EMA phase in selective dissolution studies. In selective dissolution studies, when the blend nanocomposites were treated with xylene, a solvent for EMA but not for EOC, the resulting solution was of an intense black colour. This was not the case when the blend nanocomposites were treated with hexane, a solvent for EOC. The outcomes of the selective dissolution test and the HR

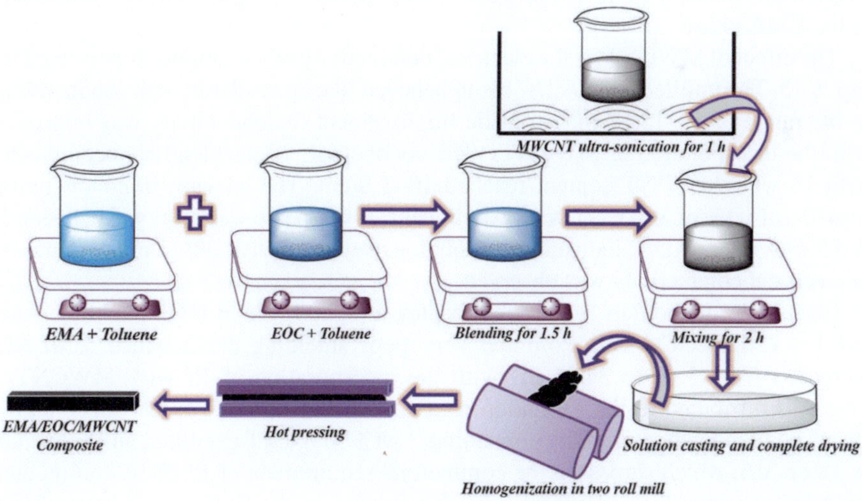

Fig. 8.11 Schematic representation of composite preparation. Reproduced with permission from [96]. Copyright 2019, Elsevier Science Ltd.

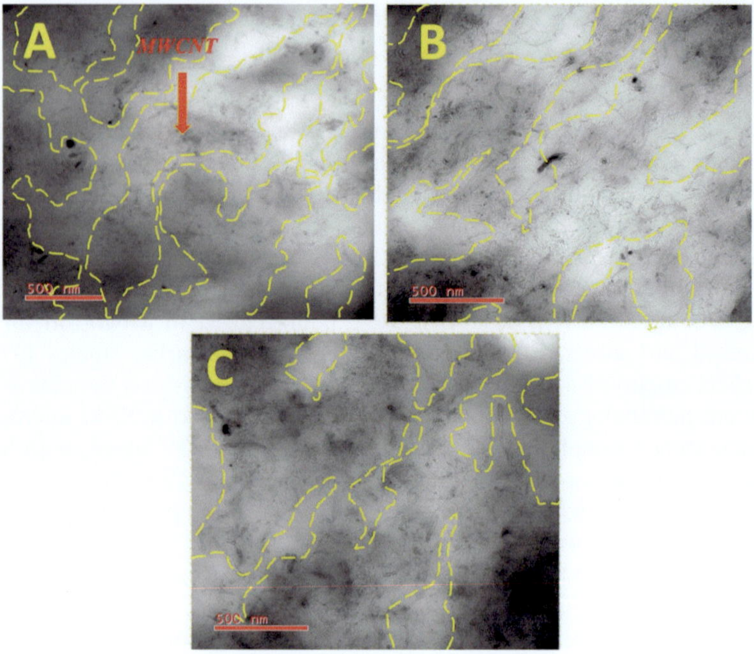

Fig. 8.12 HRTEM micrographs for EMA/EOC blend composite containing **a** 3 wt%, **b** 5 wt% and (C) 15 wt% of MWCNT. Reproduced with permission from [96]. Copyright 2019, Elsevier Science Ltd.

TEM studies clearly indicated that the filler particles were preferentially dispersed in the EMA phase.

The effect of MWCNT on the electrical conductivity of the samples is presented in Fig. 8.13. The unfilled EMA/EOC blend behaved like an insulator, with conductivity in the range of 10–15 S/cm. A drastic improvement in conductivity was observed with the incorporation of MWCNTs. The conductivity of the blend nanocomposite with 15 wt% MWCNT content reached 10–2 S/cm. The increase in conductivity from 0 to 1 wt% of MWCNT was insignificant. A sharp rise was observed between 1 and 5 wt% of MWCNT, indicating percolation. Beyond 5 wt%, only a minor increase in electrical conductivity was observed.

Electrical conductivity was found to play a pivotal role in EMI shielding. The unfilled EMA/EOC blend exhibited very poor shielding performance. EMI SE increased from 3.46 to 34.06 dB with the incorporation of 15 wt% MWCNTs. This was attributed to the formation of a conducting network of MWCNT particles. The blend nanocomposite containing 3 wt% MWCNT exhibited an SE_T value of 18.66 dB, which surpassed the commercial requirement of 15 db. A continuous percolating network of MWCNT particles formed in the EMA phase beyond the percolation threshold, leading to the predominance of the absorption mechanism in

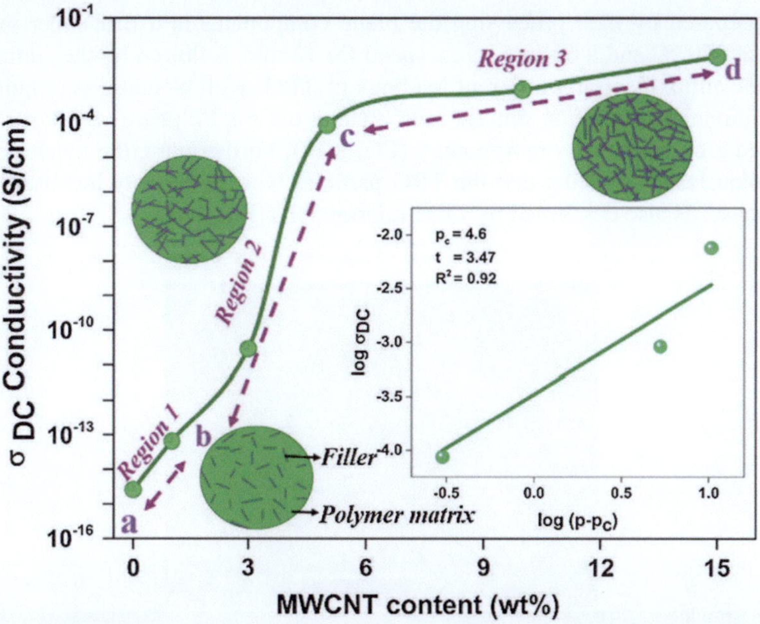

Fig. 8.13 Variation in DC conductivity with MWCNT. Inset: log σ plotted against log $(p - p_c)$ by using the least squares method. Reproduced with permission from [96]. Copyright 2019, Elsevier Science Ltd.

EMI shielding, as the particles were unevenly distributed, i.e. preferentially localised in the EMA phase.

The skin depth was found to decrease with an increase in filler loading, and the blend nanocomposite with 15 wt% MWCNT content exhibited a skin depth of approximately 0.26 nm. This indicates that a thin sample with high filler loading would be effective in attenuating EM waves.

EM SE was also found to be influenced by sample thickness, with an increase in thickness leading to a rise in SE. A sudden improvement in SE_T from 26.2 to 37.2 dB was observed when the thickness was increased from 0.5 to 2 mm.

Furthermore, only a slight reduction in SE was observed when the blend nanocomposite containing 15 wt% MWCNT was subjected to 500 cycles of bending. Thus, the shielding material remained effective even under harsh dynamic conditions.

The researchers were thus successful in fabricating a polymer blend nanocomposite with superior electrical conductivity and a low percolation threshold by selectively localising MWNTs in one of the phases of the blend.

A similar combination of high electrical conductivity and superior EMI SE at a low percolation threshold was reported by Sadeghi et al. [98] in their studies on thermally reduced graphene (TRG) reinforced co-continuous (50/50 wt/wt) polypropylene (PP)/poly(ethylene terephthalate) blends, with the TRG particles selectively localised in the thermodynamically preferred PET phase. The samples in this case were prepared by melt processing the blend components in a Brabender internal mixer at 270 °C and a 60 rpm screw speed for 12 min, followed by the addition of 0.1 wt% antioxidant and different loadings of TRG, with blending continuing for an additional 3 min. SEM studies, after etching out the PP phase with hot xylene, revealed a co-continuous morphology (Fig. 8.14). Furthermore, the xylene extract was colourless, indicating that the TRG particles were selectively localised in the PET phase, as also confirmed by TEM micrographs (Fig. 8.15).

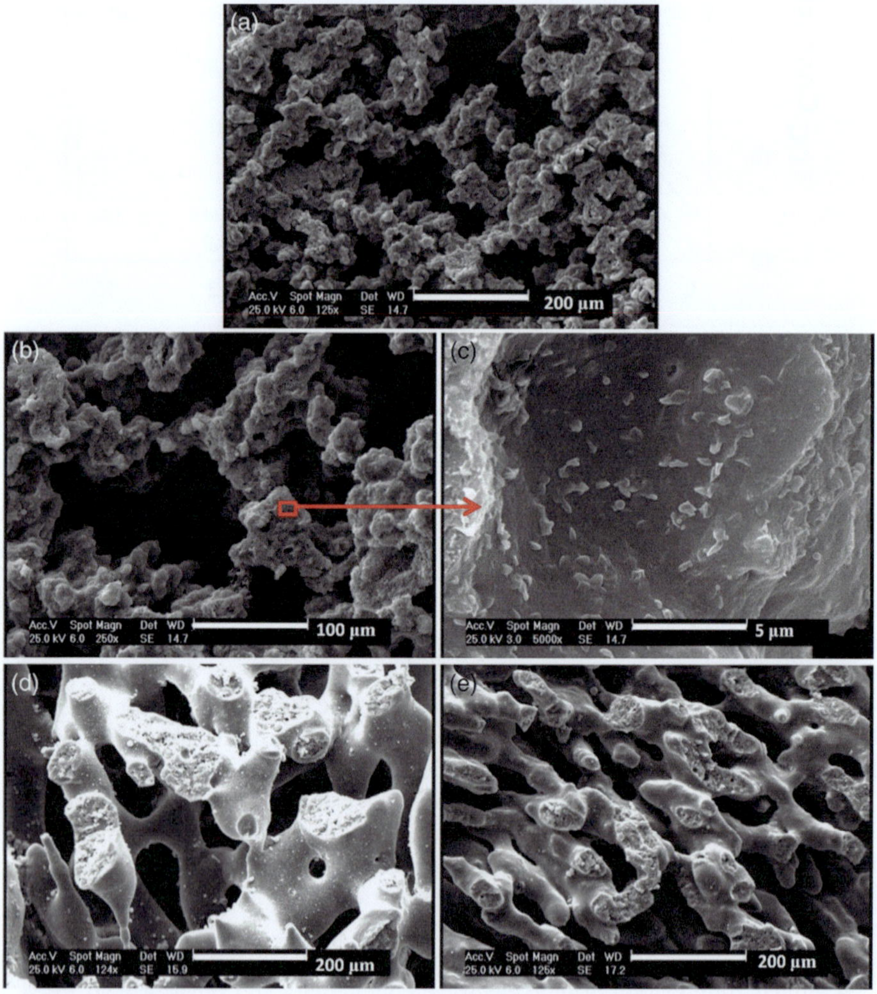

Fig. 8.14 SEM images of **a** the neat PP/PET 50/50 blend; and those containing **b** and **c** 0.22 vol%; **d** 0.44 vol%, and **e** 0.66 vol% of TRG. Reproduced with permission from [98]. Copyright 2019, Wiley Periodicals LLC

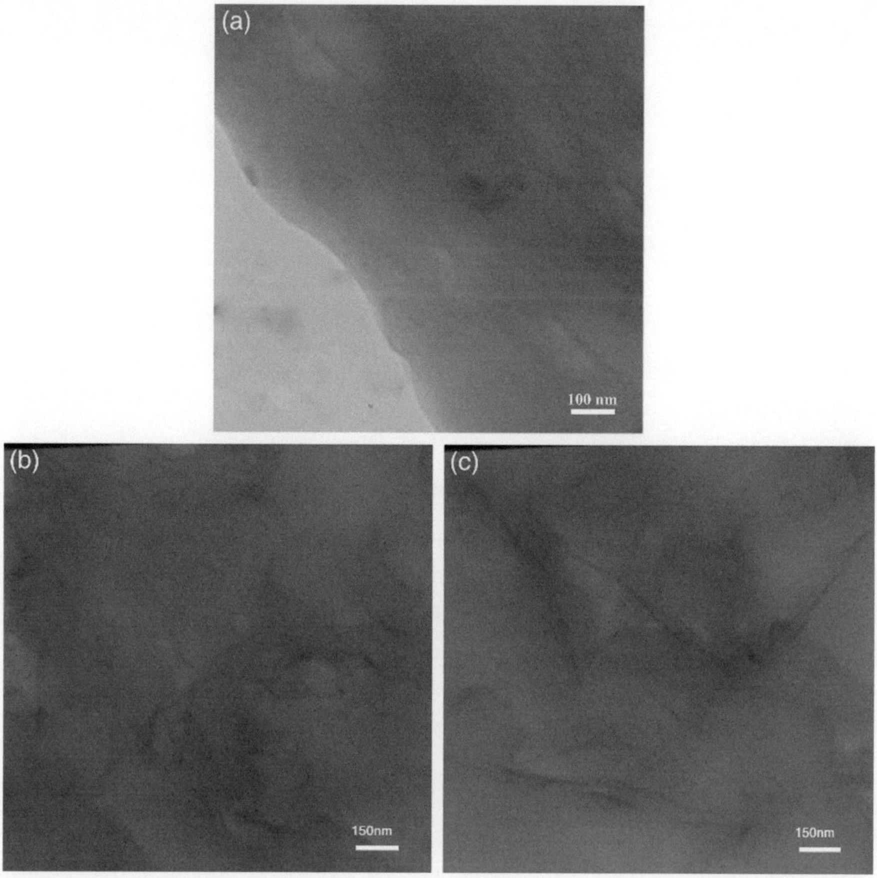

Fig. 8.15 TEM images of the PP/PET sample containing 0.44 vol% TRG: **a** from two-phase region; **b**, **c** only within PET polymer. Reproduced with permission from [98]. Copyright 2019, Wiley Periodicals LLC

The incorporation of TRG led to a significant increase in the electrical conductivity of the blends, with a percolation threshold as low as 0.08 vol% being achieved. EMI SE was found to rise with increasing TRG loading (Fig. 8.16), and depending on the frequency, an SE value in the range of 40–60 dB was attained with the addition of only 2 vol% TRG.

Researchers have also achieved a reduced percolation threshold, leading to superior EMI shielding performance at low filler loadings by selectively localising the conductive filler particles at the blend interface. In this regard, Kar et al. [99] employed a unique strategy to localise MWCNTs at the interface of (50/50 wt/wt) poly(vinylidene fluoride) (PVDF)/acrylonitrile butadiene styrene (ABS) blends by wrapping the CNTs with a mutually miscible polymer, PMMA. The PMMA-wrapped MWCNTs were produced by sonicating pristine MWCNTs in o-dichlorobenzene for

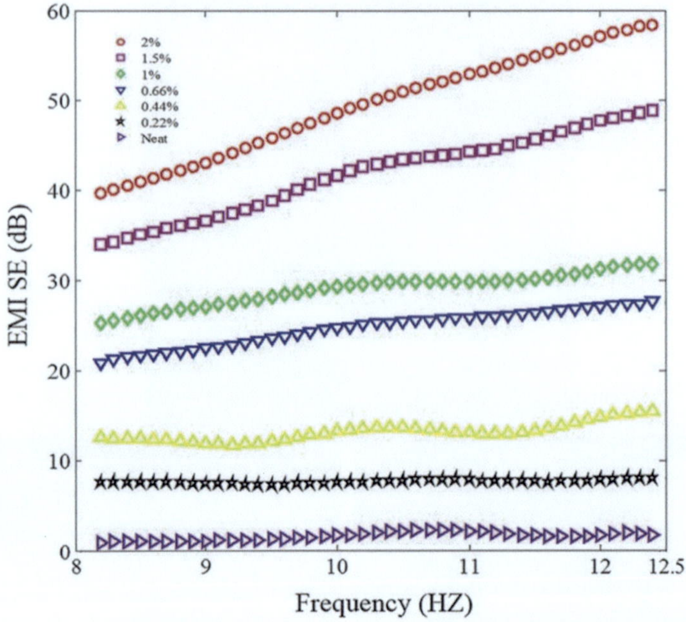

Fig. 8.16 The EMI shielding effectiveness as a function of frequency for PP/PET samples containing different volume fractions of TRG. Reproduced with permission from [98]. Copyright 2019, Wiley Periodicals LLC

20 min, followed by the addition of methyl methacrylate and azobisisobutyronitrile to the mixture. The resultant mixture was poured into a round-bottomed flask, the dissolved oxygen was removed, and the flask was sealed. The round-bottomed flask was heated with constant stirring in a pre-heated oil bath, under a nitrogen atmosphere, for 24 h. The black suspension was mixed with chloroform and dried under vacuum. The presence of PMMA on the MWCNT surface was confirmed using infrared spectroscopy. SEM studies on the blend after selectively etching out the ABS phase revealed a co-continuous morphology.

Blend nanocomposites with varying loadings of PMMA-wrapped MWCNTs were produced by melt processing in a twin-screw microcompounder at 230 °C and a screw speed of 60 rpm for 20 min. The unfilled blends and blend nanocomposites of pristine MWCNTs were also prepared under the same conditions. Selective dissolution studies showed that, unlike the blend nanocomposites of pristine MWCNTs, in which the MWCNTs were selectively localised in the PVDF phase, the blend nanocomposites of PMMA-wrapped MWCNTs had the filler particles localised at the blend interface.

Compared to the blend nanocomposites of pristine MWCNT, the blend nanocomposites of PMMA-wrapped MWCNTs exhibited a lower percolation threshold. This was attributed to the interfacial localisation of the filler particles, which facilitated the formation of an interconnected network of CNTs at a lower filler loading. Furthermore, at 3 wt% filler loading, the blend nanocomposite of PMMA-wrapped CNTs exhibited a higher electrical conductivity than the blend nanocomposite of pristine MWCNTs. This was attributed to the fact that interfacial localisation resulted in a more interconnected network at the same filler loading.

Selective localisation of the PMMA-wrapped CNTs at the interface resulted in a significant improvement in the EMI SE of the blend nanocomposites. The blend nanocomposite with 3 wt% pristine MWCNTs exhibited an SE of 24 dB, whereas the same loading of PMMA-wrapped MWCNTs achieved an SE of 32 db. Furthermore, in the blend nanocomposites of pristine MWCNTs, shielding was primarily due to reflection (83%), while in the blend nanocomposites of PMMA-wrapped CNTS, shielding was mainly due to absorption (62%). This was possibly attributed to the high dielectric constant of PMMA, which caused charge attenuation over a longer time scale. The researchers successfully enhanced the SE of conductive polymer blend nanocomposites with a co-continuous blend morphology by localising the nanoparticles at the blend interface. A similar improvement in SE due to interfacial localisation of MWCNTs was also reported by Lencar et al. [100] in their studies on MWCNT reinforced PVDF/polyethylene blends.

Researchers have frequently employed CNTs alongside magnetic nanoparticles to enhance SE [101–103]. The utilisation of magnetic nanomaterials facilitates the dissipation of incident EM waves through absorption. A typical instance of utilising hybrid nanofillers to achieve improved EMI SE can be found in the investigations by Pawar et al. [101] on polycarbonate (PC)/poly(styrene-acrylonitrile) (SAN) blends reinforced with CNTs that were grafted onto dopamine-anchored Fe_3O_4 nanoparticles. The hybrid nanofiller not only demonstrated a superior quality of dispersion compared to CNTs but also provided enhanced EMI SE. The best results stemmed from the 60/40 blend, attributed to double percolation of the conductive filler particles and the enhanced dispersion of the hybrid nanofiller. The same research group examined the effect of various magnetic nanomaterials adorned onto graphene sheets (rGO-Ni [104, 105] rGO-Fe_3O_4 [106]), used in conjunction with CNTs in PC/SAN blends. In all instances, the nanoparticles were preferentially localised within the PC phase and formed an interconnected network, leading to an improvement in electrical conductivity and EMI SE. Furthermore, the high dielectric and magnetic permeability losses augmented the attenuation of incident EM waves through absorption. A loading of 10 wt% rGO-Fe_3O_4 combined with 3 wt% CNTs proved effective in achieving a total SE of 50.7 db. The enhancement in SE through absorption was attributed to the high saturation magnetisation of Fe_3O_4.

Another strategy employed to enhance the EMI SE of conductive polymer blend nanocomposites is to foam the material. The advantages of foaming are numerous. Firstly, foaming can result in significant weight reduction, thus promoting sustainability. Secondly, incorporating gas may improve the dispersion of nanoparticles by lowering the melt viscosity [107]. Subsequent foaming can induce the orientation of filler particles within the cell wall [108], aiding the formation of percolating networks, reducing the percolation threshold, and enhancing EMI SE [84, 109–116]. Furthermore, the foam cells create sites for multiple internal reflections [5].

A typical example of the use of foaming to enhance the EMI SE of conductive polymer blend nanocomposites can be found in the work of Zhao et al. [117]. The researchers investigated the preferential distribution of two fillers in the two components of a blend and the consequent impact of this distribution on foaming and the EMI SE of the foams. PE/PVDF/ferrite (Fe_3O_4)/CNT nanocomposites containing 5 wt% Fe_3O_4, a PE:PVDF weight ratio of 1:1, and various CNT loadings were used for studies. High-density PE was used for the compositions. Nanocomposites were produced by sequential melt blending. Initially, PE, the ferrite, and the CNTs were melt processed in a torque rheometer at 50 rpm and 220 °C for 8 min to produce a conducting network. This was followed by the introduction of PVDF and blending for another 7 min. The blends were granulated and moulded into 1.4-mm-thick rectangular specimens at 220 °C and 0.25 MN pressure for 10 min. The moulded specimens were foamed by pressure-quench batch foaming using carbon dioxide as a blowing agent at a temperature of 175 °C, a pressure of 16 MPa, and a gas saturation time of 1 h.

SEM micrographs and elemental mapping of the nanocomposites are presented in Fig. 8.17a–c. The blend morphology is characterised by the island-sea type. Based on the density values and the volume ratios of the two blend components, PE is identified as the continuous phase and PVDF as the dispersed phase. The Fe_3O_4 particles and the CNTs are found to be localised in the PVDF phase and the PE phase, respectively. The localisation of Fe_3O_4 in the PVDF phase is attributed to the higher viscosity of PVDF, as noted from the complex viscosity data (Fig. 8.17d), as well as to the polarity of PVDF, which enhances its affinity for Fe_3O_4. The elemental mapping of Fe, F, and C aligns with the SEM findings.

SEM studies on the foams (Fig. 8.18) revealed closed cells of uniform sizes. The inner walls of the cells exhibited a relatively smoother appearance compared to the outer walls. This was attributed to the two separate polymer matrices and the selective distribution of the fillers in the two blend components. As the foaming temperature was close to the melting temperature of PVDF but significantly higher than that of PE, the researchers inferred that the PVDF phase was selectively foamed. In contrast, the PE phase acted as the continuous phase surrounding the PVDF cells. The distribution of PE and PVDF in the foam structure, as indicated by EDS studies and elemental mapping, aligned with the interpretation of the SEM micrographs. The Fe and F contents on the inner cell wall were relatively high, whereas the C content was elevated outside the cells.

As observed in Fig. 8.19, the nanocomposite foams exhibited higher electrical conductivities than their solid counterparts with the same CNT content. For a deeper understanding of the EMI shielding performances of the materials, specific SE values (SE_T value per unit material density per unit sample thickness) were compared. All the foams demonstrated higher specific SE values than their solid counterparts. Thus, the researchers successfully fabricated foams of blended nanocomposites with selective localisation of nanoparticles, whereby one of the blend components was selectively foamed. These foams showcased higher electrical conductivity and superior EMI shielding performance compared to their solid counterparts. However, studies

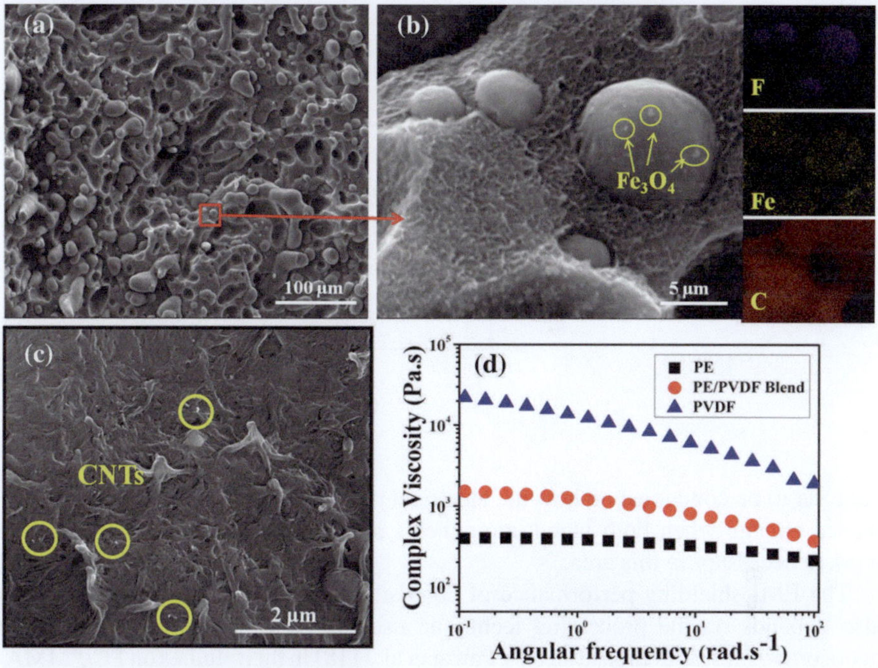

Fig. 8.17 **a–c** SEM images of PE/PVDF/Fe_3O_4/CNT nanocomposites with 2 wt% CNTs and 5 wt% Fe_3O_4. The elemental mapping of F, Fe, and C is shown in (**b**) [**a** 300×, **b**, 5000×, **c** 50,000×]. **d** Complex viscosities of PVDF, PE, and the PE/PVDF blend. Reproduced with permission from [117]. Copyright 2020, Elsevier Science Ltd.

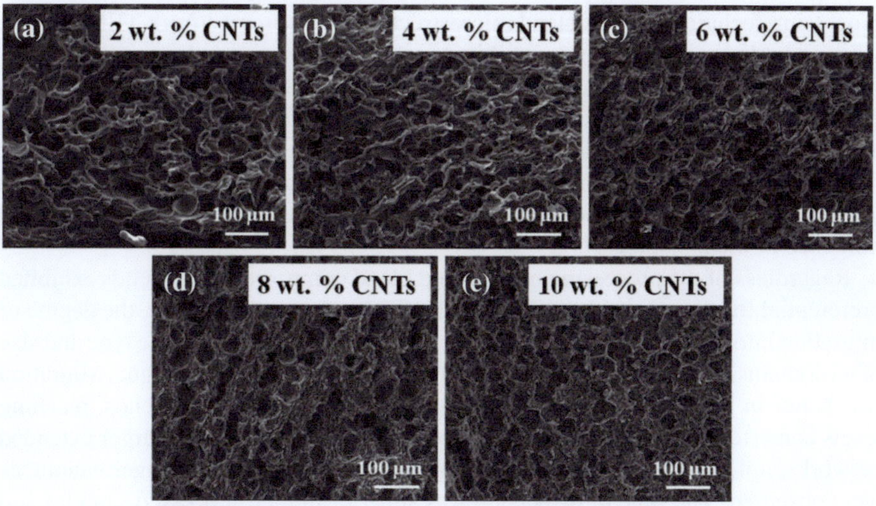

Fig. 8.18 SEM images of PE/PVDF/Fe_3O_4/CNT nanocomposite foams foamed at 175 °C, 16 MPa (200×). Reproduced with permission from [117]. Copyright 2020, Elsevier Science Ltd.

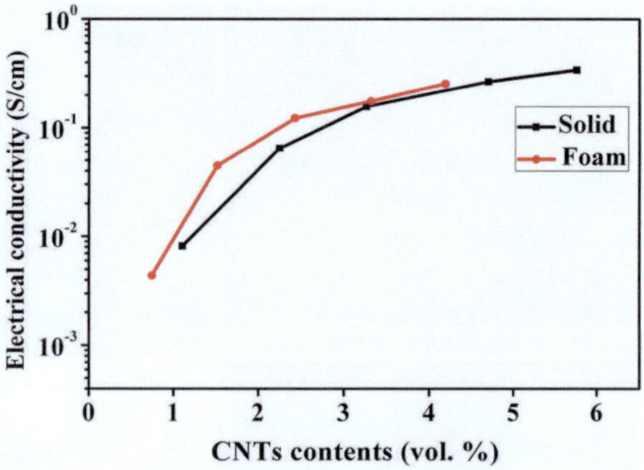

Fig. 8.19 Electrical conductivities of PE/PVDF/Fe_3O_4/CNT nanocomposites and foams with various CNT contents. Reproduced with permission from [117] Copyright 2020, Elsevier Science Ltd.

have yet to be conducted on how the electrical properties and EMI shielding performance change when both blend components are simultaneously foamed. Further work is necessary in this area.

The EMI shielding performance of conductive polymer blend nanocomposites also depends on the processing technique used for the fabrication of the blend nanocomposites, as demonstrated by Pawar et al. [118] in their studies on PC/PMMA/MWCNT blend nanocomposites prepared through solution processing and melt processing. Various compositions (PC/PMMA weight ratios of 30/70, 50/50, 60/40, and 70/30, and MWNT in the range of 1–3 wt%) were employed in the studies. In both processing techniques, island-sea, co-continuous, and phase-inverted morphologies were produced, depending on the blend composition. A significant influence of the processing technique on the blend microstructure was observed from TEM studies, wherein melt-processed blends exhibited co-continuous morphology starting at 50 wt% of PC content, whereas solution-processed blends exhibited such morphologies beginning at 30 wt% PC content. In the latter case, the blends were produced in a miscible state. Phase separation occurred later during annealing at a temperature higher than the glass transition temperature of the components. These conditions led to the restriction of phase separation, which was also favoured by the additional influence of the nanotubes.

Regardless of the processing method and blend composition, all blends exhibited preferential localisation of MWCNTs in the PC component. However, the degree of migration into this component was influenced by the PC content and the type and size of its domains, which, in turn, depended on the blend processing technique. Migration was better in melt-processed blends compared to solution-mixed blends, reaching levels consistent with the PC content. In solution-processed blends, larger extended networks, not confined within the PC component, combined with longer nanotubes were observed. The state of dispersion was superior in melt-processed samples and in those with a higher volume of accommodating PC. The processing technique significantly influenced the EMI SE of the compositions, indicating that intertube and interphase scattering played an important role in the attenuation of incident EM radiation. The highest EMI SE [−27 dB (@18 GHz)] was observed in the PC/PMMA 30/70 solution-processed blend containing 3 wt% MWCNT content. Shielding was primarily due to absorption. These blends exhibited a fine co-continuous morphology. Due to the relatively low PC content and the fine morphology, the long CNTs were not confined within the PC phase but formed segregated networks spanning the sample volume, resulting in enhanced shielding performance. The researchers thus demonstrated the influence of processing technique on the morphology and EMI shielding performance of conductive polymer blend nanocomposites.

Thus, it may be summarised that conductive polymer blend nanocomposites could be suitable for EMI shielding applications. The SE depends on the quality of dispersion of nanoparticles, the blend morphology and composition, and the processing technique used in the production of the blend nanocomposites. The key to achieving a combination of high EMI SE and a low percolation threshold is to produce a percolating network of conductive filler particles at a low filler concentration. This may be accomplished by selectively localising the filler particles in one of the blend components or at the blend interface. The EMI SE of blend nanocomposites can be enhanced by using magnetic nanoparticles in combination with carbon nanoparticles. Foaming of conductive blend nanocomposites is another strategy employed to improve SE, as the concentration of filler particles on the cell walls facilitates the formation of a percolating network. Furthermore, multiple reflections further attenuate EM radiation. Overall, an in-depth understanding of the structure-property-processing relationship of conductive blend nanocomposites and their foams is crucial for fabricating EMI shielding materials of superior performance.

8.5 Energy Storage

Besides EMI shielding applications, conductive polymer blend nanocomposites and their foams can also be utilised for energy storage applications. Batteries and supercapacitors are essential components in energy storage devices [119], and conductive polymer nanocomposites are recognised for their dielectric properties [120–122]. Conductive nanoparticles dispersed in an insulating polymeric matrix function as nanocapacitor electrodes, leading to a significant enhancement of space charge polarization, as shown in Fig. 8.20.

The potential of a conductive polymer nanocomposite for energy storage is influenced by its dielectric constant and dielectric loss behaviour [124, 125]. A high dielectric constant, coupled with a low dielectric loss, results in enhanced capacitance and minimal energy loss. The electrical percolation point is characterised by a sudden and significant rise in both dielectric constant and dielectric loss. The challenge lies in achieving a combination of high dielectric constant and low dielectric loss [126]. The increase in the dielectric constant at the percolation point has been attributed to a greater number of nanocapacitors and a reduced distance between neighbouring nanoparticles [123]. In contrast, the rise in dielectric loss has been ascribed to the formation of conducting networks in the polymer matrix (ohmic loss)

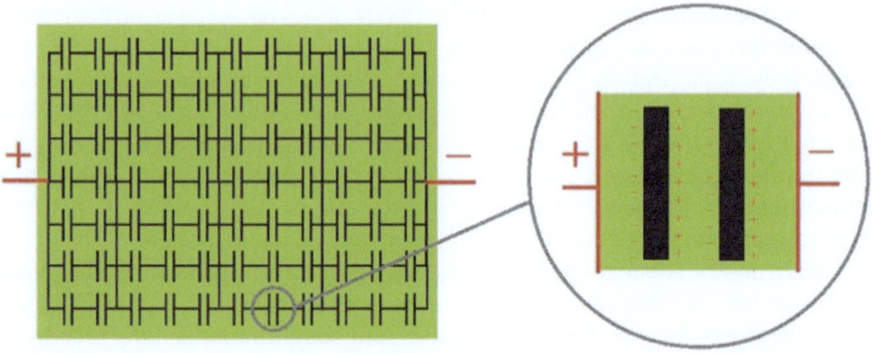

Fig. 8.20 Schematic of the formation of numerous nanocapacitors in nanocomposites. Reproduced with permission from [123]. Copyright 2014, Wiley Periodicals Ltd.

and an increased number of nanocapacitors (polarisation loss). Therefore, conductive polymer nanocomposites with filler contents at or exceeding the percolation threshold are not regarded as ideal dielectric materials [5]. Several strategies have been employed to prevent the formation of a three-dimensional conducting network, ensuring that conductive nanocomposites are suitable for energy storage applications [126]. These include the surface modification of filler particles [127–129], orientating the filler particles perpendicular to the direction of the electric field [130, 131], using hybrid nanofiller mixtures [132, 133], and foaming in such a manner that a three-dimensional conductive network is not created [134–140].

A typical example of fabricating a polymer blend nanocomposite for energy storage applications can be found in the work of Morsi et al. [141], where the researchers produced (50/50 wt/wt) Al_2O_3/TiO_2 hybrid nanofiller reinforced (50/50 wt/wt) polyethylene oxide (PEO)/chitosan (Cs) blends. The TiO_2 particles were prepared using the sol-gel process. Blend nanocomposites with various loadings (0, 0.3, 0.6, 1.2, and 2.4 wt%) of nanofiller were prepared by solution casting, as depicted schematically in Fig. 8.21.

The AC conductivity of the blended nanocomposites exhibited higher values compared to the unfilled blend. The frequency dependence of the dielectric constant (ε') and dielectric loss (ε'') is presented in Fig. 8.22. It is commonly observed that materials initially display polarization with a high value of ε', but an increase in the frequency of the applied field results in a gradual decrease in this value due to the inability of all forms of polarization [electronic (or induced) polarization, ionic polarization, orientational polarization, and interfacial polarization (space charge polarization)] [142] to keep pace with the changes in the field [143–145]. In single- phase materials like neat polymers, only the first three types of polarization are present, while space charge polarization is characteristic of multi- phase materials. The PEO/Cs-Al_2O_3/TiO_2 blended nanocomposite is expected to demonstrate this type of polarization. The observed frequency-dependent behaviour in Fig. 8.23 is attributed to the interface polarization effect or the Maxwell/ Wagner/Sillars effect [146]. In the

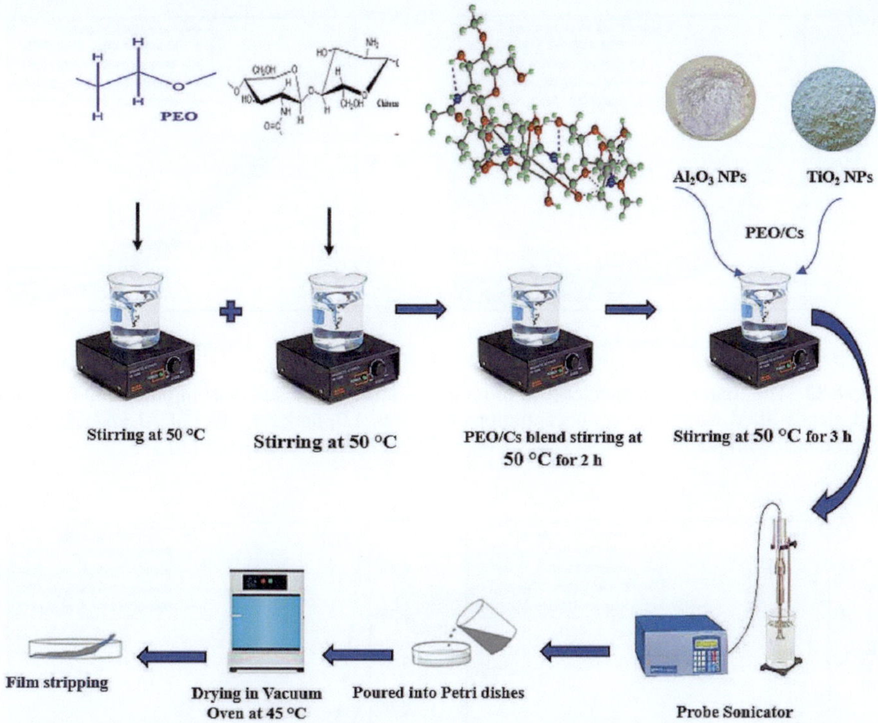

Fig. 8.21 Illustration of the step-by-step process for the preparation of PEO/Cs-Al_2O_3/TiO_2 nanocomposites. Reproduced with permission from [141]. Copyright 2024, Elsevier Science Ltd.

low- frequency regime, interface polarization is predominant, leading to a rapid reduction of both ε' and ε'' with increasing frequency. This behaviour reflects the influence of charge carriers and how they interact with the composite structure. The blended nanocomposites exhibit a prominent response at higher frequencies, which highlights the unique electrical properties introduced by the hybrid nanofillers. The significant changes in dielectric properties are ascribed to the interaction between the hybrid nanoparticles and the blend. The introduction of these nanoparticles altered the dipolar ordering of the PEO/Cs chains, leading to increased dielectric polarization. The interface polarization effect observed in the lower frequency region suggests the formation of localized charge carriers and enhanced electrical conductivity within the nanocomposites. This indicates the suitability of the blended nanocomposites for applications requiring effective charge transport, such as energy storage applications. Furthermore, the observed variation in dielectric properties with frequency and filler content indicates the potential of PEO/Cs-Al_2O_3/TiO_2 nanocomposites for tunable dielectric devices. The changes in dielectric properties with varying frequency and nanofiller concentration provide a basis for tuning the dielectric response according to requirements.

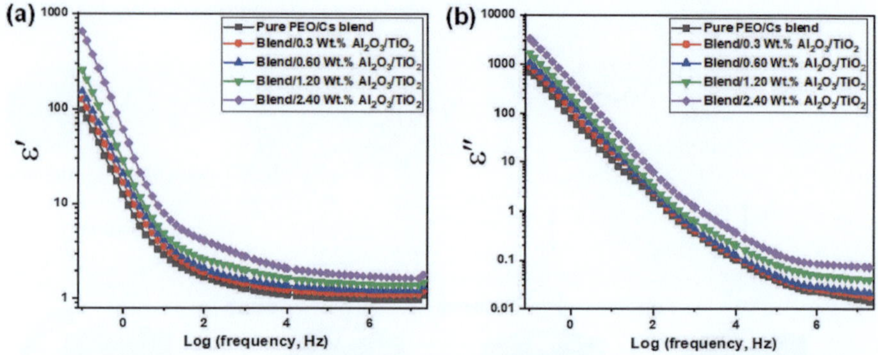

Fig. 8.22 The frequency-dependence at room temperature of **a** ε′ and **b** ε″ of pure PEO/Cs blend and blends filled with varying concentrations (0.3, 0.6, 1.2, and 2.4 wt%) of Al_2O_3/TiO_2 NPs. Reproduced with permission from [141]. Copyright 2024, Elsevier Science Ltd.

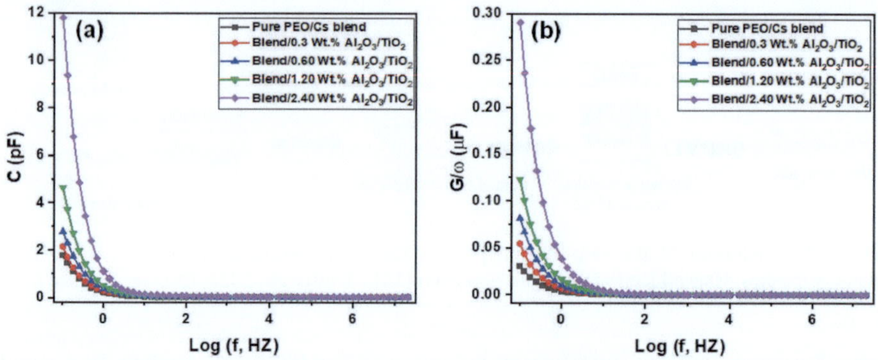

Fig. 8.23 PEO/Cs-Al_2O_3/TiO_2 nanocomposites capacitor performance at 1 V and room temperature **a** capacitance–frequency (C–f) and **b** conductance–frequency (G/ω–f) features recorded for pure PEO/Cs blend and blends filled with varying concentrations (0.3, 0.6, 1.2, and 2.4 wt%) of Al_2O_3/TiO_2 NPs. Reproduced with permission from [141]. Copyright 2024, Elsevier Science Ltd.

The researchers further investigated the performance of the blend nanocomposites in relation to capacitors, examining both the capacitance–frequency (C–f) and conductance–frequency (G/ω–f) behaviour. These studies provided an in-depth understanding of the dielectric and conductive properties of the fabricated capacitor. When the applied voltage was 1 V at room temperature, the C–f curve (Fig. 8.23a) demonstrated a remarkable exponential decrease in capacitance across the frequency spectrum up to approximately 100 Hz. Consequently, a period of stabilisation occurred in the 100 Hz–20 MHz range. The introduction of hybrid nanofillers enhanced the capacitive energy storage capacity of the blend, although there was a reduction in capacitance values due to the decrease in dielectric constant with increasing frequency, a typical characteristic of dielectric materials. The stabilisation phase in the 100 Hz–20 MHz range indicated a favourable frequency range for practical capacitor applications. Capacitors operating within this frequency range would display relatively stable and consistent capacitance values, which is crucial for maintaining reliable energy storage performance in various electronic devices. The G/ω–f curves (Fig. 8.23b) showed a consistent trend, with an increase in frequency resulting in a decrease in conductance. Higher conductance values were particularly notable in the low-frequency regime, suggesting the influence of leakage current. The reduction of conductance at higher frequencies indicated a diminished impact of leakage currents, especially in the high-frequency regime. The G/ω–f curves further enriched the understanding of capacitor behaviour. Lower conductance in the higher frequency regime is a desirable attribute of capacitors, as it signifies lower energy losses and improved efficiency.

The researchers demonstrated the potential of PEO/Cs-Al_2O_3/TiO_2 blend nanocomposites for use in energy storage devices through their extensive studies.

An example of using high permittivity ceramic nanofillers to increase the energy density of dielectric polymer-based film capacitors can be found in the work of Wang et al. [147], where the researchers incorporated polydopamine-coated barium titanate (BT) nanoparticles into a PVDF/PMMA blend. PVDF offers advantages such as high breakdown strength and a high dielectric constant. However, its substantial residual polarisation and energy loss limit its application in energy storage devices [148]. Blending with PMMA can significantly enhance the polarisation and energy storage performance of PVDF [149, 150]. The dielectric constant of the blend was expected to be further improved by incorporating a ceramic nanofiller like BT. To increase the interface coupling between the inorganic filler and the organic polymer matrix, the researchers employed polydopamine (PDA) as an interface modifier. For this purpose, BT@PDA nanofiller with a core–shell structure was synthesised, and the synthesised nanofiller was used to produce 1, 3, 5, and 7 vol% BT@PDA (4/1 wt/wt) PVDF/PMMA blend nanocomposites through solution casting. The processes for synthesising BT@PDA and fabricating the nanocomposite have been schematically presented in Fig. 8.24.

XPS (X-ray photoelectron spectroscopy) characterisation of the blend nanocomposite films revealed the presence of O–C bonds, which were formed by the combination of PDA and BT. This indicated that the BT particles were indeed coated with PDA. The Weibull distribution of the breakdown strength of the blend nanocomposite films is presented in Fig. 8.25a. The shape factor β of all compositions was greater than 10, indicating that all produced compositions had superior stability and reliability, with a breakdown strength distribution that was quite concentrated. The surface modification of the BT particles by PDA reduced defects in the films, minimising the damage caused by the filler to the reliability and stability of the films, while also slowing the decline in the breakdown strength of the films.

An increase in filler content resulted in a gradual decrease in breakdown strength (Fig. 8.25b), with the breakdown strength of the blend nanocomposite films containing 1, 3, 5, and 7 wt% BT@PDA content recorded at 430, 416, 387, and 360 MV/m, respectively.

SEM micrographs (Fig. 8.25c–d) revealed that the films exhibited no visible pores or cracks. The filler particles were uniformly distributed throughout the matrix, with

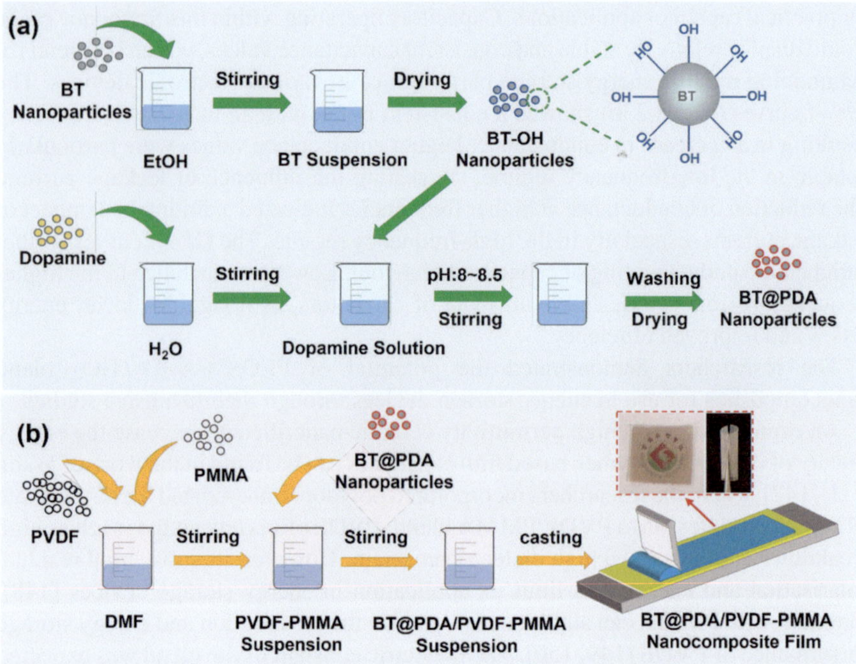

Fig. 8.24 **a** Preparation process flow of filler BT@PDA and **b** process flow of BT@PDA/PVDF-PMMA nanocomposite films. Reproduced with permission from [147]. Copyright 2022, Springer Nature

no significant agglomeration evident. This was attributed to the enhanced compatibility and dispersibility of the nanoparticles within the polymer matrix owing to the PDA coating, which improved the coupling between the filler and the polymer. When the filler content increased to 5%, some agglomeration was noted (Fig. 8.25c). As the filler content further increased to 7%, agglomeration became more pronounced (Fig. 8.25d), leading to a distortion of the local electric field and a decline in the breakdown performance of the films.

To gain a better understanding of the effect of filler concentration on electric field distortion and breakdown characteristics, finite element simulations were conducted on the breakdown process of the films (Fig. 8.25e–f). When an electric field was applied, a local weak field formed around the filler particles, which hindered the further progression of the breakdown path (Fig. 8.25e). However, when the filler concentration was high, these local weak fields overlapped to produce a sub-strong field. During the breakdown process, this led to the rapid advancement of the breakdown path, resulting in a decrease in breakdown strength (Fig. 8.25f). There was thus a good correspondence between the SEM micrographs, breakdown strength values, and finite element simulation results.

The dielectric properties of the blend nanocomposite films are presented in Fig. 8.26. An increase in filler loading resulted in a significant improvement in the

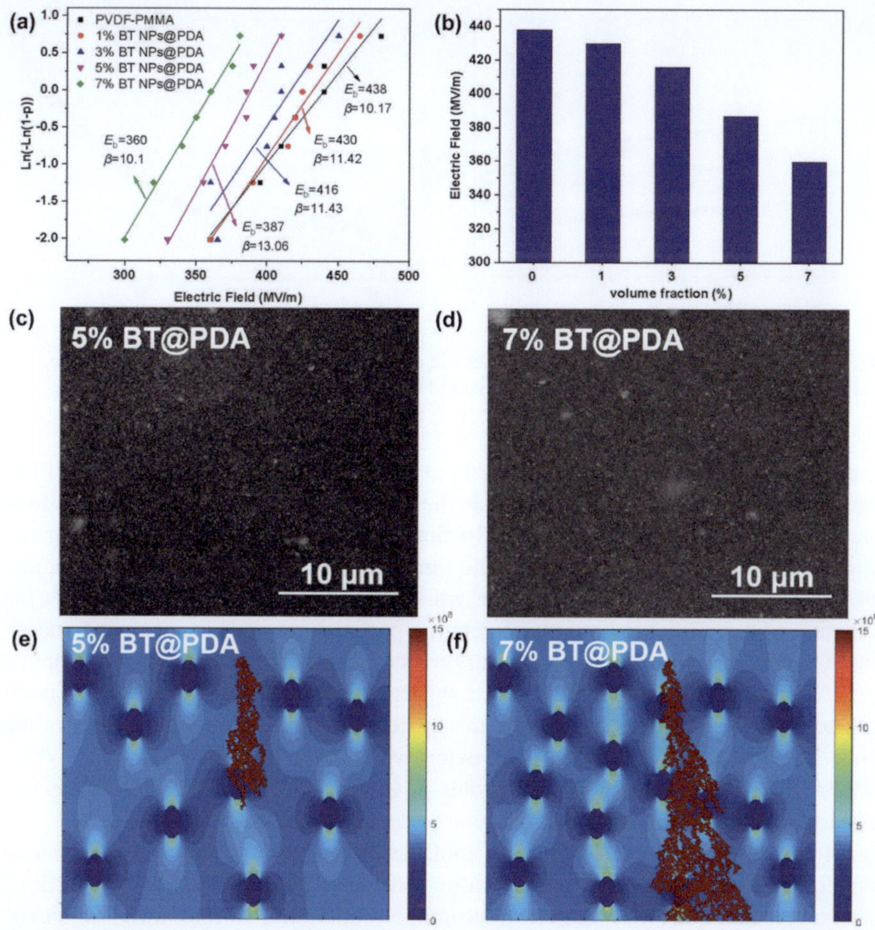

Fig. 8.25 **a** Weibull distribution of breakdown strength of nanocomposite films, **b** breakdown strength, and **c–d** SEM images. **e–f** Finite element simulation results of the thin films dielectric breakdown process. **c**, **e** 5% BT@PDA/PVDF-PMMA and **d**, **f** 7% BT@PDA/PVDF-PMMA. Reproduced with permission from [147]. Copyright 2022, Springer Nature

dielectric constant due to the high dielectric constant of BT. Changes in frequency had little effect on the dielectric constant of the blend nanocomposites, and the samples exhibited quite stable dielectric properties. All compositions showed a dielectric loss in the range of 0.025–0.15. The dielectric loss of the blend nanocomposites at a frequency of 1 kHz was lower than 0.05. Thus, coating the BT particles with PDA improved the compatibility of the filler particles with the polymer matrix, leading to superior dielectric properties.

An increase in filler content initially led to an increase in the discharge energy density Ue, followed by a decrease. When the loadings of BT@PDA were 1% and 3%, the Ue values of the nanocomposite films were 9.32 J/cm^3 and 10.46 J/cm^3,

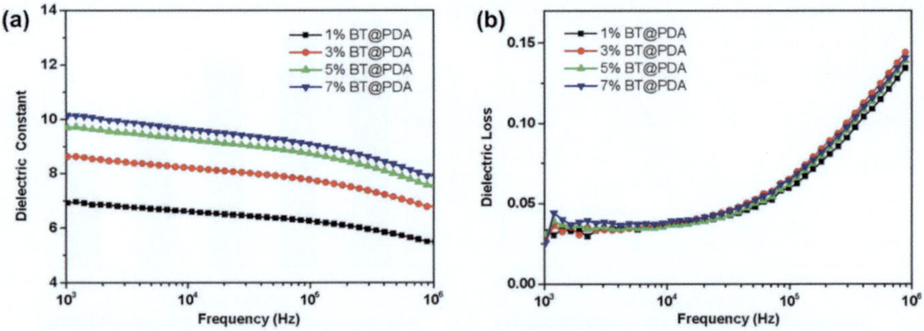

Fig. 8.26 **a** Frequency dependence of **a** dielectric constant and **b** dielectric loss of the nanocomposite films. Reproduced with permission from [147]. Copyright 2022, Springer Nature

respectively. The maximum value of Ue (11.15 J/cm^3) was achieved at a BT@PDA loading of 5%. This was 24% higher than the Ue value of the blend nanocomposite of uncoated BT particles at the same filler loading and blend composition. The discharge efficiency η of the blend nanocomposite films with BT@PDA contents of 1, 3, and 5% was all higher than 65%. However, when the filler content increased to 7%, the inclusion of a significant amount of ceramic particles resulted in increased energy loss. Ue decreased to 7.62 J/cm^3, energy loss manifested as heat, and η dropped to 52%. PDA and PMMA exhibited a synergistic effect, leading to a substantial enhancement of breakdown strength, energy density, and discharge efficiency of the films. The researchers, through their extensive studies, demonstrated that PVDF/PMMA/BT@PDA blend nanocomposites possessed superior and stable dielectric properties, high breakdown strength, energy density, and discharge efficiency.

An example of the use of carbon nanofiller in the fabrication of polymer blend nanocomposites suitable for energy storage devices can be found in the work of Alsalmah et al. [151], where the researchers conducted studies on (7/3 wt/wt) PVA/sodium alginate (NaAlg)/GO-TiO_2 hybrid nanocomposites prepared by solvent casting. The electrical and dielectric properties of the blend nanocomposites revealed that the optimal combination of properties was achieved at 8 wt% GO/TiO_2 content. The dielectric properties of all the blend nanocomposite films exhibited similar behaviour, rapidly decreasing with increasing frequencies and stabilising at higher frequency values. The maximum values of ε' and ε'' were obtained at low frequency, attributed to the influence of the metal electrode polymer contact and the fact that the dipoles had adequate time to interact with the applied field before it changed [152]. The improvement in dielectric properties following the incorporation of the GO/TiO_2 hybrid nanofiller indicated that the interactions of the nanoparticles with the blend significantly enhanced the parallel-dipolar-ordering of the C–O–C, C=O, and –OH groups. The films were found to have potential applications in energy storage devices.

Therefore, it can be summarised that conductive polymer blend nanocomposites can be utilised as energy storage devices. By selecting the appropriate blend components, filler composition, and content, the electrical and dielectric properties of these materials can be tailored. The challenge is to achieve a high dielectric constant while maintaining a low dielectric loss. This can be realised by employing various strategies, such as using hybrid fillers and controlling morphology, to prevent the formation of a percolating network of interconnected conductive filler particles.

8.6 Conclusions and Future Outlook

The application of polymer blend nanocomposites in scaffolds, flexible packaging, EMI shielding, and energy storage is well-established. However, each of these applications has a distinct set of requirements. These can be met through the appropriate selection of blend components and nanofillers, along with morphological control by tailoring composition, selecting suitable processing techniques, and optimising process parameters.

Although a significant amount of work has been done on the development of polymer blend nanocomposites for the aforementioned applications, there is a lack of knowledge in certain areas. For example, it is unclear whether it is possible to achieve the selective localisation of nHA particles in one of the blend components or at the blend interface in a scaffold, and what effect such localisation would have on the properties of the scaffold. Similarly, while the effect of foaming on the EMI shielding performance of conductive blend nanocomposites has been extensively studied, a comparative study of the properties of these foams—when one of the blend components is selectively foamed and when both components are foamed—has yet to be conducted. The use of controlled foaming to achieve superior dielectric properties in conductive polymer nanocomposites by preventing the formation of a percolating threshold is well established. However, studies on the development of foamed polymer blend nanocomposites for energy storage applications are still needed. The influence of selective localisation of filler particles on the dielectric properties of polymer blend nanocomposites also remains to be explored.

It can be concluded that a thorough understanding of the structure-property-processing relationship of polymer blend nanocomposites is essential for utilising these multiphase materials in advanced applications.

References

1. R. Banerjee, S.S. Ray, Foamability and special applications of microcellular thermoplastic polymers: a review on recent advances and future direction. Macromol. Mater. Eng. (2020). https://doi.org/10.1002/mame.202000366
2. S.S. Ray, R. Banerjee, *Foamability of Thermoplastic Polymeric Materials* (Elsevier, Amsterdam, 2021)
3. C. Shuai, W. Yang, P. Feng, S. Peng, H. Pan, Accelerated degradation of HAP/PLLA bone scaffold by PGA blending facilitates bioactivity and osteoconductivity. Bioact. Mater. **6**, 490–502 (2021). https://doi.org/10.1016/J.BIOACTMAT.2020.09.001

4. N. Bitinis, R. Verdejo, E.M. Maya, E. Espuche, P. Cassagnau, M.A. Lopez-Manchado, Physicochemical properties of organoclay filled polylactic acid/natural rubber blend bionanocomposites. Compos. Sci. Technol. **72**, 305–313 (2012). https://doi.org/10.1016/J.COMPSCITECH.2011.11.018
5. R. Banerjee, A. Gebrekrstos, J.T. Orasugh, S.S. Ray, Nanocarbon-containing polymer composite foams: a review of systems for applications in electromagnetic interference shielding, energy storage, and piezoresistive sensors. Ind. Eng. Chem. Res. **62**, 6807–6842 (2023). https://doi.org/10.1021/ACS.IECR.3C00089
6. H.-Y. Mi, X. Jing, L.-S. Turng, Fabrication of porous synthetic polymer scaffolds for tissue engineering. J. Cell. Plast. **51**, 165–196 (2015). https://doi.org/10.1177/0021955X14531002
7. D.W. Hutmacher, Scaffold design and fabrication technologies for engineering tissues - state of the art and future perspectives. J. Biomater. Sci. Polym. Ed. **12**, 107–124 (2001). https://doi.org/10.1163/156856201744489
8. H.Y. Kweon, M.K. Yoo, I.K. Park, T.H. Kim, H.C. Lee, H.S. Lee et al., A novel degradable polycaprolactone networks for tissue engineering. Biomaterials **24**, 801–808 (2003). https://doi.org/10.1016/S0142-9612(02)00370-8
9. L.G. Cima, J.P. Vacanti, C. Vacanti, D. Ingber, D. Mooney, R. Langer, Tissue engineering by cell transplantation using degradable polymer substrates. J. Biomech. Eng. **113**, 143–151 (1991). https://doi.org/10.1115/1.2891228
10. B.B. Mandal, S.C. Kundu, Osteogenic and adipogenic differentiation of rat bone marrow cells on non-mulberry and mulberry silk gland fibroin 3D scaffolds. Biomaterials **30**, 5019–5030 (2009). https://doi.org/10.1016/j.biomaterials.2009.05.064
11. B.B. Mandal, S.C. Kundu, Cell proliferation and migration in silk fibroin 3D scaffolds. Biomaterials **30**, 2956–2965 (2009). https://doi.org/10.1016/j.biomaterials.2009.02.006
12. D.W. Hutmacher, Scaffolds in tissue engineering bone and cartilage. Biomaterials **21**, 2529–2543 (2000). https://doi.org/10.1016/S0142-9612(00)00121-6
13. M. Shah Mohammadi, M.N. Bureau, S.N. Nazhat, Polylactic acid (PLA) biomedical foams for tissue engineering. Biomed. Foam. Tissue Eng. Appl. (Elsevier Ltd.) 313–334 (2014). https://doi.org/10.1533/9780857097033.2.313
14. I. Shirzaei Sani, M. Rezaei, A. Baradar Khoshfetrat, D. Razzaghi, Preparation and characterization of polycaprolactone/chitosan-g-polycaprolactone/hydroxyapatite electrospun nanocomposite scaffolds for bone tissue engineering. Int. J. Biol. Macromol. **182**, 1638–1649 (2021). https://doi.org/10.1016/J.IJBIOMAC.2021.05.163
15. P. Douglas, M. Kuhs, M. Sajjia, M. Khraisheh, G. Walker, M.N. Collins et al., Bioactive PCL matrices with a range of structural & rheological properties. React. Funct. Polym. **101**, 54–62 (2016). https://doi.org/10.1016/J.REACTFUNCTPOLYM.2016.02.004
16. P. Douglas, A.B. Albadarin, M. Sajjia, C. Mangwandi, M. Kuhs, M.N. Collins et al., Effect of poly ethylene glycol on the mechanical and thermal properties of bioactive poly(ε-caprolactone) melt extrudates for pharmaceutical applications. Int. J. Pharm. **500**, 179–186 (2016). https://doi.org/10.1016/J.IJPHARM.2016.01.036
17. K. Rezwan, Q.Z. Chen, J.J. Blaker, A.R. Boccaccini, Biodegradable and bioactive porous polymer/inorganic composite scaffolds for bone tissue engineering. Biomaterials **27**, 3413–3431 (2006). https://doi.org/10.1016/J.BIOMATERIALS.2006.01.039
18. V. Karageorgiou, D. Kaplan, Porosity of 3D biomaterial scaffolds and osteogenesis. Biomaterials **26**, 5474–5491 (2005). https://doi.org/10.1016/j.biomaterials.2005.02.002
19. M. Diba, M. Kharaziha, M.H. Fathi, M. Gholipourmalekabadi, A. Samadikuchaksaraei, Preparation and characterization of polycaprolactone/forsterite nanocomposite porous scaffolds designed for bone tissue regeneration. Compos. Sci. Technol. **72**, 716–723 (2012). https://doi.org/10.1016/J.COMPSCITECH.2012.01.023
20. G.F. Santana-Melo, B.V.M. Rodrigues, E. da Silva, R. Ricci, F.R. Marciano, T.J. Webster et al., Electrospun ultrathin PBAT/nHAp fibers influenced the in vitro and in vivo osteogenesis and improved the mechanical properties of neoformed bone. Colloids Surfaces B Biointerfaces **155**, 544–552 (2017). https://doi.org/10.1016/J.COLSURFB.2017.04.053

21. W. Shao, J. He, F. Sang, B. Ding, L. Chen, S. Cui et al., Coaxial electrospun aligned tussah silk fibroin nanostructured fiber scaffolds embedded with hydroxyapatite–tussah silk fibroin nanoparticles for bone tissue engineering. Mater. Sci. Eng. C **58**, 342–351 (2016). https://doi.org/10.1016/J.MSEC.2015.08.046
22. A. Nahanmoghadam, M. Asemani, V. Goodarzi, S. Ebrahimi-Barough, In vivo investigation of PCL/PHBV/hydroxyapatite nanocomposite scaffold in regeneration of critical-sized bone defects. Fibers Polym. **22**, 2507–2516 (2021). https://doi.org/10.1007/S12221-021-1243-Z/METRICS
23. A.K. Pal, F. Wu, M. Misra, A.K. Mohanty, Reactive extrusion of sustainable PHBV/PBAT-based nanocomposite films with organically modified nanoclay for packaging applications: compression moulding vs. cast film extrusion. Compos. Part B Eng. **198**, 108141 (2020). https://doi.org/10.1016/J.COMPOSITESB.2020.108141
24. H. Moustafa, N. El Kissi, A.I. Abou-Kandil, M.S. Abdel-Aziz, A. Dufresne, PLA/PBAT bionanocomposites with antimicrobial natural rosin for green packaging. ACS Appl. Mater. Interfaces **9**, 20132–20141 (2017). https://doi.org/10.1021/ACSAMI.7B05557/ASSET/IMAGES/LARGE/AM-2017-055578_0010.JPEG
25. H. Moustafa, C. Guizani, C. Dupont, V. Martin, M. Jeguirim, A. Dufresne, Utilization of torrefied coffee grounds as reinforcing agent to produce high-quality biodegradable PBAT composites for food packaging applications. ACS Sustain. Chem. Eng. **5**, 1906–1916 (2017). https://doi.org/10.1021/ACSSUSCHEMENG.6B02633/ASSET/IMAGES/SC-2016-026336_M001.GIF
26. D. Srinivasan, S. Nathan, T. Suresh, P.P. Lakshmana, Antimicrobial activity of certain Indian medicinal plants used in folkloric medicine. J. Ethnopharmacol. **74**, 217–220 (2001). https://doi.org/10.1016/S0378-8741(00)00345-7
27. H. Moustafa, A.M. Youssef, M.A. Nour, Investigation of morphology, mechanical, thermal and flame retardant properties of an EVA/EPDM blend by combination of organoclay with Na+-tripolyphosphate. RSC Adv. **6**, 36467–36474 (2016). https://doi.org/10.1039/C6RA03341E
28. D.O. De Castro, J. Bras, A. Gandini, N. Belgacem, Surface grafting of cellulose nanocrystals with natural antimicrobial rosin mixture using a green process. Carbohydr. Polym. **137**, 1–8 (2016). https://doi.org/10.1016/J.CARBPOL.2015.09.101
29. S. Sinha Ray, M. Okamoto, Polymer/layered silicate nanocomposites: a review from preparation to processing. Prog. Polym. Sci. **28**, 1539–1641 (2003). https://doi.org/10.1016/j.progpolymsci.2003.08.002
30. E. Makhado, S.S. Ray, Mechanical, barrier and antimicrobial properties of biodegradable poly (ε-caprolactone) nanocomposites. Adv. Sci. Eng. Med. **7**, 351–360 (2015)
31. Y. Cui, S. Kumar, B.R. Kona, D. van Houcke, Gas barrier properties of polymer/clay nanocomposites. RSC Adv. **5**, 63669–63690 (2015)
32. R. Banerjee, S.S. Ray, An overview of the recent advances in polylactide-based sustainable nanocomposites. Polym. Eng. Sci. **61**, 617–649 (2021). https://doi.org/10.1002/PEN.25623
33. M.J. Mochane, J.S. Sefadi, T.S. Motsoeneng, T.E. Mokoena, T.G. Mofokeng, T.C. Mokhena, The effect of filler localization on the properties of biopolymer blends, recent advances: a review. Polym. Compos. **41**, 2958–2979 (2020). https://doi.org/10.1002/PC.25590
34. R. Salehiyan, S.S. Ray, Influence of nanoclay localization on structure-property relationships of polylactide-based biodegradable blend nanocomposites. Macromol. Mater. Eng. **303**, 1800134 (2018). https://doi.org/10.1002/MAME.201800134
35. E. Garofalo, L. Di Maio, P. Scarfato, P. Russo, L. Incarnato, Selective localization of nanoparticles to enhance the properties of PBS/PLA nanocomposite blown films. J. Polym. Environ. **31**, 4546–4558 (2023). https://doi.org/10.1007/S10924-023-02883-1/TABLES/5
36. R. Scaffaro, A. Maio, E.F. Gulino, M. Morreale, M.F.P. La, The effects of nanoclay on the mechanical properties, carvacrol release and degradation of a PLA/PBAT blend. Materials **13**, 983 (2020). https://doi.org/10.3390/MA13040983

37. M. Dadras Chomachayi, A. Jalali-arani, F.R. Beltrán, M.U. de la Orden, U.J. Martínez, Biodegradable nanocomposites developed from PLA/PCL blends and silk fibroin nanoparticles: study on the microstructure, thermal behavior, crystallinity and performance. J. Polym. Environ. **28**, 1252–1264 (2020). https://doi.org/10.1007/S10924-020-01684-0/FIGURES/10
38. E. Jalalvandi, R.A. Majid, T. Ghanbari, H. Ilbeygi, Effects of montmorillonite (MMT) on morphological, tensile, physical barrier properties and biodegradability of polylactic acid/starch/MMT nanocomposites. J. Thermoplas. Compos. Mater. **28**, 496–509 (2013). https://doi.org/10.1177/0892705713486129
39. H.M. Alghamdi, M.M. Abutalib, A. Rajeh, M.A. Mannaa, O. Nur, E.M. Abdelrazek, Effect of the Fe_2O_3/TiO_2 nanoparticles on the structural, mechanical, electrical properties and antibacterial activity of the biodegradable chitosan/polyvinyl alcohol blend for food packaging. J. Polym. Environ. **30**, 3865–3874 (2022). https://doi.org/10.1007/S10924-022-02478-2/TABLES/2
40. S. Dadbin, M. Noferesti, M. Frounchi, Oxygen barrier LDPE/LLDPE/organoclay nanocomposite films for food packaging. Macromol. Symp. **274**, 22–27 (2008). https://doi.org/10.1002/MASY.200851404
41. P.B. Jaiswal, B.K. Pushkar, K. Maikap, P.A. Mahanwar, Abiotic aging assisted bio-oxidation and degradation of LLDPE/LDPE packaging polyethylene film by stimulated enrichment culture. Polym. Degrad. Stab. **206**, 110156 (2022). https://doi.org/10.1016/J.POLYMDEGRADSTAB.2022.110156
42. S.C. Li, B. Li, X.M. Qin, Properties of LLDPE film modified by LDPE. Polym. Plast. Technol. Eng. **51**, 307–310 (2012). https://doi.org/10.1080/03602559.2011.639833
43. B.A. Morris, Polymer blending for packaging applications, in *Multilayer Flexible Packaging*, 2nd ed (Elsevier Inc., 2016), pp. 173–204. https://doi.org/10.1016/B978-0-323-37100-1.00013-2
44. B.D.S. Deeraj, J.S. Jayan, A. Raman, A. Saritha, K. Joseph, Polymeric blends and nanocomposites for high performance EMI shielding and microwave absorbing applications. Compos. Interfaces **29**, 1505–1547 (2022). https://doi.org/10.1080/09276440.2022.2068245
45. B. Wen, M. Cao, M. Lu, W. Cao, H. Shi, J. Liu et al., Reduced graphene oxides: lightweight and high-efficiency electromagnetic interference shielding at elevated temperatures. Adv. Mater. **26**, 3484–3489 (2014). https://doi.org/10.1002/ADMA.201400108
46. J.M. Thomassin, C. Jérôme, T. Pardoen, C. Bailly, I. Huynen, C. Detrembleur, Polymer/carbon based composites as electromagnetic interference (EMI) shielding materials. Mater. Sci. Eng. R. Rep. **74**, 211–232 (2013). https://doi.org/10.1016/J.MSER.2013.06.001
47. C. Zongping Chen, C. Xu, C. Ma, W. Ren, H.-M. Cheng, Z.P. Chen et al., Lightweight and flexible graphene foam composites for high-performance electromagnetic interference shielding. Adv. Mater. **25**, 1296–1300 (2013). https://doi.org/10.1002/ADMA.201204196
48. P. Hu, J. Lyu, C. Fu, W.B. Gong, J. Liao, W. Lu et al., Multifunctional aramid nanofiber/carbon nanotube hybrid aerogel films. ACS Nano **14**, 688–697 (2020). https://doi.org/10.1021/ACSNANO.9B07459/ASSET/IMAGES/LARGE/NN9B07459_0005.JPEG
49. Z. Zeng, T. Wu, D. Han, Q. Ren, G. Siqueira, G. Nyström, Ultralight, flexible, and biomimetic nanocellulose/silver nanowire aerogels for electromagnetic interference shielding. ACS Nano **14**, 2927–2938 (2020). https://doi.org/10.1021/ACSNANO.9B07452/SUPPL_FILE/NN9B07452_SI_004.AVI
50. B.D.S. Deeraj, M.S. Mathew, J. Parameswaranpillai, K. Joseph, EMI shielding materials based on thermosetting polymers, in *Materials for Potential EMI Shielding Applications*, ed. by K. Joseph, R. Wilson, G. George (Elsevier, Amsterdam, 2020), pp. 101–110. https://doi.org/10.1016/B978-0-12-817590-3.00006-3
51. D. Wanasinghe, F. Aslani, A review on recent advancement of electromagnetic interference shielding novel metallic materials and processes. Compos. Part B Eng. **176**, 107207 (2019). https://doi.org/10.1016/J.COMPOSITESB.2019.107207
52. J.B. Anooja, K.S. Dijith, K.P. Surendran, G. Subodh, A simple strategy for flexible electromagnetic interference shielding: hybrid rGO@CB-reinforced polydimethylsiloxane. J. Alloys Compd. **807**, 151678 (2019). https://doi.org/10.1016/J.JALLCOM.2019.151678

53. N. Devi, S.S. Ray, Electromagnetic interference cognizance and potential of advanced polymer composites toward electromagnetic interference shielding: a review. Polym. Eng. Sci. **62**, 591–621 (2022). https://doi.org/10.1002/PEN.25876
54. A. Joshi, S. Datar, Carbon nanostructure composite for electromagnetic interference shielding. Pramana – J. Phys. **84**, 1099–1116 (2015). https://doi.org/10.1007/S12043-015-1005-9/FIGURES/3
55. T.K. Gupta, B.P. Singh, S.R. Dhakate, V.N. Singh, R.B. Mathur, Improved nanoindentation and microwave shielding properties of modified MWCNT reinforced polyurethane composites. J. Mater. Chem. A **1**, 9138–9149 (2013). https://doi.org/10.1039/C3TA11611E
56. D.X. Yan, P.G. Ren, H. Pang, Q. Fu, M.B. Yang, Z.M. Li, Efficient electromagnetic interference shielding of lightweight graphene/polystyrene composite. J. Mater. Chem. **22**, 18772–18774 (2012). https://doi.org/10.1039/C2JM32692B
57. G.A. Gelves, M.H. Al-Saleh, U. Sundararaj, Highly electrically conductive and high performance EMI shielding nanowire/polymer nanocomposites by miscible mixing and precipitation. J. Mater. Chem. **21**, 829–836 (2010). https://doi.org/10.1039/C0JM02546A
58. P. Kumar, F. Shahzad, S. Yu, S.M. Hong, Y.H. Kim, C.M. Koo, Large-area reduced graphene oxide thin film with excellent thermal conductivity and electromagnetic interference shielding effectiveness. Carbon N Y **94**, 494–500 (2015). https://doi.org/10.1016/J.CARBON.2015.07.032
59. F. Shahzad, P. Kumar, Y.H. Kim, S.M. Hong, C.M. Koo, Biomass-derived thermally annealed interconnected sulfur-doped graphene as a shield against electromagnetic interference. ACS Appl. Mater. Interfaces **8**, 9361–9369 (2016). https://doi.org/10.1021/ACSAMI.6B00418/ASSET/IMAGES/LARGE/AM-2016-004186_0007.JPEG
60. A. Iqbal, P. Sambyal, C. Min Koo, A. Iqbal, P. Sambyal, C.M. Koo et al., 2D MXenes for electromagnetic shielding: a review. Adv. Funct. Mater. **30**, 2000883 (2020). https://doi.org/10.1002/ADFM.202000883
61. H. Wang, L. Wu, J. Jiao, J. Zhou, Y. Xu, H. Zhang et al., Covalent interaction enhanced electromagnetic wave absorption in SiC/Co hybrid nanowires. J. Mater. Chem. A **3**, 6517–6525 (2015). https://doi.org/10.1039/C5TA00303B
62. H. Wang, G. Wang, W. Li, Q. Wang, W. Wei, Z. Jiang et al., A material with high electromagnetic radiation shielding effectiveness fabricated using multi-walled carbon nanotubes wrapped with poly(ether sulfone) in a poly(ether ether ketone) matrix. J. Mater. Chem. **22**, 21232–21237 (2012). https://doi.org/10.1039/C2JM35129C
63. C. Li, C. Zhou, J. Lv, B. Liang, R. Li, Y. Liu et al., Bio-molecule adenine building block effectively enhances electromagnetic interference shielding performance of polyimide-derived carbon foam. Carbon N Y **149**, 190–202 (2019). https://doi.org/10.1016/J.CARBON.2019.04.012
64. Y. Yuan, W. Yin, M. Yang, F. Xu, X. Zhao, J. Li et al., Lightweight, flexible and strong core-shell non-woven fabrics covered by reduced graphene oxide for high-performance electromagnetic interference shielding. Carbon N Y **130**, 59–68 (2018). https://doi.org/10.1016/J.CARBON.2017.12.122
65. E. Collett, Mueller-Stokes matrix formulation of Fresnel's equations. Am. J. Phys. **39**, 517 (2005). https://doi.org/10.1119/1.1986205
66. D. Yevick, T. Friese, F. Schmidt, A comparison of transparent boundary conditions for the Fresnel equation. J. Comput. Phys. **168**, 433–444 (2001). https://doi.org/10.1006/JCPH.2001.6708
67. S.P. Pawar, S. Kumar, S. Jain, M. Gandi, K. Chatterjee, S. Bose, Synergistic interactions between silver decorated graphene and carbon nanotubes yield flexible composites to attenuate electromagnetic radiation. Nanotechnology **28**, 025201 (2016). https://doi.org/10.1088/0957-4484/28/2/025201
68. T. Yun, H. Kim, A. Iqbal, Y. Soo Cho, G. San Lee, M.-K. Kim et al., Electromagnetic shielding of monolayer MXene assemblies. Adv. Mater. **32**, 1906769 (2020). https://doi.org/10.1002/ADMA.201906769

69. Q. Song, F. Ye, X. Yin, W. Li, H. Li, Y. Liu et al., Carbon nanotube-multilayered graphene edge plane core-shell hybrid foams for ultrahigh-performance electromagnetic-interference shielding. Adv. Mater. **29**, 1701583 (2017). https://doi.org/10.1002/ADMA.201701583
70. S. Biswas, G.P. Kar, S. Bose, Microwave absorbers designed from PVDF/SAN blends containing multiwall carbon nanotubes anchored cobalt ferrite via a pyrene derivative. J. Mater. Chem. A **3**, 12413–12426 (2015). https://doi.org/10.1039/C5TA02177D
71. R. Kumar, S.R. Dhakate, T. Gupta, P. Saini, B.P. Singh, R.B. Mathur, Effective improvement of the properties of light weight carbon foam by decoration with multi-wall carbon nanotubes. J. Mater. Chem. A **1**, 5727–5735 (2013). https://doi.org/10.1039/C3TA10604G
72. B.P. Singh, Prabha, P. Saini, T. Gupta, P. Garg, G. Kumar et al., Designing of multiwalled carbon nanotubes reinforced low density polyethylene nanocomposites for suppression of electromagnetic radiation. J. Nanoparticle Res. **13**, 7065–7074 (2011). https://doi.org/10.1007/S11051-011-0619-1/FIGURES/7
73. B.R. Kim, H.K. Lee, E. Kim, S.H. Lee, Intrinsic electromagnetic radiation shielding/absorbing characteristics of polyaniline-coated transparent thin films. Synth. Met. **160**, 1838–1842 (2010). https://doi.org/10.1016/J.SYNTHMET.2010.06.027
74. B. Zhao, M. Hamidinejad, S. Wang, P. Bai, R. Che, R. Zhang et al., Advances in electromagnetic shielding properties of composite foams. J. Mater. Chem. A **9**, 8896–8949 (2021). https://doi.org/10.1039/D1TA00417D
75. D.D.L. Chung, Materials for electromagnetic interference shielding. Mater. Chem. Phys. **255**, 123587 (2020). https://doi.org/10.1016/J.MATCHEMPHYS.2020.123587
76. S. Geetha, K.K.S. Kumar, C.R.K. Rao, M. Vijayan, D.C. Trivedi, EMI shielding: methods and materials—a review. J. Appl. Polym. Sci. **112**, 2073–2086 (2009). https://doi.org/10.1002/APP.29812
77. N. Wu, Q. Hu, R. Wei, X. Mai, N. Naik, D. Pan et al., Review on the electromagnetic interference shielding properties of carbon based materials and their novel composites: recent progress, challenges and prospects. Carbon N Y **176**, 88–105 (2021). https://doi.org/10.1016/J.CARBON.2021.01.124
78. Y. Qing, J. Su, Q. Wen, F. Luo, D. Zhu, W. Zhou, Enhanced dielectric and electromagnetic interference shielding properties of FeSiAl/Al_2O_3 ceramics by plasma spraying. J. Alloys Compd. **651**, 259–265 (2015). https://doi.org/10.1016/J.JALLCOM.2015.08.107
79. X. Hao, X. Yin, L. Zhang, L. Cheng, Dielectric, electromagnetic interference shielding and absorption properties of Si_3N_4–PyC composite ceramics. J. Mater. Sci. Technol. **29**, 249–254 (2013). https://doi.org/10.1016/J.JMST.2013.01.011
80. H. Guan, S. Liu, Y. Duan, J. Cheng, Cement based electromagnetic shielding and absorbing building materials. Cem. Concr. Compos. **28**, 468–474 (2006). https://doi.org/10.1016/J.CEMCONCOMP.2005.12.004
81. R.A. Khushnood, S. Ahmad, P. Savi, J.M. Tulliani, M. Giorcelli, G.A. Ferro, Improvement in electromagnetic interference shielding effectiveness of cement composites using carbonaceous nano/micro inerts. Constr. Build. Mater. **85**, 208–216 (2015). https://doi.org/10.1016/J.CONBUILDMAT.2015.03.069
82. Y. Wang, X. Jing, Intrinsically conducting polymers for electromagnetic interference shielding. Polym. Adv. Technol. **16**, 344–351 (2005). https://doi.org/10.1002/PAT.589
83. J. Joo, C.Y. Lee, High frequency electromagnetic interference shielding response of mixtures and multilayer films based on conducting polymers. J. Appl. Phys. **88**, 513 (2000). https://doi.org/10.1063/1.373688
84. G. Wang, J. Zhao, C. Ge, G. Zhao, C.B. Park, Nanocellular poly (ether-block-amide)/MWCNT nanocomposite films fabricated by stretching-assisted microcellular foaming for high-performance EMI shielding applications. J. Mater. Chem. C **9**, 1245–1258 (2021)
85. H. Zhang, G. Zhang, M. Tang, L. Zhou, J. Li, X. Fan et al., Synergistic effect of carbon nanotube and graphene nanoplates on the mechanical, electrical and electromagnetic interference shielding properties of polymer composites and polymer composite foams. Chem. Eng. J. **353**, 381–393 (2018). https://doi.org/10.1016/j.cej.2018.07.144

86. S. Ayub, B.H. Guan, F. Ahmad, Y.A. Oluwatobi, Z.U. Nisa, M.F. Javed et al., Graphene and iron reinforced polymer composite electromagnetic shielding applications: a review. Polymers **13**, 2580 (2021). https://doi.org/10.3390/POLYM13152580
87. S. Sankaran, K. Deshmukh, M.B. Ahamed, S.K. Khadheer Pasha, Recent advances in electromagnetic interference shielding properties of metal and carbon filler reinforced flexible polymer composites: a review. Compos. Part A Appl. Sci. Manuf. **114**, 49–71 (2018). https://doi.org/10.1016/J.COMPOSITESA.2018.08.006
88. M.H. Al-Saleh, U. Sundararaj, Electromagnetic interference (EMI) shielding effectiveness of PP/PS polymer blends containing high structure carbon black. Macromol. Mater. Eng. **293**, 621–630 (2008). https://doi.org/10.1002/MAME.200800060
89. S.P. Mahapatra, V. Sridhar, D.K. Tripathy, Impedance analysis and electromagnetic interference shielding effectiveness of conductive carbon black reinforced microcellular EPDM rubber vulcanizates. Polym. Compos. **29**, 465–472 (2008). https://doi.org/10.1002/PC.20421
90. E. Mikinka, M. Siwak, Recent advances in electromagnetic interference shielding properties of carbon-fibre-reinforced polymer composites—a topical review. J. Mater. Sci. Mater. Electron. **32**, 24585–24643 (2021). https://doi.org/10.1007/S10854-021-06900-8/FIGURES/14
91. H.G. Kim, H.J. Shin, G.C. Kim, H.J. Park, H.J. Moon, L.K. Kwac, Electromagnetic interference shielding characteristics for orientation angle and number of plies of carbon fiber reinforced plastic. Carbon Lett. **15**, 268–276 (2014). https://doi.org/10.5714/CL.2014.15.4.268
92. D. Feng, P. Liu, Q. Wang, Exploiting the piezoresistivity and EMI shielding of polyetherimide/carbon nanotube foams by tailoring their porous morphology and segregated CNT networks. Compos. Part A Appl. Sci. Manuf. **124**, 105463 (2019). https://doi.org/10.1016/J.COMPOSITESA.2019.05.031
93. M.P. Tran, J.M. Thomassin, M. Alexandre, C. Jerome, I. Huynen, C. Detrembleur, Nanocomposite foams of polypropylene and carbon nanotubes: preparation, characterization, and evaluation of their performance as EMI absorbers. Macromol. Chem. Phys. **216**, 1302–1312 (2015)
94. Y. Li, H. Shimizu, Conductive PVDF/PA6/CNTs nanocomposites fabricated by dual formation of cocontinuous and nanodispersion structures. Macromolecules **41**, 5339–5344 (2008). https://doi.org/10.1021/MA8006834/SUPPL_FILE/MA8006834-FILE001.PDF
95. O.J. Botlhoko, J. Ramontja, S.S. Ray, Thermally shocked graphene oxide-containing biocomposite for thermal management applications. RSC Adv. **7**, 33751–33756 (2017). https://doi.org/10.1039/C7RA05421A
96. R. Ravindren, S. Mondal, K. Nath, N.C. Das, Investigation of electrical conductivity and electromagnetic interference shielding effectiveness of preferentially distributed conductive filler in highly flexible polymer blends nanocomposites. Compos. Part A Appl. Sci. Manuf. **118**, 75–89 (2019). https://doi.org/10.1016/J.COMPOSITESA.2018.12.012
97. R. Ravindren, S. Mondal, P. Bhawal, S.M.N. Ali, N.C. Das, Superior electromagnetic interference shielding effectiveness and low percolation threshold through the preferential distribution of carbon black in the highly flexible polymer blend composites. Polym. Compos. **40**, 1404–1418 (2019). https://doi.org/10.1002/PC.24874
98. A. Sadeghi, R. Moeini, J.K. Yeganeh, Highly conductive PP/PET polymer blends with high electromagnetic interference shielding performances in the presence of thermally reduced graphene nanosheets prepared through melt compounding. Polym. Compos. **40**, E1461–E1469 (2019). https://doi.org/10.1002/PC.25051
99. G.P. Kar, S. Biswas, S. Bose, Simultaneous enhancement in mechanical strength, electrical conductivity, and electromagnetic shielding properties in PVDF–ABS blends containing PMMA wrapped multiwall carbon nanotubes. Phys. Chem. Chem. Phys. **17**, 14856–14865 (2015)
100. C. Lencar, S. Ramakrishnan, E. Erfanian, U. Sundararaj, The role of phase migration of carbon nanotubes in melt-mixed PVDF/PE polymer blends for high conductivity and EMI shielding applications. Molecules **27**, 933 (2022). https://doi.org/10.3390/MOLECULES27030933

101. S.P. Pawar, D.A. Marathe, K. Pattabhi, S. Bose, Electromagnetic interference shielding through MWNT grafted Fe_3O_4 nanoparticles in PC/SAN blends. J. Mater. Chem. A **3**, 656–669 (2015)
102. A.V. Menon, G. Madras, S. Bose, Magnetic alloy-MWNT heterostructure as efficient electromagnetic wave suppressors in soft nanocomposites. ChemistrySelect **2**, 7831–7844 (2017). https://doi.org/10.1002/SLCT.201700986
103. G.P. Kar, S. Biswas, R. Rohini, S. Bose, Tailoring the dispersion of multiwall carbon nanotubes in co-continuous PVDF/ABS blends to design materials with enhanced electromagnetic interference shielding. J. Mater. Chem. A **3**, 7974–7985 (2015)
104. S.P. Pawar, S. Stephen, S. Bose, V. Mittal, Tailored electrical conductivity, electromagnetic shielding and thermal transport in polymeric blends with graphene sheets decorated with nickel nanoparticles. Phys. Chem. Chem. Phys. **17**, 14922–14930 (2015). https://doi.org/10.1039/C5CP00899A
105. S.P. Pawar, M. Gandi, C. Saraf, S. Bose, Exceptional microwave absorption in soft polymeric nanocomposites facilitated by engineered nanostructures. J. Mater. Chem. C **4**, 4954–4966 (2016)
106. S.P. Pawar, M. Gandi, S. Bose, High performance electromagnetic wave absorbers derived from PC/SAN blends containing multiwall carbon nanotubes and Fe_3O_4 decorated onto graphene oxide sheets. RSC Adv. **6**, 37633–37645 (2016)
107. A. Rizvi, S.S. Bae, N.M.A. Mohamed, J.H. Lee, C.B. Park, Extensional flow resistance of 3D fiber networks in plasticized nanocomposites. Macromolecules (2019). https://doi.org/10.1021/ACS.MACROMOL.9B00885/ASSET/IMAGES/LARGE/MA9B00885_0003.JPEG
108. M. Okamoto, P. Nam, P. Maiti, T. Kotaka, T. Nakayama, M. Takada et al., Biaxial flow-induced alignment of silicate layers in polypropylene/clay nanocomposite foam. Nano Lett. **1**, 503–505 (2001)
109. G. Wang, G. Zhao, S. Wang, L. Zhang, C.B. Park, Injection-molded microcellular PLA/graphite nanocomposites with dramatically enhanced mechanical and electrical properties for ultra-efficient EMI shielding applications. J. Mater. Chem. C **6**, 6847–6859 (2018). https://doi.org/10.1039/C8TC01326H
110. H.B. Zhang, Q. Yan, W.G. Zheng, Z. He, Z.Z. Yu, Tough graphene− polymer microcellular foams for electromagnetic interference shielding. ACS Appl. Mater. Interfaces **3**, 918–924 (2011)
111. A. Ameli, P.U. Jung, C.B. Park, Electrical properties and electromagnetic interference shielding effectiveness of polypropylene/carbon fiber composite foams. Carbon N Y **60**, 379–391 (2013)
112. A. Ameli, Y. Kazemi, S. Wang, C.B. Park, P. Pötschke, Process-microstructure-electrical conductivity relationships in injection-molded polypropylene/carbon nanotube nanocomposite foams. Compos. Part A Appl. Sci. Manuf. **96**, 28–36 (2017). https://doi.org/10.1016/j.compositesa.2017.02.012
113. B. Zhao, C. Zhao, M. Hamidinejad, C. Wang, R. Li, S. Wang et al., Incorporating a microcellular structure into PVDF/graphene-nanoplatelet composites to tune their electrical conductivity and electromagnetic interference shielding properties. J. Mater. Chem. C **6**, 10292–10300 (2018). https://doi.org/10.1039/C8TC03714K
114. H. Zhang, G. Zhang, Q. Gao, M. Zong, M. Wang, J. Qin, Electrically electromagnetic interference shielding microcellular composite foams with 3D hierarchical graphene-carbon nanotube hybrids. Compos. Part A Appl. Sci. Manuf. **130**, 105773 (2020)
115. X. Fan, G. Zhang, J. Li, Z. Shang, H. Zhang, Q. Gao et al., Study on foamability and electromagnetic interference shielding effectiveness of supercritical CO_2 foaming epoxy/rubber/MWCNTs composite. Compos. Part A Appl. Sci. Manuf. **121**, 64–73 (2019). https://doi.org/10.1016/J.COMPOSITESA.2019.03.008
116. V. Eswaraiah, V. Sankaranarayanan, S. Ramaprabhu, Functionalized graphene–PVDF foam composites for EMI shielding. Macromol. Mater. Eng. **296**, 894–898 (2011). https://doi.org/10.1002/MAME.201100035

117. Y. Zhao, J. Hou, Z. Bai, Y. Yang, X. Guo, H. Cheng et al., Facile preparation of lightweight PE/PVDF/Fe_3O_4/CNTs nanocomposite foams with high conductivity for efficient electromagnetic interference shielding. Compos. Part A Appl. Sci. Manuf. **139**, 106095 (2020). https://doi.org/10.1016/J.COMPOSITESA.2020.106095
118. S.P. Pawar, P. Rzeczkowski, P. Pötschke, B. Krause, S. Bose, Does the processing method resulting in different states of an interconnected network of multiwalled carbon nanotubes in polymeric blend nanocomposites affect EMI shielding properties? ACS Omega **3**, 5771–5782 (2018). https://doi.org/10.1021/ACSOMEGA.8B00575/ASSET/IMAGES/MEDIUM/AO-2018-00575G_M001.GIF
119. Z.M. Dang, J.K. Yuan, S.H. Yao, R.J. Liao, Flexible nanodielectric materials with high permittivity for power energy storage. Adv. Mater. **25**, 6334–6365 (2013). https://doi.org/10.1002/ADMA.201301752
120. J.Y. Kim, T. Kim, J.W. Suk, H. Chou, J.H. Jang, J.H. Lee et al., Enhanced dielectric performance in polymer composite films with carbon nanotube-reduced graphene oxide hybrid filler. Small **10**, 3405–3411 (2014). https://doi.org/10.1002/SMLL.201400363
121. Z.M. Dang, M.S. Zheng, J.W. Zha, 1D/2D carbon nanomaterial-polymer dielectric composites with high permittivity for power energy storage applications. Small **12**, 1688–1701 (2016). https://doi.org/10.1002/SMLL.201503193
122. R. Simoes, J. Silva, R. Vaia, V. Sencadas, P. Costa, J. Gomes et al., Low percolation transitions in carbon nanotube networks dispersed in a polymer matrix:dielectric properties, simulations and experiments. Nanotechnology **20**, 035703 (2008). https://doi.org/10.1088/0957-4484/20/3/035703
123. N. Yousefi, X. Sun, X. Lin, X. Shen, J. Jia, B. Zhang et al., Highly aligned graphene/polymer nanocomposites with excellent dielectric properties for high-performance electromagnetic interference shielding. Adv. Mater. **26**, 5480–5487 (2014). https://doi.org/10.1002/ADMA.201305293
124. K. Deshmukh, M.B. Ahamed, K.K. Sadasivuni, D. Ponnamma, M.A.A. AlMaadeed, S.K. Khadheer Pasha et al., Graphene oxide reinforced poly (4-styrenesulfonic acid)/polyvinyl alcohol blend composites with enhanced dielectric properties for portable and flexible electronics. Mater. Chem. Phys. **186**, 188–201 (2017). https://doi.org/10.1016/J.MATCHEMPHYS.2016.10.044
125. X. Xu, C. Yang, J. Yang, T. Huang, N. Zhang, Y. Wang et al., Excellent dielectric properties of poly(vinylidene fluoride) composites based on partially reduced graphene oxide. Compos. Part B Eng. **109**, 91–100 (2017). https://doi.org/10.1016/J.COMPOSITESB.2016.10.056
126. M.H. Al-Saleh, Carbon-based polymer nanocomposites as dielectric energy storage materials. Nanotechnology **30**, 062001 (2018). https://doi.org/10.1088/1361-6528/AAF12C
127. Y. Zhao, L. Luo, H. Tang, Z. Zhou, G.X. Chen, Q. Li, Preparation of high-k composites with low dielectric loss based on the double-layer coaxial structure of inorganic/polymer. J. Appl. Polym. Sci. **135**, 46299 (2018). https://doi.org/10.1002/APP.46299
128. M. Yang, H. Zhao, D. He, J. Bai, Constructing a continuous amorphous carbon interlayer to enhance dielectric performance of carbon nanotubes/polyvinylidene fluoride nanocomposites. Carbon N Y **116**, 94–102 (2017). https://doi.org/10.1016/J.CARBON.2017.01.105
129. C. Yang, Y. Lin, C.W. Nan, Modified carbon nanotube composites with high dielectric constant, low dielectric loss and large energy density. Carbon N Y **47**, 1096–1101 (2009). https://doi.org/10.1016/J.CARBON.2008.12.037
130. Q. Guan, L. Yuan, Y. Zhang, A. Gu, G. Liang, Improving the mechanical, thermal, dielectric and flame retardancy properties of cyanate ester with the encapsulated epoxy resin-penetrated aligned carbon nanotube bundle. Compos. Part B Eng. **123**, 81–91 (2017). https://doi.org/10.1016/J.COMPOSITESB.2017.05.029
131. X. Xia, J. Hao, Y. Wang, Z. Zhong, G.J. Weng, Theory of electrical conductivity and dielectric permittivity of highly aligned graphene-based nanocomposites. J. Phys. Condens. Matter **29**, 205702 (2017). https://doi.org/10.1088/1361-648X/AA68EC

132. M. Arjmand, S. Sadeghi, M. Khajehpour, U. Sundararaj, Carbon nanotube/graphene nanoribbon/polyvinylidene fluoride hybrid nanocomposites: Rheological and dielectric properties. J. Phys. Chem. C **121**, 169–181 (2017). https://doi.org/10.1021/ACS.JPCC.6B10741/ASSET/IMAGES/MEDIUM/JP-2016-10741K_0009.GIF
133. S. Luo, S. Yu, R. Sun, C.P. Wong, Nano Ag-deposited $BaTiO_3$ hybrid particles as fillers for polymeric dielectric composites: toward high dielectric constant and suppressed Loss. ACS Appl. Mater. Interfaces **6**, 176–182 (2014). https://doi.org/10.1021/AM404556C/SUPPL_FILE/AM404556C_SI_001.PDF
134. A. Ameli, M. Nofar, C.B. Park, P. Pötschke, G. Rizvi, Polypropylene/carbon nanotube nano/microcellular structures with high dielectric permittivity, low dielectric loss, and low percolation threshold. Carbon N Y **71**, 206–217 (2014). https://doi.org/10.1016/J.CARBON.2014.01.031
135. M. Hamidinejad, B. Zhao, R.K.M. Chu, N. Moghimian, H.E. Naguib, T. Filleter et al., Ultralight microcellular polymer-graphene nanoplatelet foams with enhanced dielectric performance. ACS Appl. Mater. Interfaces **10**, 19987–19998 (2018). https://doi.org/10.1021/ACSAMI.8B03777/ASSET/IMAGES/MEDIUM/AM-2018-037779_0010.GIF
136. B. Zhao, M. Hamidinejad, C. Zhao, R. Li, S. Wang, Y. Kazemi et al., A versatile foaming platform to fabricate polymer/carbon composites with high dielectric permittivity and ultra-low dielectric loss. J. Mater. Chem. A **7**, 133–140 (2018). https://doi.org/10.1039/C8TA05556D
137. F. Jia, C. Liu, B. Yang, A. Lee, L. Wu, H. Ning, Preparation of novel graphene/silicone rubber nanocomposite dielectric foams. Polymers **14**, 3273 (2022). https://doi.org/10.3390/POLYM14163273
138. M. Arjmand, M. Mahmoodi, S. Park, U. Sundararaj, Impact of foaming on the broadband dielectric properties of multi-walled carbon nanotube/polystyrene composites. J. Cell. Plast. **50**, 551–562 (2014). https://doi.org/10.1177/0021955X14539778
139. J. Zhu, J. Shen, S. Guo, C.B. Park, Poly (vinylidene fluoride)-based microcellular dielectrics filled with polydopamine coated carbon nanotubes for achieving high permittivity and ultralow dielectric loss. Compos. Part A Appl. Sci. Manuf. **163**, 107222 (2022). https://doi.org/10.1016/J.COMPOSITESA.2022.107222
140. X. Zhang, P. Li, P. Gong, Z. Xie, B. Jin, C.B. Park et al., Supercritical carbon dioxide foaming for ultra-low dielectric loss perfluorinated foam. J. CO_2 Util. **65**, 102226 (2022). https://doi.org/10.1016/J.JCOU.2022.102226
141. M.A. Morsi, G.M. Asnag, A.S. Assran, R. Alwafi, A.E. Tarabiah, N.A. Alshehri et al., Reinforced PEO/Cs polymers blend with Al_2O_3/TiO_2 hybrid nanofillers: nanocomposites for optoelectronics and energy storage. J. Energy Storage **88**, 111554 (2024). https://doi.org/10.1016/J.EST.2024.111554
142. S.R. Alharbi, M. Alhassan, O. Jalled, S. Wageh, A. Saeed, Structural characterizations and electrical conduction mechanism of $CaBi_2Nb_2O_9$ single-phase nanocrystallites synthesized via sucrose-assisted sol–gel combustion method. J. Mater. Sci. **53**, 11584–11594 (2018). https://doi.org/10.1007/S10853-018-2458-2/FIGURES/7
143. M.A. Gabal, A.A. Al-Juaid, S. El-Rashed, M.A. Hussein, Y.M. Al Angari, A. Saeed, Structural, thermal, magnetic and electrical properties of polyaniline/$CoFe_2O_4$ nano-composites with special reference to the dye removal capability. J. Inorg. Organomet. Polym. Mater. **29**, 2197–2213 (2019). https://doi.org/10.1007/S10904-019-01179-Z/TABLES/4
144. E.A. Al-Harthy, Y.M. Al Angari, A. Salam, M. Awad, A. Al-Juaid, A.A. Saeed et al., Synthesis, characterization and dye removal capability of conducting polypyrrole/$Mn_{0.8}Zn_{0.2}Fe_2O_4$/graphite oxide ternary composites. Catalysts **12**, 1624 (2022). https://doi.org/10.3390/CATAL12121624
145. M.A. Gabal, E.A. Al-Harthy, Y.M. Al Angari, A. Awad, A.A. Al-Juaid, M.A. Hussein et al., Synthesis, structural, magnetic and high-frequency electrical properties of $Mn_{0.8}Zn_{0.2}Fe_2O_4$/polypyrrole core–shell composite using waste batteries. J. Inorg. Organomet. Polym. Mater. **32**, 1975–1987 (2022). https://doi.org/10.1007/S10904-022-02241-Z/FIGURES/11

146. E.M. Abdelrazek, A.M. Abdelghany, A.E. Tarabiah, H.M. Zidan, AC conductivity and dielectric characteristics of PVA/PVP nanocomposite filled with MWCNTs. J. Mater. Sci. Mater. Electron. **30**, 15521–15533 (2019). https://doi.org/10.1007/S10854-019-01929-2/FIGURES/15
147. Y. Wang, D. Yang, M.M. Hessien, K. Du, M.M. Ibrahim, Y. Su et al., Flexible barium titanate@polydopamine/polyvinylidene fluoride/polymethyl methacrylate nanocomposite films with high performance energy storage. Adv. Compos. Hybrid Mater. **5**, 2106–2115 (2022). https://doi.org/10.1007/S42114-022-00552-W/FIGURES/5
148. Prateek, V.K. Thakur, R.K. Gupta, Recent progress on ferroelectric polymer-based nanocomposites for high energy density capacitors: synthesis, dielectric properties, and future aspects. Chem. Rev. **116**, 260–317 (2016)
149. X. Zhang, Y. Jiang, R. Gao, X. Li, Z. Shen, B.W. Li et al., Tuning ferroelectricity of polymer blends for flexible electrical energy storage applications. Sci. China Mater. **64**, 1642–1652 (2021). https://doi.org/10.1007/s40843-020-1588-3
150. Q. Chi, Y. Zhou, C. Yin, Y. Zhang, C. Zhang, T. Zhang et al., A blended binary composite of poly(vinylidene fluoride) and poly(methyl methacrylate) exhibiting excellent energy storage performances. J. Mater. Chem. C **7**, 14148–14158 (2019). https://doi.org/10.1039/C9TC04695J
151. H.A. Alsalmah, S.J. Almehmadi, M.O. Farea, Q.A. Alsulami, H. Albalawi, A. Rajeh, Hybrid GO/TiO_2 nanoparticles reinforced NaAlg/PVA blend: nanocomposites for high-performance energy storage devices. Polym. Adv. Technol. **34**, 2831–2840 (2023). https://doi.org/10.1002/PAT.6105
152. A.M. Abdelghany, A.H. Oraby, M.O. Farea, Influence of green synthesized gold nanoparticles on the structural, optical, electrical and dielectric properties of (PVP/SA) blend. Phys. B Condens. Matter. **560**, 162–173 (2019). https://doi.org/10.1016/J.PHYSB.2019.02.029

Index

S. S. Ray and R. Banerjee, *Polymer Blend Nanocomposites*, Springer Series in Materials Science 354, https://doi.org/10.1007/978-3-032-00309-6

MIX
Papier aus verantwortungsvollen Quellen
Paper from responsible sources
FSC® C105338
FSC
www.fsc.org

If you have any concerns about our products,
you can contact us on
ProductSafety@springernature.com

In case Publisher is established outside the EU,
the EU authorized representative is:
Springer Nature Customer Service Center GmbH
Europaplatz 3, 69115 Heidelberg, Germany

Printed by Libri Plureos GmbH
in Hamburg, Germany